Microchip
Fabrication

Other Reference Books of Interest by McGraw-Hill

Handbooks

BENSON • *Television Engineering Handbook*

CHEN • *Computer Engineering Handbook*

COOMBS • *Printed Circuits Handbook*

DI GIACOMO • *Digital Bus Handbook*

DI GIACOMO • *VLSI Handbook*

FINK AND CHRISTIANSEN • *Electronics Engineers' Handbook*

HARPER • *Electronic Packaging and Interconnection Handbook*

JURAN AND GRYNA • *Juran's Quality Control Handbook*

RORABAUGH • *Digital Filter Designer's Handbook*

SERGENT AND HARPER • *Hybrid Microelectronic Handbook*

TUMA • *Engineering Mathematics Handbook*

WAYNANT • *Electro-Optics Handbook*

WILLIAMS AND TAYLOR • *Electronic Filter Design Handbook*

Other

ANTOGNETTI • *Power Integrated Circuits*

ANTOGNETTI AND MASSOBRIO • *Semiconductor Device Modeling with SPICE*

BEST • *Phase-Locked Loops*

BUCHANAN • *CMOS / TTL Digital Systems Design*

BUCHANAN • *BiCMOS / CMOS Systems Design*

BYERS • *Printed Circuit Board Design with Microcomputers*

ELLIOTT • *Integrated Circuits Fabrication Technology*

HECHT • *The Laser Guidebook*

KIELKOWSKI • *Inside SPICE*

SMITH • *Thin-Film Deposition*

SZE • *VLSI Technology*

TSIVIDIS • *Mixed Analog Digital VLSI Devices and Technology: An Introduction*

TSUI • *LSI / VLSI Testability Design*

WOBSCHALL • *Circuit Design for Electronic Instrumentation*

WYATT • *Electro-Optical System Design*

To order or receive additional information on these or any other McGraw-Hill titles, please call 1-800-822-8158 in the United States. In other countries, contact your local McGraw-Hill representative.

KEY=WM16XXA

Microchip Fabrication

A Practical Guide to Semiconductor Processing

Peter Van Zant

Third Edition

McGraw-Hill

New York   San Francisco   Washington, D.C.   Auckland   Bogotá
Caracas   Lisbon   London   Madrid   Mexico City   Milan
Montreal   New Delhi   San Juan   Singapore
Sydney   Tokyo   Toronto

Library of Congress Cataloging-in-Publication Data

Van Zant, Peter.
   Microchip fabrication : a practical guide to semiconductor
processing / Peter Van Zant.—3rd ed.
      p.   cm.
   Includes index.
   ISBN 0-07-067250-4 (hardcover)
   1. Semiconductors—Design and construction. I. Title.
IN PROCESS
621.3815'2—dc20   96-15729
                                                CIP

McGraw-Hill

*A Division of The **McGraw·Hill** Companies*

ISBN 0-07-067250-4

*The sponsoring editor for this book was Steve Chapman, the editing
supervisor was Fred Bernardi, and the production supervisor was
Pamela Pelton. It was set in Century Schoolbook by Dina E. John of
McGraw-Hill's Professional Book Group composition unit.*

Printed and bound by R. R. Donnelley & Sons Company.

 This book is printed on recycled, acid-free paper containing a
minimum of 50% recycled, de-inked fiber.

To the wonderfully creative, dedicated, and hard-playing band of chip pioneers who created the world's most advanced manufacturing process, transforming Silicon Valley into the Florence of the twentieth century. And to the spirit of Marchetti's and the Wagon Wheel.

Contents

Preface
to the Third Edition

The first edition of *Microchip Fabrication* came off press in 1984. Feature sizes were still in the micron range, a circuit could be fabricated with a few dozen major steps, and a wafer-fabrication facility could be built for about $10 million. We knew that Moore's law was racing along like some out-of-control train. Since 1984 it has crashed through the micron feature size barrier, the multimillion transistor chip, and the gigabuck ($1 billion) wafer-fabrication facility. These pressures have driven wafer fabrication into a dizzying pace of changes.

The pressures tormenting the industry also show up in this edition. It has been our goal to write *Microchip Fabrication* in a way such that fabrication workers or suppliers to wafer-fabrication operations would have a clear, concise first text with which to launch their careers. Exploding new processes and technologies, each more complicated and demanding, have placed new challenges on achieving our goals. For any given process, there are numbers of options and endless variations on those options. *Microchip Fabrication* covers the basics of processing and reflects current manufacturing practices. It is our intention to give the reader a thorough grounding in the problems and current solutions, thus allowing self-education as the industry continues its relentless march to the outer fringes of technology.

As in the second edition, this third edition is organized around the principal processing technologies. Where there are shared processes (such as tube furnaces in diffusion and CVD), the process is described once and referenced in the other sections. This approach has created a lot of forward and backward references. We hope the confusion is tolerable.

We would like to thank the following industry folks who graciously shared their information and experience for this update: Anne Miller

(Semiconductor Services), Don Keenan (Siltec Corporation), Roger Brennan (Solecon), Jerry Scheely (Technical Instruments Company), Mike Wright (Wright Associates), Gene Sullivan (Dryden Engineering), Jim Reynolds (Reynolds Consulting), Don Levinthal (Semiconductor International), Adam Japko (Solid State Technology), Bill Moffet and Jim Reynolds (Yield Engineering Systems), and John Price (In-Spec).

Also, thanks and gratitude to my ever-supportive wife, Mary DeWitt.

Peter Van Zant

1

The Semiconductor Industry

Overview

In this chapter you will be introduced to the history, development, and trends of the semiconductor industry. The major product types and transistor-building structures and different levels of integration will be explained.

Objectives

Upon completion of this chapter you should be able to:

1. Describe the difference between discrete devices and integrated circuits.
2. Write the definitions of the terms *solid-state, planar processing,* and N- and P-type semiconducting materials.
3. List the four major stages of semiconductor processing.
4. Explain the *integration scale* and the implications of processing circuits at different levels of integration.
5. List the major process and device trends in semiconductor processing.

Birth of an Industry

Significant technological breakthroughs usually result from seeking a solution to a problem and they add to a body of existing knowledge.

The development of solid-state electronics followed this pattern, building on the limitations of early computers and leading to the development of the transistor in 1947.

One problem occurred prior to the 1890 U.S. census. It was predicted that the 1890 census would take 10 years to condense and correlate by manual tabulation. This length of time was unacceptable and led the U.S. Census Bureau to sponsor a contest for a new method of processing the data. The winner was Herman Hollerith, who invented a mechanical tabulating machine that completed the census data crunching in an amazing six weeks.

His machine was a mechanical system driven by an electrical motor that introduced the punch card. Hollerith went on to form the Tabulating Machine Company, which eventually evolved into the International Business Machines Corporation (IBM). This tabulating machine was one of the earliest computers, a machine that was designed to perform automatic calculations on data. The technology of computers continued to progress in the 1930s with the development of computers using electromechanical switches.

During World War II, intelligent and fast machines to crack secret codes were developed. One result of these efforts was the world's first electronic computer, the Electronic Numeric Integrator and Calculator, or ENIAC for short. The ENIAC was first demonstrated at the Moore School of Engineering in Pennsylvania in 1947.

ENIAC hardly fits the modern picture of a computer. It occupied some 3000 ft^2, weighed 50 tons, generated large quantities of heat, required the services of a small power station, and cost \$400,000 in 1940 dollars. ENIAC needed 19,000 vacuum tubes along with thousands of resistors and capacitors (Fig. 1.1). However crude this early computer now seems, it was the system that gave birth to the electronics era with all its famous progeny. The revolution it launched eventually led to the end of the industrial age and the dawn of the information age in the mid 1970s. In that decade more

Size, ft	30 × 50
Weight, tons	30
Vacuum Tubes	18,000
Resistors	70,000
Capacitors	10,000
Switches	6000
Power Requirements, W	150,000
Cost (in 1940)	\$400,000

Figure 1.1   Eniac statistics. (*Foundations of Computector Technology, J. G. Giarratano, Howard W. Sams & Co., Indianapolis, Ind., 1983.*)

workers became employed in the processing of information than in manufacturing.

However, many inventions have been made along the way from the monstrous ENIAC to laptop computers. A measure of this progress was the duplication in the mid 1970s of the entire electronics of the ENIAC on a $\frac{3}{8}$-in^2 piece of silicon, drawing less power than a light bulb and costing less than $20!

The Solid-State Transistor

Vacuum tubes were the primary electronic devices or components of electrical products from their invention in the early 1900s until the 1950s. A vacuum tube consists of three elements: two electrodes separated by a grid in a glass enclosure (Fig. 1.2). Inside the enclosure is a vacuum, required to prevent the elements from burning up and to allow the easy transfer of electrons.

Tubes perform two important electrical functions, switching and amplification. Switching refers to the ability of an electrical device to turn a current on and off. Amplification is a little more complicated. It is the ability of a device to receive a small signal (or current) and amplify it while retaining its electrical characteristics.

However, vacuum tubes suffer from a number of drawbacks. They are bulky, prone to loose connections and vacuum leaks, and fragile; they require a relatively large amount of power to operate; and their elements deteriorate rather rapidly. One of the major drawbacks to ENIAC and other tube-based computers was a limited operating time due to tube burnout.

These problems formed the impetus leading many laboratories around the country to seek a replacement for the vacuum tube. That effort came to fruition on December 23, 1947, when three Bell

Figure 1.2   Vacuum tube.

Figure 1.3   The first transistor.

Laboratory scientists demonstrated an electrical amplifier made from the semiconducting material germanium (Fig. 1.3). This device offered the electrical functioning of a vacuum tube, but with the advantages of solid state: no vacuum, small size, light in weight, low power requirements, and long lifetime. First named a transfer resistor, the new device soon became known as the *transistor.*

The three scientists, John Bardeen, Walter Brattin, and William Shockley, were awarded the 1956 Nobel Prize in physics for the invention of the transistor.

Solid-State Devices

The transistor, like the vacuum tube, is a three-element device. But instead of electrons "jumping" between the elements in a vacuum, they cross electrical *junctions* formed in the solid-state material. These junctions are the interfaces between regions in the semiconducting material that are either rich in electrons or holes (Fig. 1.4). The electron-rich regions are called N type (for negative) and the hole-rich regions are called P type (for positive). The electrical junctions are more formally called NP or PN junctions. The concept of holes is explained in Chap. 3 and junctions are explained in Chap. 12.

Solid-state technology is also used to create diodes, resistors, and capacitors. Diodes are two-element devices that function in a circuit as a switch. Resistors are monoelement devices that serve to limit cur-

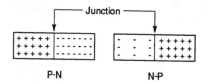

Figure 1.4   P-N and N-P junctions.

rent flow. Capacitors are two-element devices that store charge in a circuit. In some circuits the technology is used to create fuses. Refer to Chap. 14 for an explanation of these concepts and an explanation of how these devices work.

Discrete devices

The announcement in 1947 of the first solid-state transistor launched the solid-state electronics age and the microelectronics industry. The industry is divided into two major segments, the semiconductor industry and the electronics industry. The electronics industry designs and produces a vast number of semiconductor-device-based products, from consumer electronics to computers. It also includes the manufacturers of printed circuit boards.

The semiconductor segment is composed of two major subsegments. One comprises the firms that actually make semiconductor solid state-devices and circuits, using a manufacturing process called *wafer fabrication*. The other subsegment is the cadre of companies that develop and supply the special equipment, materials, and services required for wafer fabrication.

The semiconductor industry was in full swing by the early 1950s supplying devices for transistor radios and transistor-based computers. In this early stage the semiconductor industry produced separate transistors, diodes, resistors, and capacitors (Fig. 1.5), known as *discrete devices,* because they contained only one device per chip. Most discrete devices have less demanding operational and fabrication requirements than integrated circuits. In general, discrete devices are not considered leading edge products. Yet they are required in most sophisticated circuits, and traditionally have accounted for 20 to 30 percent of the dollar volume of all semiconductor devices sold.

Integrated circuits (ICs)

The dominance of discrete devices in solid-state circuits came to an end in 1959. In that year, Jack Kilby, a new engineer at Texas Instruments in Dallas, formed a complete circuit on a single piece of

Figure 1.5 Solid-state discrete devices.

the semiconducting material germanium. His invention combined several transistors, diodes, and capacitors (five components total), and used the natural resistance of the germanium chip (called "a bar" by Texas Instruments) as a circuit resistor. This invention was the *integrated circuit,* the first successful integration of a complete circuit in and on the same piece of a semiconducting substrate. Figure 1.6 is a drawing of the Kilby circuit. Note that the devices are connected with individual wires.

The Kilby circuit did not have the form that is prevalent today. Robert Noyce, then at Fairchild Camera, created the final piece of the puzzle. Earlier, Jean Horni, also at Fairchild Camera, had combined a diffused junction transistor on a silicon chip to produce a solid-state transistor with a flat profile (Fig. 1.7). The flattened profile was the outcome of taking advantage of the easily formed natural oxide of silicon, which also happened to be a dielectric (electrical insulator). Horni's transistor used a layer of evaporated aluminum that was patterned into the proper shape to serve as wiring for the device. This technique is called *planar technology.*

Noyce applied this technique to "wire" individual devices together on the silicon wafer surface. The Noyce integrated circuit became *the* model for all integrated circuits. The techniques used not only met

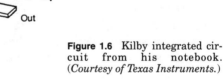

Figure 1.6 Kilby integrated circuit from his notebook. (*Courtesy of Texas Instruments.*)

Figure 1.7 Horni "teardrop" transistor.

the needs of that era, but contained the seeds for the miniaturization and cost-effective manufacturing techniques that still drive the industry. Kilby and Noyce shared the patent for the integrated circuit.

Starting with the simple Kilby IC, the industry has never looked back. Each year the number of devices in a single circuit has increased. In 1964 Gordon Moore, a founder of Intel, predicted that integrated circuit density would double every year. This prediction became known as Moore's law and has proven very accurate (Fig. 1.8).

The industry keeps track of these integration levels by a scale (Fig. 1.9) ranging from small-scale integration (SSI) to ultra-large-scale integration (ULSI). ULSI chips are sometimes referred to as very very large-scale integration ((VVLSI). The popular press calls these newest products *megachips*.

In addition to the integration scale, memory circuits are identified by the number of memory bits contained in the circuit (e.g., a four-meg memory chip can store four million bits of memory). Logic circuits are often rated by their number of gates. A *gate* is the basic operational component of a logic circuit.

Figure 1.8 Growth in IC density [*After Moore, "VLSI, What Does the Future Hold?", Electronic News, 42(14) (1980).*]

Level	Abbreviation	# Components per Chip
Small Scale Integration	SSI	2 - 50
Medium Scale Integration	MSI	50 - 5000
Large Scale Integration	LSI	5000 - 100,000
Very Large Scale Integration	VLSI	Over 100,000 - 1,000,000
Ultra Large Scale Integration	ULSI	> 1,000,000

Figure 1.9 IC integration table.

Figure 1.10   Stages of semiconductor manufacturing.

Stages of Manufacturing

Solid-state devices are manufactured in four distinct stages (Fig. 1.10). They are material preparation, crystal growth and wafer preparation, wafer fabrication, and packaging.

In the first stage, material preparation (see Chap. 2), the raw semiconducting materials are mined and purified to meet semiconductor standards. For silicon, the starting material is sand (Fig. 1.11).

In stage two the material is formed into wafers (Fig. 1.12) in a process called crystal growth and wafer preparation (Chap. 3). Wafers (also called slices) are thin disks of the semiconducting material. This stage is performed by either a separate department within the semiconductor company or vendors outside the company. Figure 1.11 shows the conversion of sand to semiconductor grade silicon. The industry also makes devices and circuits from germanium and compounds of different semiconductor materials.

It is in stage three, wafer fabrication, that the devices or integrated circuits are actually formed in and on the wafer surface (Fig. 1.13). Up to several thousand identical devices are formed on each wafer,

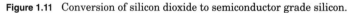

Figure 1.11   Conversion of silicon dioxide to semiconductor grade silicon.

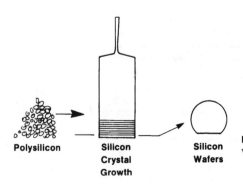

Figure 1.12   Crystal growth and wafer preparation.

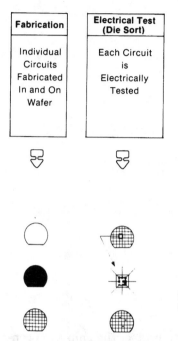

Fabrication	Electrical Test (Die Sort)
Individual Circuits Fabricated In and On Wafer	Each Circuit is Electrically Tested

Figure 1.13 Wafer fabrication (and electrical test).

although 200 to 300 are more common. The area on the wafer occupied by the discrete device or integrated circuit is called a chip or die. The wafer fabrication process is also called *fabrication, fab, chip fabrication,* or *microchip fabrication.*

Chips are fabricated by both *merchant* and *captive* producers. Merchant suppliers manufacture chips to sell on the open market. Captive producers, whose final products are computers, communications systems, and the like, produce chips in-house for their own products. Some firms produce chips for in-house use and also sell on the open market, and others produce specialty chips in-house and buy others on the open market. During the 1980s a greater percentage of chips being fabricated in captive fab areas became the trend. Also, during the 1980s a number of firms started supplying chips but contracted out the wafer fabrication. They are known as *fabless chip suppliers.*

After wafer fabrication, the chips on the wafer are complete but untested. Next comes an electrical test (called *wafer sort*) of every chip to identify those that meet customer specifications. Wafer sort may be the last step in wafer fabrication or the first step in the *packaging* process. Packaging (Fig. 1.14) is the series of processes that separates the wafer into individual die and places them into protective packages.

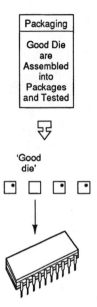

Figure 1.14 Packaging stage.

A protective chip package is necessary to protect the chip from contamination and abuse and to provide a durable and substantial electrical lead system, which will connect the chip onto a printed circuit board or directly into an electronic product. Packaging takes place in a different department of the semiconductor company and quite often in an offshore plant.

While most chips are packaged in individual packages, a growing percentage are being incorporated into hybrid circuits or multichip modules (MCMs) or mounted directly on printed circuit boards (chip-on-board or COB). An integrated circuit comprises entirely solid-state devices contained on one chip. A hybrid circuit combines semiconductor devices (discretes and ICs), with thick or thin film resistors and conductors, and other electrical components on a ceramic substrate. These techniques are explained in Chap. 18.

Process and Product Trends

Since 1947, the semiconductor industry has seen the continuous development of new and improved processes. These process improvements have led to more highly integrated and reliable circuits that have, in turn, fueled the continuing electronics revolution. These process improvements fall into two broad categories, process and structural. Manufacturing processes and price structure improvements result in the fabrication of the devices and circuits in ever higher density, quantity, and reliability. Structural improvements are new device designs that yield greater circuit performance, power control, and reliability.

The advancement of chip density from SSI to ULSI has been accompanied by two trends, increasing chip size and decreasing the size of the individual circuit components. Discrete and SSI chips average about 100 mils (0.1 in) on a side. ULSI chips are about 500–750 mil (0.5–0.75 in) per side. Increasing die size also pushed the drive to larger wafer diameters. The 1 in diameter wafers of the 1960s have given way to 200- and 300-mm (8- and 12-in) wafers.

Although integrated circuit chip size has grown as more and more devices are crowded into the circuits, the higher densities would not be possible without a commensurate decrease in the size of each part of individual devices, known as the device *feature size*. This decrease has been brought about by dramatic improvements in the imaging process, known as *photolithography,* and the trend to multilayers of conductors. The decrease in feature size over a twenty-year span is shown in Fig. 1.15.

Projections for these interdependent factors—feature size, die size, wafer diameter, number of interconnect levels, and package pin count (I/O, for in/out connections)—have been made by the Semiconductor Industry Association (SIA) through the year 2007. Their "road map" is shown in Fig. 1.16.

An analogy used to explain these trends is the layout of a neighborhood of single family homes. The density of the neighborhood is a func-

Figure 1.15 Decreasing image feature size. (*After Wolf and Tauber, "Silicon Processing for the VLSI Era."*)

DRAM Production	1995	1998	2001	2004	2007
Feature Size (nm)	350	250	180	120	100
Bits (DRAM)	64M	256M	1G	4G	16G
Processing Cost ($/cm^2)	$3.90	$3.80	$3.70	$3.60	$3.50
Chip Size (mm^2)					
Logic	400	600	800	1000	1250
DRAM	200	320	500	700	1000
Defect Density (1/cm^2)	0.05	0.03	0.01	0.004	0.002
Logic Interconnect Levels	4-5	5	5-6	6	6-7

Figure 1.16 SIA lithography roadmap. (*Courtesy Semiconductor Industry Association*)

tion of the house size, lot size, and width of the streets. Increasing the density can be accomplished by building more houses, which results in a larger neighborhood. Or the houses can be put closer together by reducing the lot and street size. Another possibility is to reduce the size of the individual houses. However, at some point the streets cannot be reduced anymore or they won't be wide enough for autos, and at some point the houses cannot be reduced in size and still function as dwelling units.

The same types of options take place in chip design. The individual parts of the devices can be reduced in size (houses) and moved closer together (lot and street size). A denser chip is the result.

There are several benefits to the reduction of feature size and the attendant increase in circuit density. At the circuit performance level there is an increase in circuit speed. With shorter distances to travel and with the individual devices occupying less space, information can be put into and gotten out of the chip in less time. Anyone who has waited for a personal computer to perform a simple operation can appreciate the effect of faster performance. These same density improvements result in a chip or circuit that requires less power to operate. The small power station required to run the ENIAC has given way to powerful laptop computers that run on a set of batteries.

A companion trend has been an increase in die size. The benefits of increasing chip density also apply to larger circuits. Making the individual chip components smaller and closer together *and* creating a larger chip yields faster and more powerful circuits.

Perhaps the most significant effect of these process and product improvements is the cost of the chips. Figure 1.17 shows the year-by-year drop in memory chips. The curves are typical for any maturing

.035¢

.025¢   4K

                    ∶ 256K

.015¢   16K

        64K

1980 '81 '82 '83 '84 '85 '86 '87

Figure 1.17 Price of chips per bit of memory.

product. Prices start high, but as the technology is mastered and manufacturing efficiencies increase, the prices drop and eventually become stable. These chip prices have constantly declined even as the performance of the chips has increased. However, the economics of mega-chip manufacturing may actually drive the cost of chips up (Fig. 1.18 and Chap. 15). The two factors, increased performance and lower cost, have driven the switch from conventional electronic components to solid-state devices. An example is the automobile. A modern auto carries hundreds of dollars of on-board computing power. That same amount of power would have cost some $3000 in a 1970 auto. Considering that a 1970s car cost about $3000, this is a remarkable achievement. Even more impressive is the personal computer. Today,

Figure 1.18 Rising cost of transistors with decreasing feature size and decreasing manufacturing efficiency (X). (*Courtesy Semiconductor International, June 1994*)

CHIP USES	1996
Computer	48.0%
Consumer	21.2%
Telecom	14.7%
Industrial	9.8%
Automotive	4.4%
Military	1.9%
Total	100.0%

Figure 1.19 Semiconductor chip uses. (*Courtesy In-Stat-1995 SEMI ISS seminar*)

a moderate priced desktop computer can deliver the same power as an IBM mainframe manufactured in 1970.

The availability of cheap chips has led to their inclusion in many different products. Major industry use of chips is shown in Fig. 1.19.

To offset the decline in chip prices, the semiconductor industry has continually had to make dramatic improvements in processing technology, including vendor-supplied materials and equipment. The fabrication processes have become more numerous as the use of automation is becoming an increasing requirement in many fields (see Chap. 15).

Overall, the semiconductor industry has experienced continual growth. From its birth in the 1950s it is now approaching worldwide sales of $2 trillion a year, supported by a supplier industry of over $1 trillion. Millions of chips are supplied by factories located throughout the world. Interestingly, even as the industry shows signs of maturing it is still growing faster than other "mature" industries, indicating that microchips are nowhere near the technical and performance limits (Fig. 1.20).

KEY INDUSTRY	% Growth[*]
Automobiles	1.3%
Pharmaceuticals	1.5%
Aircraft	2.1%
Computers/Peripherals	2.5%
GDP	3.0%
Telecom-Equipment	3.7%
Semiconductors/Elec. Comp.	9.3%

[*] Compound Annual Growth Rate 1987-1994

Figure 1.20 Key industry growth rates. (*Courtesy U.S. Industrial Outlook 1994, SEMI ISS 1994*)

Year	1993	1996	1999	2002	2005
DRAM Capacity	16Mb	64Mb	256Mb	1Gb	4Gb
Feature Size (micron)	0.5	0.35	0.25	0.18	0.08
Volumes of Enc. Britannics	1/4/ Vol	1 Vol.	4 Vol.	16 Vol.	2 Sets

Figure 1.21   Future DRAM capacity. (*Source:* BusinessWeek, *July 4, 1994*)

An example of increasing chip power is shown in Fig. 1.21, which indicates the number of volumes of the Encyclopedia Britannica that can be stored on larger capacity dynamic random access memory (DRAM) chips.

Five Decades of Progress

The history of the semiconductor industry is one of continual developments and advances. Each new level of process expertise and material refinement has made possible newer and more powerful circuits. The first five decades of the semiconductor era have seen an emerging industry rise to world dominance.

The development decade (1951–1960)

While the tremendous advantages of solid-state electronics were recognized early, the advances possible from miniaturization were not realized until two decades later. During the 1950s engineers set to work and defined many of the basic processes and materials still used today.

The Bell transistor came to be known as the *point contact* transistor because the surface contacts were made by springing two thin wires onto the semiconductor surface. This system was unreliable and quickly gave way to the *alloy junction* transistor (Fig. 1.22). In this

Figure 1.22   Alloy junction transistor.

Figure 1.23 Grown junction transistor.

process the contacts were alloyed to the surface. Another early technique was the *grown* junction (Fig. 1.23), where the required semiconductor junction was grown into the semiconductor crystal during the crystal-growing process. These advances all came out of Bell Labs.

These transistors were the bipolar type (see Chap. 14), which was the dominate transistor construction technique well into the 1970s. The term *bipolar* refers to a transistor structure that operates on both negative and positive currents. The other major method of building a solid-state transistor is the field effect transistor (FET). William Shockley published the operational basics of a FET in 1951. These transistors operate with only one type of current and are also called *unipolar* devices. The FET came to the marketplace in volume in a structure known as the metal oxide semiconductor (MOS) transistor.

William Shockley and Bell Labs get much of the credit for the spread of semiconductor technology. Shockley left Bell Labs in 1955 and formed Shockley Laboratories in Palo Alto, California. While his company did not survive, it established semiconductor manufacturing on the West Coast and started what eventually became known as Silicon Valley. Bell Labs helped the fledgling industry by licensing its semiconductor discoveries to a host of companies.

The early semiconductor devices were made of germanium. Texas instruments changed that with the introduction of the first silicon transistor in 1954. The issue over which material would dominate was settled in 1956 and 1957 by two more developments from Bell Labs, diffused junctions and oxide masking. *Diffused junctions* were an important part of the early mesa transistor (Fig. 1.24) brought to the market place in silicon by Texas Instruments. The *mesa* transis-

Figure 1.24 Mesa transistor with diffused emitter and base.

Wafer

Wafer and
Oxide Layer

Patterned
Oxide Layer

Dopants

Doping of Wafer
through Patterned
Oxide Layer

Metal Conducting
Layer

Figure 1.25 Basics of silicon planar processing.

tor was named for its mesa-like shape, which allowed electrical contacts to the subsurface N- and P-type regions.

It was the development of *oxide masking* that ushered in the Silicon Age. Silicon dioxide (SiO_2) grows uniformly on silicon and having a similar index of expansion, allows high temperature processing without warping. Silicon dioxide is a dielectric material which allows it to function on the silicon surface as an insulator. Additionally, SiO_2 is an effective block to the dopants that form the N and P regions in silicon.

The net effect of these advances was planar technology (Fig. 1.25), introduced by Fairchild Camera in 1960. With the aforementioned techniques it was possible to form (using diffusion) and protect (because of silicon dioxide) junctions during and after the manufacturing process. Also, the development of oxide masking allowed all of the transistor formation to take place on the top surface of the wafer (Fig. 1.26), that is, in one plane. It was this process that set the stage for Robert Noyce's development of thin film wiring.

Bipolar Transistor

Epitaxial Layer

Wafer

Figure 1.26 Double diffused bipolar transistor formed in epitaxial layer.

Bell Labs developed the technique of forming transistors into a high purity layer of semiconducting material deposited on top of the wafer. Called an epitaxial layer, this invention allowed higher-speed devices and provided a scheme for the closer packing of components in a bipolar circuit.

The semiconductor industry developed in high technology centers off Route 128 around Boston, in Silicon Valley in Northern California, and at Texas Instruments in Dallas, Texas. These areas maintained their dominance until the 1970s when the process technology moved from laboratory-type processes to equipment-based ones.

The 1950s were indeed the golden age of semiconductor development. During this incredibly short time almost all of the basic processes and materials were discovered. The decade opened with the manufacture of small volumes of crude devices in germanium and ended with the first integrated circuit and silicon firmly established as the material for the semiconductor of the future.

The processing decade (1961–1970)

The 1960s were the decade of the process engineer and start-up companies. The trend of chip price erosion was well established in the 1950s. In the 1960s many new chip manufacturers entered the market, while established firms solidified their positions. The field got crowded, which only served to create more downward price pressure. It fell to the process engineers to develop high-yield volume processes to crank out the needed chips. At the same time, the technology spread as engineers brought the technology from company to company.

During this period, process engineers were aided by a growing number of vendors. In the 1950s much of the equipment was made or modified in house for semiconductor use. Likewise, most of the chemicals were purchased from outside vendors at industrial grade levels and "cleaned up" in house. By the 1960s the number of fab areas had grown sufficiently and processes were approaching a level of commonalty to attract semiconductor specialty suppliers.

Kodak corporation was in the game early with its line of KPR, KMER, and KTFR negative photoresists. The Shipley Company introduced its line of positive resists and Waycoat developed its line of negative photoresists, the first based on a synthetic polymer. On the equipment front, BTU and Thermco sold diffusion and oxidation tube furnaces, and K & S and Kasper supplied contact aligners critical for the photomasking operation. Three traditional chemical giants, Merck, Monsanto, and Dow Corning, helped fuel the semiconductor industry growth by becoming suppliers of polished silicon wafers.

Many of the key company players of the 1950s formed new companies in the 1960s. Robert Noyce left Fairchild to found Intel with

Andrew Grove and Gordon Moore. Charles Sporck also left Fairchild to join National Semiconductor, helping it into a major position in the industry. Signetics became the first company dedicated exclusively to the fabrication of ICs. However, the price dropping had a purging effect and some companies gave up fabrication operations. It was rare that a fabrication area was closed down for good, though. Usually a buyer with a new circuit idea was waiting in the wings to start production in the closed facility.

Price dropping was accelerated by the development of a plastic package for silicon devices in 1963. Also in that year, RCA announced the development of the insulated field effect transistor (IFET), which paved the way for the MOS industry. RCA also pioneered the first complementary MOS (CMOS) circuits.

During this period the dominant marketing strategy was to design a new chip and ride the sales curve up before the competition could bring out a competing design. At the same time, in fabrication areas process engineers were working to keep up with higher yield, shrinking feature size, and higher volume requirements.

In the 1970s semiconductor manufacturing changed from processing small batches in laboratories to high-volume manufacturing. The dependence of manufacturing costs and corporate profit on volume was established.

The production decade (1971–1980)

At the start of the 1970s, the semiconductor industry was manufacturing ICs primarily at the medium-scale integration (MSI) level. The move to profitable, high-yield large-scale integration (LSI) devices was being somewhat hampered by mask-caused defects and the damage inflicted on the wafers by contact aligners. In fact, all of the existing processes in one way or another represented barriers to the volume production of high-level circuits.

The defect and aligner mask problems were solved by the development of the first practical projection aligner by Perkin and Elmer Company. This decade also saw the improvement of clean-room construction and operation because of the introduction of ion implantation machines and E-beam machines for high-quality mask generation. Continuous spin/bake and develop/bake systems came into universal use, and lithography mask steppers used in mask making began to show up in fabrication areas for wafer imaging.

During this decade, equipment changed from manual to automatic operation with on-board solid-state controllers and sequencers. Button-to-button operation became the guiding principle of equipment suppliers. The move from operator control to automatic control of the processes increased both wafer throughput and uniformity.

Reliance on equipment-based processes drove the wafer fabrication of the industry to new heights, up to a level of some $10 billion a year in sales.

Once the processes produced actual pieces of equipment, the stage was set for dissemination throughout the world. Along with the process improvements came a more detailed understanding of the physics of solid-state devices, allowing the mastery of the technology by student engineers worldwide.

The automation decade (1981–1990)

Marketplace pressures and trends resulted in automatic machines that, for the most part, would load the wafers, process them through predetermined steps, and return them to transfer carriers.

These advances were needed for the production of advanced chips. A characteristic of the industry is the wide distribution of chip sophistication. While leading edge chips have feature sizes at the submicron level, a surprising number of "older" circuits still have a significant market life. These "older" circuits do not require the sophistication and process control of submicron ULSI circuits and their lower profit margins cannot support the major expense of fabrication area upgrades. This decade opened with a wide variety of processing modes, from the manual processes of the early 1970s to the most sophisticated, almost operator-free, fabrication areas.

With most of the individual processes automated, the focus in the 1980s was on elimination of operators from the fabrication areas and automation of material delivery and movement. The elimination of operators is required by the fact that humans are one of the major sources of contamination in the process. Even elaborate gowning procedures cannot control particulate contamination generated by people moving about the wafers. These issues are examined in more detail in Chap. 4. The automation of individual processes challenged the industry to develop methods for transferring wafers among the various machines. This aspect has been the focus of the automation decade, with the fabled "lights out" fabrication area being the goal.

While most industries were moving toward standardization of manufacturing processes, the opposite happened in the semiconductor industry. It is true that most wafer fabrication processes became better characterized and controlled. But at the same time various trends pushed manufacturers into designing more complicated chips. The new designs in turn presented new manufacturing challenges that led to the development of new processes.

Global semiconductor manufacturing became driven by a proliferation of international cross licensing and joint venture agreements.

Another trend was in chip manufacturer and supplier codevelopment agreements, which were forced by the special needs of modern chip fabrication plants and the ever increasing cost of development. The 1980s saw the rise of Japanese semiconductor producers, followed closely by the "Four Tigers": Hong Kong, Taiwan, Singapore, and Korea.

The production era (1991–?)

Throughout the 1970s and 1980s, the 1-μm feature size barrier loomed as both opportunity and challenge. The opportunity was a new era of megachips with vastly increased speeds and memory. The challenge was found in the limitations of conventional lithography, which were made worse by increased surface steps, additional layers, and necessary wafer diameter increases, to mention a few. In the early 1990s, the 1-μm barrier was crossed in laboratories and 50 percent of fab lines[1] were working at the micron or submicron level.

While feature size drives the technical advances, the increasing number of process steps (Fig. 1.27) drives changes in manufacturing strategies. Although the chip industry continues to pump out new and more powerful chips, the industry is maturing. Early on the profit strategy was to ride the innovation curve. That meant always being first (or close to first) with the latest and greatest chip that could be sold with enough profit to pay for research and development (R&D) and finance new designs. The profit potential of this strategy overcame manufacturing yield problems and lower efficiencies.

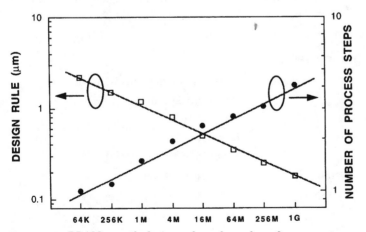

Figure 1.27 DRAM growth design rule and number of process steps. (*SEMI 1995 ISS Conference*)

But the spread of the technology (resulting in increased competition) and improvements in process control moved the industry to greater emphasis on production issues. These issues are primarily productivity factors: automation, cost control, process characterization and control, and worker efficiency.

Strategies to control costs have included detailed analysis of equipment cost of ownership, new fabrication layouts (such as cluster tools), robotic automation, wafer isolation technology (WIT), computer integrated manufacturing (CIM), sophisticated statistical process control, use of advanced metrology instruments, just-in-time inventory schemes, and the like (see Chap. 15).

Technical driving factors, feature size reduction, wafer diameter increases, and yield improvement all have physical or statistical limits. But productivity improvement, which incorporates many of these factors (see Chap. 15), is the source of continuing profit. The pressures are enormous. Wafer fabrication facilities are approaching the gigabuck ($1,000,000,000) level and equipment and process development are equally expensive. In the era beyond the 0.35-μm feature size, new techniques such as x-ray and deep ultraviolet (UV) (DUV) lithography or a huge improvement in conventional lithography techniques will require extensive and expensive development.

This is not to say that technical advancement is finished. On the contrary, many of the processes that will be in use in 10 years are as yet unknown or in the very primitive stages of development. What has changed is that the improvements are becoming evolutionary instead of revolutionary. Engineers are learning how to wring every bit of productivity out of processes before looking for a big technological jump to solve problems. This is another sign of a maturing industry.

Key Concepts and Terms

Capacitor	Integration scale and circuit density
Chip size	Planar processing
Diode	Price erosion
Discrete device	Resistor
Feature size	Semiconducting material
Four stages of semiconductor manufacturing	Solid-state devices
Hybrid circuits	Transistor
Integrated circuit	Wafer

Review Questions

1. List the four types of discrete devices.

2. Describe the advantages of solid-state devices over vacuum tubes.

3. A VLSI circuit has more components than a ULSI circuit (true or false).

4. Describe the difference between a hybrid and integrated circuit.

5. State the stage of processing that uses wafers.

6. State the stage of processing that uses chips.

7. List two other terms used to describe chips.

8. Describe what is meant by the term feature size.

9. Name two other terms used for feature size.

10. Describe the functions of a semiconductor package.

Reference

1. Daniel Rose, "Feast and Famine in Materials," *Semiconductor Equipment and Materials International,* Industry Strategy Symposium, Popple Beach, CA, January 8–11, 1994.

Chapter

2

Semiconductor Materials and Process Chemicals

Overview

The unique functions of semiconductor devices and circuits are the result of the electrical and physical properties of semiconducting materials. In the first part of this chapter, the basic concepts of atoms, electrical classification of solids, and intrinsic and doped semiconductors are explained. In the second part, elementary chemistry as it applies to process chemicals is detailed.

Objectives

Upon completion of this chapter you should be able to

1. Identify the parts of an atom.
2. Name the two unique properties of a doped semiconductor.
3. List at least three semiconducting materials.
4. Explain the advantages and disadvantages of gallium arsenide as compared to silicon.
5. Explain the difference in composition and electrical functioning of N- and P-type semiconducting materials.
6. Describe the properties of resistivity and resistance.
7. Identify the differences between acids, alkalis, and solvents.
8. List the four states of nature.
9. Give the definitions of an atom, a molecule, and an ion.
10. Explain four or more basic chemical safety handling rules.

Atomic Structure

The Bohr atom

The understanding of semiconductor materials requires a basic knowledge of atomic structure. Atoms are the building blocks of the physical universe. Everything in the universe is made from the 96 stable and 12 unstable materials known as elements. Each element has a different atomic structure. The different structures give rise to the different properties of the elements.

The concept of an atom is arrived at by a mental process of division. It goes like this: Consider a small piece of gold. Now start dividing that piece of gold into smaller and smaller pieces. At some point you will have a final small piece that still exhibits the properties of gold. This final small piece is the atom of gold.

Dividing that last piece further will yield the three parts that compose individual atoms. They are called subatomic particles, specifically, protons, neutrons, and electrons. These subatomic particles each have their own properties. A particular combination and structure of the subatomic particles is required to form the gold atom. The basic structure of the atom most used to understand physical, chemical, and electrical differences among different elements was first proposed by the famous physicist Niels Bohr (Fig. 2.1).

The Bohr atomic model has the positively charged protons and neutral neutrons located together in the nucleus of the atom. The negatively charged electrons move in defined orbits about the nucleus, similar to the movement of the planets about the sun. There is an attractive force between the positively charged protons and the negatively charged electrons. However, this force is balanced by the out-

e⁻ = ELECTRON
+ = PROTON
N = NEUTRON
O = UNFILLED ELECTRON POSITION

Figure 2.1 Bohr atom model.

ward centrifugal force of the electrons moving in their orbits. The net result is a structurally stable atom.

Each orbit has a maximum number of positions available for electrons. In some atoms not all of the positions are filled. The unfilled position is called a *hole*. When a particular electron orbit is filled to the maximum, additional electrons must go into the next outer orbit.

The periodic table

The elements differ from each other in the number of electrons, protons, and neutrons in their atoms. Fortunately, nature combined the subatomic particles in an orderly fashion. An examination of some of the rules governing atomic structure is helpful in understanding the properties of semiconducting materials and process chemicals. Atoms (and therefore the elements) range from the simplest, hydrogen (with one electron), to the most complicated, Lawrencium (with 103 electrons). The rules of atomic structure are as follows:

1. In each atom there are an equal number of protons and electrons.

2. Each element contains a specific number of protons, and no two elements have the same number of protons.

3. Elements with the same number of outer orbit electrons have similar properties.

4. Elements are stable with a filled outer ring or with eight electrons in the outer ring. These atoms tend to be more chemically stable than atoms with partially filled rings.

5. Atoms seek to combine with other atoms to create the stable condition of full orbits or eight electrons in their outer rings.

Hydrogen consists of only one proton in the nucleus and only one electron. This arrangement illustrates the first rule of atomic structure. Rule 2 leads to the assignment of numbers to each of the elements. Known as the *atomic number*, it is equal to the number of protons (and therefore electrons) in the atom. For example, hydrogen has one proton in its nucleus while the oxygen atom has eight. The basic reference for the elements is the periodic table (Fig. 2.2). The periodic table has a box for each element, which is identified by two letters. The atomic number is in the upper lefthand corner of the box. For example, calcium (Ca) has the atomic number 20, and so we know immediately that calcium has 20 protons in its nucleus and 20 electrons in its orbital system.

Figure 2.3 shows the atomic structure of elements 1, hydrogen; 3, lithium; and 11, sodium. When constructing the diagrams several

Figure 2.2 Periodic table of elements.

Atomic number 1: hydrogen

Atomic number 3: lithium

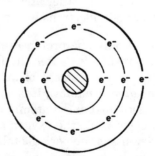

Atomic number 11: sodium

Figure 2.3 Atomic structures of hydrogen, lithium, and sodium.

rules were observed in the placement of the electrons in their proper orbits The rule is that each orbit (n) can hold $2n^2$ electrons. Solution of the math for orbit 1 dictates that the first electron orbit can hold only two electrons. This rule forces the third electron of lithium into the second ring. The rule limits the number of electrons in the second ring to 8 and that of the third ring to 18. So, when constructing the diagram of the sodium atom, with 11 protons and electrons, the first two orbits take up 10 electrons, leaving the eleventh in the third ring.

These three atoms have a commonality. Each has an outer ring with only one electron in it. This illustrates rule 3. This rule is reflected in the periodic table. Note that hydrogen, lithium, and sodium appear on the table in a vertical column labeled with Roman numeral one (I). The column number represents the number of electrons in the outer ring, and all of the elements in each column share similar properties. It is no accident that the three best electrical conductors (copper, silver, and gold) all appear in the same column (Ib) (Fig. 2.4).

The last two rules of atomic structure are relevant to the understanding of semiconductors. Rules 4 and 5 influence the creation of N- and P-type semiconductor materials as explained in the section on doped semiconductors.

Electrical Conduction

Conductors

An important property of many materials is the ability to conduct electricity or support an electrical current flow. An electrical current is simply a flow of electrons. Electrical conduction takes place in elements and materials where the attractive hold by protons on the outer-ring electrons is relatively weak. In such a material, these elec-

Figure 2.4 The three best electrical conductors.

trons can be easily moved and set up an electrical current. This condition exists in most metals.

The property of materials to conduct electricity is measured by a factor known as *conductivity*. The higher the conductivity, the better the conductor. Conducting ability is also measured by the reciprocal of the conductivity, which is called *resistivity*. The lower the resistivity of a material, the better the conducting ability.

$$C = 1/\rho$$

where C = conductivity
ρ = resistivity in ohms-centimeter ($\Omega \cdot$ cm)

Dielectrics

At the opposite end of the conductivity scale are materials that exhibit a large attractive force between the nucleus and the orbiting electrons. The net effect is a great deal of resistance to the movement of electrons. These materials, which have low conductivity and high resistivity, are known as *dielectrics*. In electrical circuits and products, dielectric materials such as glass are used as insulators.

An electrical factor related to the degree of conductivity (and resistivity) of a material is the electrical resistance of a specific volume of the material. The *resistance* is a factor of the resistivity and dimensions of the material. Resistance to electrical flow is measured in ohms, as illustrated in Fig. 2.5. The formula defines the electrical resistance of a specific volume of a specific material (in this illustration the volume is a rectangular bar with dimensions D, L, and W). The relationship is analogous to density and weight, *density* being a material property and *weight* the force exerted by a specific volume of the material.

R = Resistance
L = Length
W = Width
D = Depth
A = Cross sectional area = WxD
ρ = Resistivity of material

$$R = \rho\frac{L}{A}$$

$$R = \rho\frac{L}{WxD}$$

Figure 2.5   Resistance of rectangular bar.

Electric current flow is analogous to water flowing in a hose. For a given hose diameter and water pressure only a given amount of water will flow out of the hose. The resistance to flow can be reduced by increasing the hose diameter, shortening the hose, and/or increasing the pressure. In an electrical system the electron flow can be increased by increasing the cross section of the material, shortening the length of the piece, increasing the voltage (analogous to pressure), and/or decreasing the resistivity of the material.

Intrinsic semiconductors

Semiconducting materials, as the name implies, are materials that have some natural electrical conducting ability. There are two elemental semiconductors, silicon and germanium, both found in column IV (Fig. 2.6) of the periodic table. In addition, there are some tens of material compounds (a *compound* is a material containing two or more chemically bound elements) that also exhibit semiconducting properties. These compounds come from elements, such as gallium

Figure 2.6 Semiconductor materials.

arsenide and gallium phosphide, found in columns III and V. Other compounds come from elements in columns II and VI.

The term *intrinsic* refers to these materials in their purified state, that is, not contaminated with impurities or other substances.

Doped semiconductors

Semiconducting materials, in their intrinsic state, are not useful in solid-state devices. However, through a process called *doping,* specific elements can be introduced into intrinsic semiconductor materials. These elements increase the conductivity of the intrinsic semiconductor material. The doped material displays two unique properties that are the basis of solid-state electronics. The two properties are

1. Precise resistivity control through doping

2. Electron and hole conduction

Resistivity of doped semiconductors

Metals have a conductivity range of 10^4 to 10^6 $\Omega \cdot$ cm. The implications of this limited range are illustrated by an examination of the resistor represented in Fig. 2.5. Given a specific metal with a specific resistivity, the only way to change the resistance of a given volume is to change the dimensions. In a semiconducting material the resistivity can be changed, giving another degree of freedom in the design of the resistor. Semiconductor resistivity can be extended over a range of 10^{-3} to 10^3 by the addition of dopant atoms.

Semiconducting materials can be doped into a useful resistivity range by elements that make the material either electron rich (N type) or hole rich (P type). Figure 2.7 shows the relationship of the doping level to the resistivity of silicon. The x axis is labeled "the carrier concentration" because the electrons or holes in the material are called carriers. Note that there are two curves, N type and P type, because different amounts of energy are required to move an electron or a hole through the material. As the curves indicate, it takes less of a concentration of N-type dopants than P-type dopants to create a given resistivity in silicon. Another way to express this phenomenon is that it takes less energy to move an electron than to move a hole.

It takes only 0.000001 to 0.1 percent of a dopant to bring a semiconductor material into a useful resistivity range. This property of semiconductors allows the creation of regions of very precise resistivity values in the material.

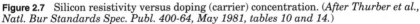

Figure 2.7 Silicon resistivity versus doping (carrier) concentration. (*After Thurber et al., Natl. Bur Standards Spec. Publ. 400-64, May 1981, tables 10 and 14.*)

Electron and hole conduction

Another limitation of a metal conductor is that it conducts electricity only through the movement of electrons. Metals are permanently N type. Semiconductors can be made either N- and P-type by doping with specific dopant elements. N and P type semiconductors can conduct electricity by either electrons or holes. Before examining the conduction mechanism, it is instructive to examine the creation of free (or extra) electrons or holes in a semiconductor structure.

To understand the concept of N-type semiconductors, consider a piece of silicon (Si) doped with a very small amount of arsenic (As), as shown in Fig. 2.8. Assuming even mixing, each of the arsenic atoms would be surrounded by silicon atoms. Applying rule 5 from the section entitled "the periodic table" that atoms attempt to stabilize by having eight electrons in their outer ring, the arsenic atom is shown sharing four electrons from its neighboring silicon atoms. However, arsenic is from column V, which means it has five electrons in its outer ring. The net result is that four electrons from the arsenic atom pair up with electrons from the silicon atoms, leaving one left over. This one electron is available for electrical conduction. Considering that a crystal of silicon has millions of atoms per cubic centimeter, a huge number of electrons are available to conduct an electrical current. Doping silicon with the elements arsenic, phosphorus, and antimony creates N-type conditions.

An understanding of P-type material is approached in the same manner (Fig. 2.9). The difference is that only boron, from column III of the periodic table, is used to make P-type silicon. When mixed into the silicon it, too, borrows electrons from silicon atoms. However, having only three outer electrons, there is a place in boron's outer ring that is not filled by an electron. It is this unfilled position that is defined as a hole.

Within a doped semiconductor material there is a great deal of activity. Electrons are constantly moving and holes are constantly being created. The electrons are attracted to holes, leaving unfilled

Figure 2.8 N-type doping of silicon with arsenic.

Figure 2.9 P-type doping of silicon with boron.

Figure 2.10 Electron conduction in N-type semiconductor material.

positions, which create more holes. How the holes contribute to electrical conduction is illustrated in Fig. 2.10. When a voltage is applied across a piece of conducting or semiconducting material the negative electrons move towards the positive pole of the voltage source, such as a battery.

In P-type material (Fig. 2.11) an electron will move toward the positive pole by jumping into a hole along the direction of route t_1. Of course, when it leaves its position, it leaves a new hole. As it continues towards the positive pole, it creates a succession of holes. The effect to someone measuring this process with a current meter is that the material is supporting a positive current, when actually it is a negative current moving in the opposite direction. This phenomenon is called *hole flow* and is unique to semiconducting materials.

The dopants that create a P-type conductivity in a semiconductor material are called *acceptors*. Dopants that create N-type conditions are called *donors*. An easy way to keep these terms straight is that acceptor has a *p* and donor is spelled with an *n*.

Figure 2.11 Hole conduction in P-type semiconductor material.

Classification	Electrons	Examples	Conductivity
1. Conductor	Free to Move	Gold Copper Silver	10^4–10^6 /ohm-cm
2. Insulator (Dielectric)	Bound	Glass Plastic	10^{-22}–10^{-10} /ohm-cm
3. Semiconductor a. Intrinsic	Some Available	Germanium Silicon III-IV	10^{-9}–10^3 /ohm-cm
b. Doped	Controlled Amount Available	N-type Semiconductor P-type Semiconductor	

Figure 2.12 Electrical classification of materials.

	N-TYPE	P-TYPE
1. Conduction	Electrons	Holes
2. Polarity	Negative	Positive
3. Dopant Term	Donor	Acceptor
4. Doping Elements in Silicon	Arsenic Phosphorus Antimony	Boron

Figure 2.13 Characteristics of doped semiconductors.

The electrical characteristics of conductors, insulators, and semiconductors are summarized in Fig. 2.12. The particular characteristics of doped semiconductors are summarized in Fig. 2.13.

N- and P-type conditions are also created in germanium and compound semiconductors with specific dopant elements.

Carrier mobility

It was mentioned in the section called "Electron and hole conduction" that it takes less energy to move an electron than a hole through a piece of semiconducting material. In a circuit we are interested in both the energy required to move these carriers (holes and electrons) and the speed at which they move. The speed of movement is called the carrier mobility, with holes having a lower mobility than electrons. This factor is an important consideration in selecting a particular semiconducting material for a circuit.

Semiconductor Production Materials

Germanium and silicon

Germanium and silicon are the two elemental semiconductors. The first transistor was made with germanium as were the initial devices of the solid-state era. However, germanium presents problems in processing and in device performance. Its 937°C melting point limits high-temperature processing. More importantly, its lack of a naturally occurring oxide leaves the surface prone to electrical leakage.

The development of silicon–silicon dioxide planar processing solved the leakage problem of integrated circuits, flattened the surface profile of the circuits, and allowed higher-temperature processing due to its 1415°C melting point. Consequently, silicon represents over 90 percent of the wafers processed worldwide.

Compound semiconducting materials

There are many semiconducting compounds formed from elements listed in columns III and IV or II and VI of the periodic table. Of these compounds, the ones used most often in commercial semiconductor devices are gallium arsenide (GaAs) and gallium arsenide phosphide (GaAsP). These compounds have special properties.[1] Unlike silicon or germanium, diodes made from GaAs and GaAsP give off visible light and laser light when activated with an electrical current. These compounds are used to make the light-emitting diodes (LEDs) in electronic panel displays.

An important property of gallium arsenide is its high carrier mobility. This property allows a gallium arsenide device to react to high-frequency microwaves and effectively switch them into electrical currents in communications systems. In a silicon device, the carriers move more slowly. This means that when stimulated by a microwave, a silicon device cannot react to a wave quickly enough before the next wave impinges upon the device.

This same property, carrier mobility, is the basis for the excitement over gallium arsenide transistors and ICs. Devices of GaAs operate two to three times faster than comparable silicon devices and are the devices of choice for superfast computers and real-time control circuits such as airplane controls.

GaAs also offers two other significant advantages over silicon. It has a natural resistance to radiation-caused leakage. This phenomenon is common to all semiconductors. Radiation, such as that found in space, causes holes and electrons to form in semiconductor materials. These electrons and holes give rise to unwanted currents that can cause the device or circuit to malfunction or cease functioning. Devices that can perform in a radiation environment are known as *radiation hardened*. GaAs is naturally radiation hardened.

Also, GaAs is semi-insulating. In an integrated circuit this property minimizes leakage between adjacent devices, allowing a higher packing density, which in turn results in a faster circuit since the holes and electrons travel shorter distances. In silicon circuits special isolating structures must be built into the surface to control surface leakage. These structures take up valuable space and reduce the density of the circuit.

These advantages make GaAs the basic material for most supercomputers and in communication systems requiring high-speed switching. Despite all of the advantages, GaAs is not expected to replace silicon as the mainstream semiconducting material, because of its tradeoffs in favor of performance versus processing difficulty. While GaAs circuits are very fast, the majority of electronic products do not require their level of speed. GaAs, like germanium, does not possess a natural oxide. To compensate for the lack of a naturally grown oxide, layers of dielectrics must be deposited on the GaAs, which leads to longer processing and lower yields. Also, half of the atoms in GaAs are arsenic, a element very dangerous to human beings. Unfortunately, the arsenic evaporates from the compound at normal process temperatures, requiring the addition of suppression layers (caps) or pressurized process chambers. These steps lengthen processing time and add to cost.

The evaporation of arsenic also occurs during the crystal-growing stage, resulting in nonuniform crystals and wafers that are very prone to breakage during fab processing. Also, the production of large diameter GaAs wafers has lagged behind that of silicon (see Chap. 3). Despite its problems, gallium arsenide is an important semiconducting material and will continue to increase in use. It will probably have a major influence on computer performance of the future. A comparison of the major semiconducting production materials and silicon dioxide is presented in Fig. 2.14.

Other compound semiconductors

The speed of GaAs circuits is due to a parameter called *band gap*, the amount of energy required to move an electron from one shell (orbit) to another. The smaller the band gap, the lower the energy required, and, therefore, the faster a device can switch signals.

There are a number of compounds that offer this advantage for special device use. They are composed of elements from columns II and VI, III and V, or IV and VI of the periodic table.

	Ge	Si	GaAs	SiO$_2$
Atomic Weight	72.6	28.09	144.63	60.08
Atoms/cm^3 or Molecules	4.42×10^{22}	5.00×10^{22}	2.21×10^{22}	2.3×10^{22}
Crystal Structure	Diamond	Diamond	Zinc-Blends	Amorphous
Atoms/Unit Cell	8	8	8	—
Density	5.32	2.33	5.65	2.27
Energy Gap	0.67	1.11	1.40	8 (approx.)
Dielectric Constant	16.3	11.7	12.0	3.9
Melting Point (°C)	937°	1415°	1238°	1700° (approx.)
Breakdown Field (V/)	8 (approx.)	30 (approx.)	35 (approx.)	600 (approx.)
Linear Coefficient of Thermal Expansion $\frac{\Delta L}{LT} \quad \frac{1}{C}$	5.8×10^{-6}	2.5×10^{-6}	5.9×10^{-6}	0.5×10^{-6}

Figure 2.14 Physical properties of semiconductor materials.

Silicon germanium. Communication devices are being made in layers of silicon germanium, SiGe, by vacuum chemical vapor deposition processes. The combination increases transistor speeds to levels allowing ultrafast radios and personal communication devices.[2]

Process Chemicals

It should be fairly obvious that a lot of processing is required to change the raw semiconducting materials into useful devices. The majority of these processes use chemicals. In fact, microchip fabrication is primarily a chemical process, or more correctly, a series of chemical processes. Up to 20 percent of all process steps are cleaning or wafer surface preparation.[3]

Great quantities of acids, bases, solvents, and water are consumed by a semiconductor plant. This is partly due to the extremely high purity and special formulations required of the chemicals for precise and clean processing. Larger wafers and higher cleanliness requirements necessitate more automated cleaning stations and, therefore, the cost of removing spent chemicals is rising. Process chemicals can account for up to 40 percent of all chip manufacturing costs. (The cleanliness requirements for semiconductor process chemicals are explored in Chap. 4. Specific chemicals and their properties are detailed in the process chapters.)

Molecules, Compounds, and Mixtures

Earlier, the basic structure of matter was explained by use of the Bohr atomic model. This model is used to explain the structural differences of the elements that make up all the materials in the physical universe. But it is obvious that the universe contains more than 103 (the number of elements) types of matter. However, it is true that the great number of materials found in nature, and made by humans, are all based on those 103 elements.

The basic unit of nonelemental material is the molecule. For example, the basic unit of water is a molecule composed of two hydrogen atoms and one oxygen atom. The proliferation of materials comes about from the ability of atoms to bond together to form molecules.

It is inconvenient to draw diagrams such as in Fig. 2.15 every time we want to designate a molecule. The more common practice is to write the molecular formula. For water it is the familiar H_2O. This formula tells us exactly the elements and their number in the material. Chemists use the more precise term *compound* to describe different combinations of elements. Thus H_2O (water), NaCl (sodium chloride or salt), H_2O_2 (hydrogen peroxide), and As_2O_3 (arsine) are all different compounds composed of aggregates of individual molecules.

Some elements combine into *diatomic molecules*. A diatomic molecule is one composed of two atoms of the same element. The familiar process gases, oxygen, nitrogen, and hydrogen in their natural states, are all composed of diatomic molecules. Thus their formulas are O_2, N_2, and H_2.

Materials also come in two other forms, mixtures and solutions. Mixtures are composed of two or more substances, but the substances

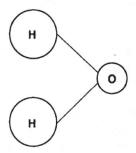

Figure 2.15 Diagram of water molecule.

retain their individual properties. A mixture of salt and pepper is the classic example. *Solutions* are composed of a solid dissolved in a liquid. In the liquid the solids are interspersed, with the solution taking on unique properties. However, the substances in a solution do not form into a new molecule. Salt water is an example of a solution. It can be separated back into its original parts: salt and water.

Ions

The term *ion* or ionic is used often in connection with semiconductor processing. This term refers to any atom or molecule with an unbalanced charge that exists in a material. An ion is designated by the chemical symbol of the element or molecule followed by a superscripted positive or negative sign (e.g., Na^+, Cl^-). One of the serious contamination problems in chip processing is mobile ionic contamination, such as that caused by sodium (Na^+). The problem comes from the positive charge carried by the sodium when it gets into the semiconductor material or device. In some processes, such as ion implantation, it is necessary to create an ion, such as boron (B^+), to accomplish the process.

States of Matter

Matter is found in four different states. They are solid, liquid, gas, and plasma (Fig. 2.16).

Solids, liquids, and gases

- Solids are defined as having definite shape and volume that are retained under normal conditions of temperature and pressure.

- Liquids have definite volume but a variable shape. A liter of water will take the shape of any container it is put in.

- Gases have neither definite shape nor volume. They, too, will take the shape of any container; but unlike liquids, they can expand or can be compressed to entirely fill the container.

```
• Solid
• Liquid
• Gas
• Plasma
```

Figure 2.16 Four states of nature.

The state of a particular material is related to its pressure and temperature. Temperature is a measure of the total energy incorporated in the material. We know that water can exist in all three states (ice, liquid water, and steam or water vapor) simply by changing the temperature (and/or pressure). The influence of pressure is more complicated and beyond the scope of this text.

Plasma state

The fourth state of nature is plasma. A star is an example of a plasma state. It certainly does not meet the definition of a solid, liquid, or gas. A plasma state is defined as a high energy collection of ionized atoms or molecules. Plasma states can be induced in process gases by the application of high energy radio frequency (RF) fields. They are used in semiconductor technology to cause chemical reactions in gas mixtures. One of their advantages is that energy can be delivered at a lower temperature than in conventional thermal systems, such as convection heating ovens.

Properties of Matter

All materials can be differentiated by their chemical compositions and the properties that arise from those compositions. In this section, several key properties of semiconductor materials and process chemicals are defined.

Temperature

The temperature of a chemical exerts great influence on its reactions with other chemicals. Additionally, safe use of some chemicals requires knowledge and control of their temperatures. Three temperature scales are used to express the temperature of a material: Fahrenheit, Celsius, and Kelvin (Fig. 2.17).

The Fahrenheit scale was developed by Gabriel Fahrenheit, a German physicist, using a water and salt solution. He assigned to the solution's freezing temperature the value of zero degrees Fahrenheit (0°F). Unfortunately, the freezing temperature of pure water is more useful, and we have ended up with the Fahrenheit scale having a water freezing point at 32°F and a boiling point of 212°F, with 180 degrees between the two points.

The Centigrade is more popular in scientific endeavors. It more sensibly sets the freezing point of water at 0°C and boiling at 100°C. Note that there is exactly 100°C between the two points. This means that a one-degree change in temperature as measured on the Celsius scale requires more energy than a one-degree change on the Fahrenheit scale.

Figure 2.17 Temperature scales.

The third is the Kelvin scale. It uses the same scaling factor as the Centrigrade scale but is based on absolute zero. *Absolute zero* is the theoretical temperature at which all atomic motion would cease. This value corresponds to $-273°C$. On the Kelvin scale water freezes at 273 K and boils at 373 K.

Density, specific gravity, and vapor density

An important property of matter is density. When we say that something is *dense,* we refer to its mass or weight per unit volume. A cork has a lower density than an equal volume of iron. Density is expressed as the weight, in grams, per cubic centimeter (cm^3) of the material. Water is the standard with 1 cm^3 (at 4°C) weighing 1 g. The densities of other substances are expressed as a ratio of their density to that of a comparable volume of water. Silicon has a density of 2.3. Therefore a piece of silicon 1 cm^3 in volume will weigh 2.3 g.

Specific gravity is a term used to refer to the density of liquids and gases at 4°C. It is the ratio of the density of the substance to that of water. Gasoline has a specific gravity of 0.75, which means it is 75 percent as dense as water.

Vapor density is a density measurement of gases under certain conditions of temperature and pressure. The reference is air, with 1 cm^3 having an assigned density of 1. Hydrogen has a vapor density of 0.60, which makes it 60 percent the density of a similar volume of air. A container of hydrogen will weigh 60 percent less than a similar container of air.

Pressure and vacuum

Another important aspect of matter is pressure. Pressure, as a property, is usually used in connection with liquids and gases. It is defined as the force per unit area exerted against the surface of the container.

It is the gas pressure in a cylinder that forces the gas out into a process chamber. All process machines using gases must have gauges to measure and control the pressure.

Pressures are expressed in pounds per square inch of area (psia), atmospheres, or torr. An atmosphere (atm) is the pressure exerted by the atmosphere surrounding the earth at a specific temperature. Thus a high-pressure oxidation system operated at 5 atm contains a pressure 5 times that of the atmosphere.

One atmosphere of air has a pressure of 14.7 psia. Pressures inside gas tanks are measured in psig or pounds per square inch gauge. This means that the gauge reading is absolute; it does not include the pressure of the outside atmosphere.

Vacuum is also a term and condition encountered in semiconducting processing. Vacuum is actually a condition of low pressure. Generally pressures below standard atmospheric pressures are referred to as vacuum. But a vacuum condition is measured in units of pressure.

Low pressures tend to be expressed in torr. The unit is named after the Italian scientist, Torricelli, who made many of the early discoveries in the field of gases and their properties. A torr is defined as the equivalent of 1 mm of mercury in a pressure-measuring device known as a *manometer.*

Imagine the effect on the column of mercury in the manometer in Fig. 2.18a of increasing the pressure above atmospheric pressure. As the pressure goes up, it pushes down the mercury in the dish and raises the mercury in the column. Now imagine what happens as air is extracted from the system (Fig. 2.18b) below atmospheric pressure, creating a vacuum. As long as there are any gas molecules or atoms in the manometer, some small pressure will be exerted on the mercury in the dish and the mercury in the column will rise some small but finite amount. The amount of the rise as measured in millimeters (mm) is relative to the pressure, or in this case the vacuum.

Figure 2.18 Pressure vacuum measurement.

Vacuum systems for evaporation, sputtering, and ion implantation are operated at vacuums (pressures) of 10^{-6} to 10^{-9} torr. Translated into a vacuum system containing a simple manometer, this means that the column of mercury would rise only 0.000000001 (1×10^{-9}) to 0.000001 (1×10^{-6}) mm, a *very* small height. In actual practice, a mercury manometer cannot measure these extremely low pressures. Other, more sensitive gauges are used.

Acids, Alkalis, and Solvents

Acids and alkalis

Semiconductor processing requires large amounts of liquid chemicals to etch, clean, and rinse the wafers and packages. These chemicals are divided by chemists into three major classifications.[4]

- Acids
- Alkalis
- Solvents

Acids and alkalis differ from each other due to the presence of specific ions in the liquid. Acids contain *hydrogen ions* while alkalis (also called bases) contain *hydroxyl ions*. An examination of the water molecule explains the differences.

The chemical formula for water normally is written as H_2O. It can also be written in the form HOH. When separated into its parts, we find that water is made up of a positively charged hydrogen ion (H^+) and a negatively charged hydroxide ion (OH^-). When water is mixed with other elements, either the hydrogen or hydroxyl ion combines with other substances (Fig. 2.19). Liquids that contain the hydrogen ion are called acids. Liquids that contain the hydroxyl ion are called alkalis or bases. Acids and bases are commonly found in the home; lemon juice and vinegar are acids, and ammonia and baking soda in a solution of water are bases.

```
H₂O = HOH = H⁺  +  OH⁻
                |       |
              Acids   Bases
                |       |

              HCl     NaOH
              HNO₃    KOH
              HF      NH₄OH
```

Figure 2.19 Acid and base solutions.

Acids are further divided into two categories, organic and inorganic. Organic acids are those that contain hydrocarbons, while inorganic acids do not. Sulfonic acid is an organic acid, and hydrogen fluoride (HF) is an inorganic acid.

The strength and reactivity of acids and bases are measured by the pH scale (Fig. 2.20). This scale ranges from 0 to 14, with 7 being a neutral point. Water is neutral, neither an acid nor a base; therefore, it has a pH of 7. Strong acids, such as sulfuric acid (H_2SO_4), have low pH values of 0 to 3. Strong bases, such as sodium hydroxide (NaOH), have pH values greater than 7. Both acids and bases are reactive with skin and other chemicals and should be stored and handled with all of the prescribed safety precautions.

Solvents

Solvents are liquids that do not ionize; they are neutral on the pH scale. Water is a solvent; in fact, it is the solvent with the greatest ability to dissolve other substances. Alcohol and acetone are commonly used solvents in the wafer-fabrication process. Most of the solvents in fabrication processing are volatile, flammable, or combustible. It is important to use them in properly exhausted stations and observe prescribed precautions in their storage and use.

Safety Issues

The storage, use, and disposal of chemicals, as well as electrical and other hazards, are present in semiconductor process areas. Companies

Figure 2.20   pH scale.

can address these risks by developing employee knowledge, skill, and awareness through training programs and safety inspections.

The material safety data sheet

For every chemical brought into a manufacturing site, the supplier must provide a form called the material safety data sheet (MSDS). It is required by the federal Occupational, Safety, and Health Act (OSHA). Also called OSHA Form 20, the MSDS contains chemical, storage, health, first aid, and use information about the chemical. Under current regulations, MSDSs must be filed on the site and be available to employees.

Key Concepts and Terms

Acids, alkalis, solvents	Intrinsic semiconductor
Bohr atomic model	Ion
Carrier mobility	MSDS
Conductivity and resistivity	Molecule and compound
Conductor	N type and P type
Dielectric	Periodic table of the elements
Dopant	pH scale
Doped semiconductor	Pressure and vacuum
Electron, proton, neutron	Silicon and germanium
Element	Solid, liquid, gas
Gallium arsenide	Temperature scales
Hole	III-V compounds

Review Questions

1. Describe the electrical differences between a conductor, a dielectric, and a semiconductor.

2. Why are doped semiconductors required for solid-state devices?

3. Which has a higher resistivity, a metal or an intrinsic semiconductor?

4. Give two reasons silicon is the most common semiconducting material.

5. Name one element that will make silicon N type.

6. A P-type semiconductor exhibits a _____ (negative or positive) current.

7. Indicate which classification (acid, base, solvent) is described below.

   a. Contains OH⁻ ions

   b. Contains H⁺ ions

   c. Is neutral on the pH scale

8. Acids have a pH between _____ and _____ .

9. Bases have a pH between _____ and _____ .

10. What is the pH of water?

References

1. R. E. Williams, *Gallium Arsenide Processing Techniques*, Artech House, Inc., Dedham, MA, 1984.
2. "Industry News," *Semiconductor International*, March 1994, p. 22.
3. R. Allen et al., "MNST Wafer Cleaning," *Solid State Technology*, January 1994, p. 61.
4. P. Van Zant, *Safety First Manual*, Semiconductor Services, Redwood City, Calif., 1984, Module 4.

3

Crystal Growth and Wafer Preparation

Overview

In this chapter, the conversion of the intrinsic semiconducting materials into raw wafers and the steps required to produce wafer fabrication-grade polished wafers are explained.

Objectives

1. Explain the difference between crystalline and noncrystalline materials.
2. Describe the conditions that characterize a polycrystalline and a single-crystal material.
3. Draw a diagram of the two major wafer crystal orientations used in semiconductor processing.
4. Explain the Czochralski, the float-zone, and the liquid crystal-encapsulated Czochralski methods of crystal growing.
5. Draw a flow diagram of the wafer-preparation process.
6. Explain the use and meaning of the flats ground on wafers.
7. Describe the benefits in the wafer-fabrication process that come from edge-rounded wafers.
8. Describe the benefits in the wafer-fabrication process that come from flat and damage-free wafer surfaces.

Semiconductor Materials Preparation

High-performance solid-state devices require extremely clean process-ing materials, including the starting wafer. The first stage of semicon-ductor manufacturing is the extraction and purification of the raw semiconductor material(s) from the earth. The material must meet electronic or semiconductor specifications, which generally means that impurities have been reduced to no more than several parts per million (ppm) or less.

Actual purification starts with a chemical reaction. For silicon, it is the conversion of the ore to a silicon-bearing gas such as silicon tetra-chloride or trichlorosilane. Solid contaminants, such as other metals, are left behind in the ore remains. The silicon-bearing gas is then reacted with hydrogen (Fig. 3.1) to produce semiconductor-grade sili-con. The silicon produced has a crystalline structure, which is a nec-essary material requirement for making semiconducting devices.

Crystalline Materials

Different materials differ from one another in many ways. One way is the organization of the atoms in the material. In some materials, such as silicon and germanium, the atoms occupy very definite positions relative to each other. These positions are repeated thoughout the material. Such materials are called *crystals*. Materials without a defi-nite periodic arrangement of their atoms are called noncrystalline or *amorphous*. Plastics are examples of amorphous materials.

Unit Cells

There are actually two levels of atomic organization possible for crys-talline materials. First is the organization of the individual atoms. The atoms in a crystal arrange themselves at specific points in a structure known as a *unit cell*. The unit cell is the first level of organi-zation in a crystal. The unit cell structure is repeated everywhere in the crystal.

Another term used to reference crystal structures is *lattice*. A crys-talline material is said to have a specific lattice structure, and the atoms are located at specific points in the lattice structure.

$$2SiHCl_3 \text{ (gas)} + 3H_2 \text{ (gas)} \rightarrow 2Si \text{ (solid)} + 6HCl \text{ (gas)}$$

Figure 3.1 Hydrogen reduction of trichlorosilane.

The number of atoms, relative positions, and binding energies between the atoms in the unit cell give rise to many of the characteristics of the material. Each crystalline material has a unique unit cell. Silicon atoms have 16 atoms arranged into a diamond structure (Fig. 3.2). GaAs crystals have 18 atoms in a unit cell configuration called a *zinc blende* (Fig. 3.3).

Polycrystals and Single Crystals

The second level of organization within a crystal is related to the organization of the unit cells. In intrinsic semiconductors, the unit cells are *not* in a regular arrangement to each other. The situation is similar to a disorderly pile of sugar cubes, with each cube representing a unit cell. A material with such an arrangement has a polycrystal structure. The second level of organization occurs when the unit cells (sugar cubes) are all neatly and regularly arranged relative to each of the others (Fig. 3.4). Materials thus arranged have a single- (or mono-) crystal structure.

Single-crystal materials have more uniform and predictable properties than polycrystalline materials. The structure allows a uniform and predictable electron flow in semiconductors. At the end of the fabrication process, crystal uniformity is essential for separating the wafer into die with nonragged edges (see Chap. 18).

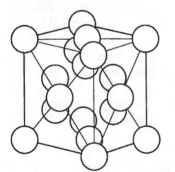

Figure 3.2 Unit cell of silicon.

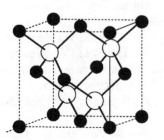

Figure 3.3 GaAs crystal structure.

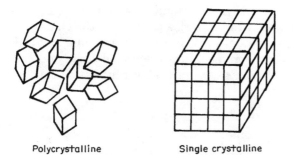

Polycrystalline   Single crystalline

Figure 3.4 Poly- and single-crystal structures.

Crystal Orientation

Besides the requirement of a single-crystal structure for a wafer, there is the requirement of a specific crystal orientation. This concept can be visualized by considering slicing the single-crystal block in Fig. 3.4. Slicing it through the vertical plane would expose one plane, while slicing it corner to corner would expose a different plane.

Wafers can be cut from the crystal at different angles, with each angle representing a different plane. Each of these slices through the crystal would also cut through each of the unit cells, exposing a specific plane within each cell. Each plane is unique, differing in atom count and binding energies between the atoms. These differences result in different chemical, electrical, and physical properties associated with each plane. A specific crystal plane must be specified for the wafers.

The planes are identified by a series of three numbers known as *Miller Indices*. Calculation of Miller indices is an exercise in geometry beyond the scope of this book. However, the reader should know that each set of Miller indices identifies the location of a specific plane in the crystal. Figure 3.5 shows two simple cubic unit cells nestled into the origin of an *XYZ* coordinate system.

The two most common orientations used for silicon wafers are the <100> and the <111> planes. The plane designations are verbalized as the one-oh-oh plane and the one-one-one plane. The brackets indicate that the three numbers are Miller indices.

<100> oriented wafers are used for fabricating metal oxide semiconductor (MOS) devices and circuits, while the <111> oriented wafers are used for bipolar devices and circuits. GaAs wafers are also cut along the <100> planes of the crystal. In actual practice, the large crystals are grown along the specific crystal planes as explained in the following section on crystal growth.

Note that the <100> plane in Fig. 3.6 has a square shape, while the <111> plane is triangular in shape. These orientations are

Figure 3.5 Crystal planes.

Figure 3.6 Wafer orientation indicators.

revealed when wafers are broken as shown in Fig. 3.6. The <100> wafers break into quarters or with right-angle (90°) breaks. The <111> wafers break into triangular pieces.

The crystal orientation can be determined by etching, x-ray diffraction, or light reflection (see section on crystal orientation, conductivity, and resistivity check).

Crystal Growth

Semiconductor wafers are cut from large crystals of the semiconducting material. These crystals, also called ingots, are grown from chunks of the intrinsic material that have a polycrystal structure and are undoped. The process of converting the polycrystal chunks to a large crystal of single-crystal structure, with the correct orientation, and containing the proper amount of dopant, is called *crystal growing*.

Three different methods are used to grow crystals, the Czochralski, liquid crystal-encapsulated czochralski, and float-zone techniques.

Czochralski (CZ) method

The majority of silicon crystals are grown by the CZ method (Fig. 3.7). The equipment consists of a quartz (silica) crucible heated by surrounding coils that carry radio-frequency (RF) waves. The crucible is loaded with polycrystalline chunks of the semiconductor material and small amounts of dopant. The dopant material is selected to create either an N-type or a P-type crystal. First the polycrystalline chunks and dopants are heated to the liquid state (Fig. 3.8). Next a seed crystal is positioned to just touch the surface of the liquid material (called

Figure 3.7 Czochralski crystal-growing system.

Figure 3.8 Crystal growth from a seed.

the *melt*). The seed is a small crystal that has the same orientation as the desired finished larger crystal. Seeds can be produced by chemical vapor techniques. In practice, they are pieces of previously grown crystals and are reused.

Crystal growth starts as the seed is slowly raised above the melt. The surface tension between the seed and the melt causes a thin film of the melt to adhere to the seed and cool. During the cooling, the atoms in the melted semiconductor material orient themselves to the crystal structure of the seed. The net effect is that the crystal orientation of the seed is propagated in the growing crystal. The dopant atoms in the melt become incorporated into the growing crystal, creating an N- or P-type crystal.

To achieve doping uniformity, crystal perfection, and diameter control, the seed and crucible (along with the pull rate) are rotated in opposite directions during the entire crystal-growing process. Process control requires a complicated feedback system integrating the parameters of rotational speed, pull speeds, and melt temperature. The crystal is pulled in three sections. First a thin neck is formed, followed by the body of the crystal, and ending with a blunt tail. The CZ method is capable of producing crystals several feet in length and with diameters up to 12 or more in.

Liquid crystal-encapsulated Czochralski (LEC) method

LEC crystal growing[1] is used for the growing of gallium arsenide crystals. It is essentially the same as the standard CZ process, but with a major modification for gallium arsenide. The modification is required because of the evaporative property of the arsenic in the melt. At the crystal-growing temperature, the gallium and arsenic react and the arsenic can evaporate, resulting in a nonuniform crystal.

Two solutions to the problem are used. One is to pressurize the crystal-growing chamber to suppress the evaporation of the arsenic. The other is the LEC process (Fig. 3.9). LEC uses a layer of boron trioxide (B_2O_3) floating on top of the melt to suppress the arsenic evaporation. In this method, a pressure of about one atm is required in the chamber.

Float-zone method

Float-zone crystal growth is one of several processes explained in this text that were developed early in the history of the technology and are still used for special needs. A drawback to the CZ method is the inclusion of oxygen from the crucible into the crystal. For some

Seed

Crystal

Boron
Trioxide
Layer

Gallium Arsenide
Melt

Figure 3.9 LEC system of crystal growth.

devices, higher levels of oxygen are intolerable. For these special cases, the crystal might be grown by the float-zone technique, which produces a lower oxygen-content crystal.

Float-zone crystal growth (Fig. 3.10) requires a bar of the polysilicon and dopants that have been cast in a mold. The seed is fused to one end of the bar and the assemblage placed in the crystal grower. Conversion of the bar to a single-crystal orientation starts when an RF coil heats the interface region of the bar and seed. The coil is then moved along the axis of the bar, heating it to the liquid point, a small section at a time. Within each molten region, the atoms align to the orientation started at the seed end. Thus the entire bar is converted to a single crystal with the orientation of the starting seed.

Float-zone crystal growing cannot produce the large diameters possible with the CZ process and the crystals have a higher dislocation density. But the absence of a silica crucible yields higher-purity crystals with lower oxygen content. Lower oxygen yields crystals with higher purity that are used in power thyristors and rectifiers. The two methods are compared in Fig. 3.11.

Crystal and Wafer Quality

Semiconductor devices require a high degree of crystal perfection. But even with the most sophisticated techniques, a perfect crystal is unobtainable. The imperfections, called *crystal defects,* result in process problems by causing uneven silicon dioxide film growth, poor epitaxial film depositions, uneven doping layers in the wafer, and other problems. In finished devices, the crystal defects cause unwanted current leakage and may prevent the devices from operating at required voltages. There are three major categories of crystal defects:

Figure 3.10 Float-zone crystal-growing system.

PARAMETER	CZ	FLOAT ZONE
Large Crystal	Yes	Difficult
Cost	Lower	
Dislocations	$0 - 10^4/cm^2$	$10^3 - 10^5/cm^2$
Resistivity	Up to 100 ohm-cm	2000 ohm-cm Max.
Radial Resistivity	5 – 10%	5 – 10%
Oxygen Content	$10^{16} - 10^{18}$ atoms/cm³	0 – Very Low

Figure 3.11 Comparison of CZ and float crystal-growing methods.

1. Point defects

2. Dislocations

3. Growth defects

Point defects

Point defects come in two varieties. One comes when contaminants in the crystal become jammed in the crystal structure, causing strain.

Figure 3.12 Vacancy crystal defect.

The second is known as a *vacancy*. In this situation, there is an atom missing from a location in the structure (Fig. 3.12).

Vacancies are a natural phenomenon that occur in every crystal. Unfortunately, vacancies occur whenever a crystal or wafer is heated and cooled, such as in the fabrication process. The minimization of vacancies is one of the driving forces behind the desire for low-temperature processing.

Dislocations

Dislocations are a misplacement of the unit cells in a single crystal. They can be imagined as an orderly pile of sugar cubes with one of the cubes slightly out of alignment with the others.

Dislocations occur from growth conditions and lattice strain in the crystal. They also occur in wafers from physical abuse during the fabrication process. A chip or abrasion of the wafer edge serves as a lattice strain site that can generate a line of dislocations that progresses into the wafer interior with each subsequent high-temperature processing of the wafer. Wafer dislocations are revealed by a special etch of the surface. A typical wafer has a density of 200 to 1000 dislocations per square centimeter.

Etched dislocations appear on the surface of the wafer in shapes indicative of their crystal orientation. <111> wafers etch into triangular dislocations and <100> wafers show "squarish" etch pits (Fig 3.6).

Growth defects

During crystal growth, certain conditions can result in structural defects. One is *slip*, which refers to the slippage of the crystal along crystal planes (Fig. 3.13). Another problem is *twinning*. This is a situation where the crystal grows in two different directions from the same interface. Both of these defects are cause for rejection of the crystal.

Figure 3.13   Crystal slip.

Wafer Preparation

End cropping

After removal from the crystal grower, the crystal goes through a series of steps that result in the finished wafer. First is the cropping off of the tapered crystal ends with a saw.

Diameter grinding

Despite the efforts made to control the crystal diameter during growth, there is a variation over the length of the crystal (Fig. 3.14). This variation is unacceptable for the final wafers. In the fabrication process, the wafers will be stored and handled in a variety of holders that require a consistent specific diameter. Poor diameter control can result in warped and broken wafers.

Diameter grinding is a mechanical operation performed in a centerless grinder. This machine grinds the crystal to the correct diameter without the necessity of clamping it into a lathe-type grinder with a fixed center point.

Crystal orientation, conductivity, and resistivity check

Before the crystal is submitted to the wafer-preparation steps, it is necessary to determine if it meets orientation and resistivity specifications. The crystal orientation (Fig 3.15) is determined by either x-ray diffraction or collimated light refraction. In both methods, an end of the crystal

Before   After

Figure 3.14   Crystal diameter grinding.

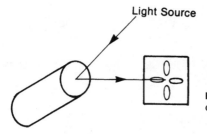

Figure 3.15 Crystal orientation determination.

is etched or polished to remove saw damage. Next, the crystal is mounted in the refraction apparatus and the x rays or collimated light are
reflected off the crystal surface onto a photographic plate (x rays) or
screen (collimated light). The pattern formed on the plate or screen is
indicative of the crystal plane (orientation) of the grown crystal. The
pattern shown in Fig 3.15 is representative of a <100> orientation.

Most crystals are grown several degrees off the major <111> or
<100> plane. This off-orientation is required for several reasons, par-
ticularly ion implantation processing. The reasons are covered in the
applicable process chapters.

During the orientation measurement, the crystal is sitting in a
layer of glue or wax on a cutting block. The crystal is glued and positioned on the block to ensure that the wafers are cut from the crystal
at the correct orientation.

Since each crystal is doped, an important electrical check is conductivity type (N or P) to ensure that the right dopant type was used. A
hot-point probe connected to a polarity meter is used to generate
holes or electrons (depending on the type) in the crystal. The conductivity type is displayed on the meter.

The amount of dopant put into the crystal is determined by a resistivity measurement using a four-point probe. See Chap. 13 for a
description of this measurement technique. The curves presented in
Chap. 2 (Fig. 2.7) show the relationship between resistivity and doping concentration for N- and P-type silicon.

The resistivity is checked along the axis of the crystal due to dopant
variation during the growing process. This variation results in wafers
that fall into several resistivity specification ranges. Later in the
process, the wafers will be grouped by resistivity range to meet customer specifications.

Flat grinding

Once the crystal is oriented on the cutting block, a flat is ground
along the axis (Fig. 3.16). This flat will show up on each of the wafers

Figure 3.16 Crystal flat grinding.

and is called the *major flat*. The position of the flat is along one of the major crystal planes, as determined by the orientation check.

In the fabrication process, the flat functions as a visual reference to the orientation of the wafer. It is used to place the first pattern mask on the wafer so that the orientation of the chips is always to a major crystal plane.

On most crystals, there is a second, smaller, secondary flat ground on the edge. The position of the secondary flat to the major flat is a code that tells the wafer-fabrication department both the orientation and conductivity type of the wafer. The code is shown in Fig. 3.17.

Wafer slicing

The wafers are sliced from the crystal with the use of diamond-coated, inside-diameter saws (Fig. 3.18). These saws are thin circular

Figure 3.17 Wafer flat locations.

Figure 3.18 Inside-diameter saw wafer slicing.

sheets of steel with a hole cut out of the center. The inside of the hole is the cutting edge and is coated with diamonds. An inside-diameter saw is rigid but thin. This reduces the kerf (cutting width) size, which in turn prevents the waste of sizable amounts of the crystal by the slicing process.

For use, the saw is tightened onto a circular frame much like a musical drumhead and is secured around the circumference of the drum. This type of saw gives the rigidity required to produce thin wafers with flat and parallel surfaces. A typical 150-mm-diameter wafer will be about 25 mils in diameter.

Rough polish

The surface of a semiconductor wafer has to be free of irregularities and saw damage and be absolutely flat. The first requirements come from the very small dimensions of the surface and subsurface layers making up the device. They have dimensions in the 0.5- to 2-μm range. To get an idea of the relative dimensions of a semiconductor device, imagine the cross section in Fig. 3.19 is as tall as a house wall, about 8 ft. On that scale, the working layers on the top of the wafer would all exist within the top 1 or 2 in of the wall.

Flatness is an absolute requirement for small-dimension patterning (see Chap. 11). The advanced patterning processes project the required pattern image onto the wafer surface. If the surface is not flat, the projected image will be distorted just as a film image will be out of focus on a nonflat screen.

The flatting and polishing process proceeds in two steps, rough polish and chemical-mechanical polishing (Fig. 3.20). Rough polishing is

Figure 3.19 Cross section of MOS transistor.

Figure 3.20 Abrasive and chemical-mechanical surface polishing.

a conventional, abrasive, slurry-lapping process, but fine-tuned to semiconductor requirements. A primary purpose of the rough polish is to remove the surface damage leftover from the wafer-slicing process.

Chemical-mechanical polishing (CMP)

The final polishing step is a combination of chemical etching and mechanical buffing. The wafers are mounted on rotating holders and lowered onto a pad surface rotating in the opposite direction. The pad material is generally a cast and sliced polyurethane with a filler or a urethane-coated felt. A slurry of silica (glass) suspended in a mild etchant, such as potassium or ammonium hydroxide, is dropped onto the polishing pad.

The alkaline slurry chemically grows a thin layer of silicon dioxide on the wafer surface. The buffing action of the pad mechanically removes the oxide in a continuous action. High points on the wafer surface are removed until an extremely flat surface is achieved. If a semiconductor wafer surface was extended to 10,000 ft (the length of an airport runway), it would vary only about ± 2 in over its entire length.

Achieving this extreme flatness requires the specification and control of the polishing time, the pressure on the wafer and pad, the speed of rotation, the slurry particle size, the slurry feed rate, the chemistry (pH) of the slurry, and the pad material and conditions.

Chemical-mechanical polishing (CMP) is one of the techniques developed by the industry that has allowed production of larger wafers. CMP is also used to planarize wafers in the wafer-fabrication process. In this application, wafers with a variegated top surface are flattened to allow accurate pattern imaging. (See Chap. 10 for a detailed explanation of this use of CMP.)

Backside processing. In most cases, only the front side of the wafer goes through the extensive chemical-mechanical polishing. The backs may be left rough or may be etched to a bright appearance. For some device use, the backs may receive a special process to induce crystal damage, called *backside damage*. Backside damage causes the growth of dislocations that radiate up into the wafer. These dislocations can act as a trap of mobile ionic contamination introduced into the wafer during the fabrication process. The trapping phenomenon is called *gettering*.

Edge grinding. Edge grinding is a mechanical process that leaves the wafer with a rounded edge (Fig. 3.21). This rounding minimizes edge

Before   After

Figure 3.21   Wafer edge grinding.

chipping and damage during fabrication that can result in wafer breakage or serve as the nucleus for dislocation lines.

Wafer evaluation

Before packing, the wafers (or samples) are checked for a number of parameters as specified by the customer. Figure 3.22 illustrates a typical wafer specification for a semiconductor wafer. Of primary concern are surface problems such as particulates, stain, and haze. These problems are detected by the use of high-intensity lights or automated inspection machines.

Oxidation

Silicon wafers may be oxidized before shipment to the customer. The silicon dioxide layer serves to protect the wafer surface from scratches and contamination during shipping. Most companies start the wafer-fabrication process with an oxidation step, and buying the wafers with an oxide layer saves a manufacturing step. (Oxidation processes are explained in Chap. 7.)

Packaging

While much effort goes into producing a high-quality and clean wafer, the quality can be lost during shipment to the customer, or worse, from the packaging method itself. Therefore, there is a very stringent requirement for clean and protective packaging. The packaging materials are of non-static, non-particle-generating materials, and the equipment and operators are grounded to drain off static charges that

Typical 200 mm Wafer Specification	
Diameter	200mm
Diam. Tolerance	0.25mm
Thickness	725 +/- μ
Crystal Orientation	<100> +/- 1 degree
Resistivity	2.7-4 Ω-cm
Res. Gradient	10%
Oxygen	25-29 ppma
Oxygen Gradient	5%
Carbon	0.3 ppma

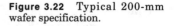

Figure 3.22   Typical 200-mm wafer specification.

attract small particles to the wafers. Wafer packaging takes place in clean rooms (see Chap. 4).

Increasing wafer diameters

The evolution of higher-density and larger-size chips has required the delivery of larger-diameter wafers. Starting with 1-in-diameter wafers in the 1960s, the industry now produces the 200-mm-diameter wafers that are considered the state of the art in the 1990s. The Semiconductor Industry Association's *Technology Roadmap* predicts that by 2007 wafer fabrication will require 400-mm-diameter wafers in production quantities.

Epitaxial silicon wafers

Despite the high quality of starting wafers, they are not adequate for the formation of complimentary metal-oxide semiconductor (CMOS) devices, which require a layer of epitaxially deposited silicon. Most major wafer suppliers have the capability to add the epitaxial layer to the wafer before delivery to the customer. (This device technology is discussed in Chap. 12 and 16.)

Key Concepts and Terms

Backside damage	Point defects
Chemical-mechanical polishing	Poly crystalline
Crystal	<100> plane
Crystal dislocations	<111> plane
Crystal growing	Seed crystal
Crystal orientation	Single crystal
Czochralski	Slip
Edge rounding	Unit cell
Float zone	Vacancy
Growth defects	Wafer flat code
LEC	Wafer flats
Melt	Wafer slicing

Review Questions

1. In a polycrystal structure, the atoms are not arranged (true/false).

2. In a single-crystal structure, the unit cells are not arranged (true/false).

3. Draw a cubic unit cell and identify the <100> plane.

4. <111> oriented wafers are used for _____ (bipolar, MOS) devices.

5. What is the orientation of a semiconductor crystal if the seed has a <100> orientation.

6. Draw a diagram of a CZ crystal grower and identify all the major parts.

7. During crystal growth, the molten material is changed from single-crystal structure to polycrystal structure (true/false).

8. Why are wafers edge rounded?

9. Draw a flow diagram of the wafer-preparation process.

10. Give two reasons why semiconductor wafers require a flat surface.

References

1. R. E. Williams, *Gallium Arsenide Processing Techniques,* Artech House, Inc., Dedham, MA, 1984, p. 37.
2. "Example Wafer Specifications," SILTEC Corporation, Salem, OR, 1996.

4

Contamination Control

Overview

In this chapter, the effects of contamination on device processing, device performance, and device reliability are detailed along with the types and sources of contamination found in a fabrication area. Clean-room strategies and major contamination-control procedures are explained.

Objectives

Upon completion of this chapter, you should be able to:

1. Identify the three major effects of contamination on semiconductor devices and processing.

2. List the major sources of contamination in a fabrication area.

3. Define the "class number" of a clean room.

4. List the particle density of class 100, 10, and 1 fabrication areas.

5. Describe the role of positive pressure, air showers, and adhesive mats in maintaining cleanliness levels.

6. List at least three techniques used to minimize contamination from fabrication personnel.

7. Identify the three contaminants present in "normal" water and their control in semiconductor plants.

8. Describe the differences between normal industrial chemicals and semiconductor-grade chemicals.

9. Name two problems associated with high static levels and two methods of static control.

The Problem

Semiconductor devices are very vulnerable to many types of contaminants. They fall into four major classes:

1. Particles

2. Metallic ions

3. Chemicals

4. Bacteria

Particles. Semiconductor devices, especially dense integrated circuits, are vulnerable to all kinds of contamination. The sensitivity is due to the small feature sizes and the thinness of deposited layers on the wafer surface. These dimensions are in the micron range. A micron or micrometer (μm) is very small. A centimeter contains 10,000 μm. Another way to envision a micron is that a human hair is about 100 μm in diameter (Fig. 4.1). The small physical dimensions of the devices make them very vulnerable to particulate contamination in the air, coming from workers, generated by the equipment, and present in processing chemicals (Fig. 4.2). As the feature size and films become smaller (Fig. 4.3), the allowable particle size must be controlled to smaller dimensions.

A rule of thumb is that the particle size must be 10 times smaller than the minimum feature size.[1] A 0.3-μm feature size device is vulnerable to 0.03-μm-diameter particles. Particles that locate in a critical part of the device and destroy its functioning are called *killer*

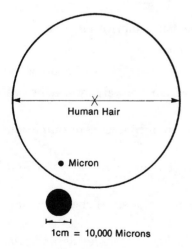

Figure 4.1 Relative size of one micron.

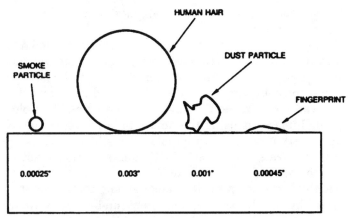

Figure 4.2 Relative size of contamination. (*Hybrid Microcircuit Technology Handbook*)

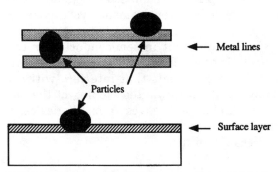

Figure 4.3 Relative size of airborne particles and wafer dimensions.

defects. Killer defects also include crystal defects and other process-induced problems. On any wafer there are a number of particles, some of which are the killer variety, but others, located in less sensitive areas, are harmless. Current state-of-the-art processing seeks to reduce the number of particles on a 200-mm wafer to no more than 10 particles, 0.3 μm in size.[2]

Metallic ions. In Chap. 2, it was established that semiconductor devices are fabricated through our ability to create the two conductivity types in wafers, control the resistivity level through doping, and create N-P junctions in the surface of the devices. These three properties are achieved by the purposeful introduction of specific dopants into the crystal and into the wafer. These desired effects are achieved with very small amounts of the dopants. Unfortunately, it only takes a small amount of certain electrically active contaminants in the wafer to alter electrical characteristics, changing device performance and reliability parameters.

The contaminants causing these types of problems are known as mobile ionic contaminants (MICs). They are atoms of metals that exist in the material in an ionic form. Further, these metallic ions are highly mobile in semiconductor materials. This mobility means that the metallic ions can move inside the device, even after passing electrical testing and shipping, causing the device to fail. Unfortunately, the metals (Fig. 4.4) that cause these problems in silicon devices are present in most chemicals. On wafers, MIC contamination must be less than 10^{10} atoms per square centimeter.[3]

Sodium is usually the most prevalent mobile ionic contaminant in most untreated chemicals and is the most mobile in silicon. Consequently, control of sodium is a prime goal in silicon processing. The MIC problem is most serious in MOS devices, a fact that has led some chemical suppliers to develop MOS or low-sodium-grade chemicals. These labels refer to low-mobile ionic contaminant levels.

Chemicals. The third major contaminant in semiconductor process areas is unwanted chemicals. Process chemicals and process water can be contaminated with trace chemicals that interfere with the wafer processing. They may result in unwanted etching of the surface, create compounds that cannot be removed from the device, or cause nonuniform processes. Chlorine is such a contaminant and is rigorously controlled in process chemicals.

Bacteria. Bacteria are the fourth major contaminant class. Bacteria are organisms that grow in water systems and on surfaces not cleaned regularly. Bacteria, once on the device, act as particulate con-

Trace Metal Impurity	Parts per billion (ppb)
Sodium	50
Potassium	50
Iron	50
Copper	60
Nickel	60
Aluminum	60
Manganese	60
Lead	60
Zinc	60
Chlorides	1000

Figure 4.4 Example resist stripper trace metal contents (*EKC Technology—830 Photoresist stripper*)

tamination or may contribute unwanted metallic ions to the device surface.

Contamination-caused problems

The four types of contaminants affect devices and device processing in three specific performance areas. They are:

1. Device processing yield
2. Device performance
3. Device reliability

Device processing yield

Device processing in a contaminated environment can cause a multitude of problems. Contamination may change the dimensions of device parts, change the cleanliness of the surfaces, and/or cause pitted layers. Within the fabrication process are a number of quality checks and inspections specifically designed to detect contaminated wafers. High levels of contamination result in fewer wafers completing the process and higher costs.

Device performance

A more serious problem is related to small pieces of contamination that may escape the in-process quality checks. Whereas the wafer may appear to be clean, undetected particulates, unwanted chemicals, and/or high levels of mobile ionic contaminants in the wafer can change the electrical performance of the devices. This problem usually shows up at the postfabrication electrical test.

Device reliability

Device reliability is the most insidious of the contamination failures. Small amounts of metallic contaminants can get into the wafer during processing and not be detected during normal device testing. However, in the field, these contaminants can travel inside the device and end up in electrically sensitive areas, causing failure. This failure mode is a primary concern of the space and defense industries.

In the rest of this chapter, the sources, nature, and control of the types of contamination that affect semiconductor devices are identified. With the advent of LSI level circuits in the 1970s, the control of contamination became essential to the industry. Since that time a great deal of knowledge about and control of contamination has been learned. Contamination control is now a discipline of its own and is one of the critical technologies that has to be mastered to profitably produce solid-state devices.

The remarks about contamination control in this chapter are requirements for fabrication areas, mask-making areas, and areas in which semiconductor equipment and materials are manufactured.

Contamination Sources

General sources

Contamination in a clean room is defined as anything that interferes with the production of the product and/or its performance. The stringent requirements of solid-state devices define levels of cleanliness that far exceed those of almost any other industry. Literally everything that comes in contact with the product during manufacture is a potential source of contamination. The major sources are:

1. Air

2. The production facility

3. Clean-room personnel

4. Process water

5. Process chemicals

6. Process gases

7. Static charge

Each source produces specific types and levels of contamination and requires special controls to render it acceptable in the clean room.

Air

Normal air is so laden with contaminants that it must be treated before entering a clean room. A major problem is airborne particles, referred to as *particulates* or *aerosols*. Normal air contains copious amounts of small dust and particles, as illustrated in Fig. 4.5. A major problem with small particles is that they "float" and remain in the air for long periods of time. These particles are called aerosols. Air cleanliness levels in clean rooms are identified by the particulate diameters and their density in the air.

Air quality is designated by the *class number* of the air in the area as defined in Federal Standard 209E.[4] This standard designates air

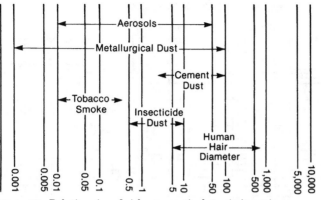

Figure 4.5 Relative size of airborne particulates (microns).

quality in the two categories of particle size and density. The class number of an area is defined as the number of particles 0.5 μm or larger in a cubic foot of air. The air in a typical city, filled with smoke, smog, and fumes, can contain up to 5 million particles per cubic foot, which is a class number of 5 million.

Figure 4.6 shows the relationship between particulate diameter and density as defined by Federal Standard 209E. Figure 4.7 lists the class numbers and associated particle size for various environments. Federal Standard 209E specifies cleanliness levels down to class 1 levels. While 209E defines class numbers at the 0.5-μm particle size, successful wafer fabrication processing requires even tighter controls. Engineers strive to achieve reduction of 0.3-μm particles in class 10 and 1 environments. Increasing chip density and decreasing feature size dictate a corresponding decrease in the class number of the processing environment. Specifications proposed by Semetech call for 64-Mbit memory process areas at class 0.1 and 256-Mbit air levels at class 0.01.[5]

Clean Air Strategies

The design of a clean room is integral to its ability to produce contamination-free wafers. A major consideration in the design is the maintenance of clean air in the process areas. Automation is also an important clean-room strategy for the reduction of contamination. This issue is explored in the equipment section and in Chap. 15. Four distinct clean-room design strategies are used:

1. Clean work stations
2. Tunnel design

Proposed Federal Standard 209E

Airborne Particulate Cleanliness Classes for Clean Rooms and Clean Zones

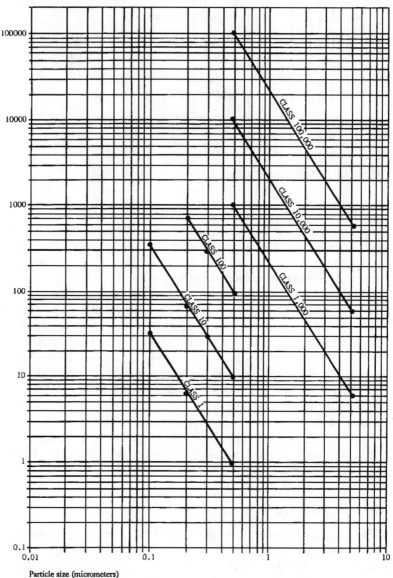

Particle size (micrometers)

Class limits in particles per cubic foot of size equal to or greater than particle sizes shown*

* The class limit particle concentrations shown are defined for class purposes only and do not
necessarily represent the size distribution to be found in any particular situation.

Figure 4.6 Air cleanliness classes standard 209E.

Environment	Class Number	Maximum particle size (micron)
Projected-256 Mbit	0.01	<<0.1
Mini environment	0.1	<0.1
ULSI Fab	1	0.1
VLSI Fab	10	0.3
VLF station	100	0.5
Assembly area	1000-10,000	0.5
House room	100000	
Outdoors	>500,000	

Figure 4.7 Typical class numbers for various environments.

3. Total clean room

4. Minienvironments

Clean-room work station strategy

Air-filtered production areas first became a need in the space industry. It was found that even single specks of dust could cause a satel-
lite to fail. Consequently, NASA and its contractors constructed
assembly rooms that featured ceiling and wall filters. They also developed much of the basic understanding of contamination control, from
operator gowning to clean materials.

The semiconductor industry adopted many of these techniques. One
problem, however, was that the cleanliness levels achieved in the
small rooms could not be maintained when the design was expanded
to larger fabrication areas with more production workers. The early
semiconductor industry evolved from a ceiling and wall filter concept
to a clean-room work station strategy. This strategy focused on individual work stations with air filters and nonshedding materials.
Outside the work stations, the wafers were stored and moved in covered boxes.

The fabrication area consisted of a large room with the work stations (called *hoods*) arranged in rows so that the wafers could move
sequentially through the process, never being exposed to dirty air.
The filters in these clean hoods are of a type called a high-efficiency
particulate attenuation (HEPA) filter. These filters are constructed of
fragile fibers with many small holes and folded into the filter holder
in an accordion design (Fig. 4.8). The high density of small holes and
large area of the filter medium allow the passage of large volumes of
air at low velocity. The low velocity contributes to the cleanliness of
the hood by not causing air currents. The low velocity is also necessary for operator comfort. A typical airflow is 90 to 100 ft/min.[6] HEPA

Figure 4.8   Hepa filter.

Figure 4.9   Cross section of VLF hood.

filters have a filtering efficiency of 99.99 percent plus and are the filters of choice in all three of the clean-room strategies.

Typically, a clean hood (Fig. 4.9) has a HEPA filter mounted in the top. Air is drawn from the room through a prefilter by a fan and forced through the HEPA filter. The air leaves the filter in a laminar pattern and at the work surface turns and exits the hood. A shield directs the exiting air over the wafers in the hood. The formal name for the work station is a vertical laminar flow (VLF) station. The term VLF is derived from the laminar nature of the airflow. Some work stations are designed with the HEPA filter in the back of the work surface. These stations are called horizontal laminar flow (HLF) hoods.

Both types of stations keep the wafers clean in two ways. First is the filtered air inside the hood. The second cleaning action is the slight positive pressure built up in the station. This pressure prevents airborne dirt from operators and from the aisle areas from entering the hood.

A special design of VLF hood is required for wet chemical processing stations (Fig. 4.10). These stations must be connected to an exhaust system to remove the chemical fumes. The fumes can pose both safety and contamination problems. In this design, care must be taken to balance the VLF air and the exhaust to maintain the

Figure 4.10 Cross section of a VLF-fume-exhaust hood.

required class number in the station. Cleanliness also requires that wafers be stored only in the cleaner front area of the work surface.

The clean-station strategy also applies to equipment in modern wafer fabrication. Each tool must be fitted with a VLF or HLF hood to keep the wafers clean during the loading and unloading steps.

Tunnel concept

As more critical particulate control became necessary, it was noted that the VLF hood approach had several drawbacks. Chief among them was the vulnerability to contamination from the many personnel moving about in the room. People entering and exiting the fabrication area had the potential of contaminating all the process stations in the area.

This particular problem is solved by dividing the fabrication area into separate tunnels or bays (Fig. 4.11). Instead of the individual VLF hoods, HEPA filters are built into the ceilings and serve the same purpose. The wafers are kept clean by the filtered air from the ceiling filters and less vulnerable to personnel-generated contamination because of fewer workers in the immediate vicinity. On the downside, tunnel arrangements cost more to construct and are less versatile when the process changes.

Total clean-room strategy

Developments in clean-room design and filtering technology have allowed an evolution back to the open fabrication area (Fig. 4.12). In

Figure 4.11 Cross section of clean-room tunnel.

Figure 4.12 Cross section of laminar flow clean room. (*Courtesy of Semiconductor International.*)

the latest version, air filtering is accomplished by HEPA filters in the ceiling with returns in the floor to give a continuous flow of clean air. The work stations are tabletops with perforations to allow the filtered air to pass through the tabletop uninterrupted.

Modern clean-room technology can produce fabrication areas that can perform at Class 10 or Class 1 conditions. An important clean-room parameter is recovery. This parameter is the time it takes for the filters to return the area to acceptable conditions after a shift start, personnel break, or other disturbance. In a Class 1 facility, the air may turn over every 6 seconds.[7] These rooms are often referred to as "world-class fabs."

In the evolution of clean rooms the trend has been to isolate the wafer from contamination sources. VLF hoods isolate the wafer from the room air, and tunnels isolate the wafer from excessive personnel exposure. The advent of CMOS ICs increased the number of process steps and the need to include more process stations in the clean room. These larger rooms (and tunnels) bring with them the potential of contamination from the sheer volume of the air and the increased number of operators. These changes, along with high equipment costs, have driven the cost of clean rooms to the $1 billion level.

Micro- and minienvironments

Projections in the mid-1980s showed increasing clean-room costs with diminishing returns on effectiveness. The concept of isolating the wafer in as small an environment as possible became the new direction. This concept was already in place with steppers and other process tools that had built-in clean *microenvironments* for wafer loading and unloading (Fig. 4.13).

The challenge was to link process microenvironments with a system such that the wafer was never exposed to the room air. In the mid-1980s Hewlett Packard developed a critical link with the invention of the *standard mechanical interface* (SMIF).[8] With SMIF, the traditional run box is replaced with a wafer enclosure (*microenvironment*) that can be pressurized with air or nitrogen to keep out room air. This approach took on the general name of *wafer isolation technology* (WIT), also called a *minienvironment system*. There are three parts to the system: the wafer box or pod for transporting the wafers, the isolated microenvironment at the tool, and a mechanism for extracting and loading the wafers. The wafer *pod* features a mechanical interface that allows direct connection to the microenvironment of the process tool (Fig. 4.14). Wafers may be loaded directly from and to the pod onto the tool-wafer system by dedicated handlers. Another approach is to remove and load the cassette from the pod to the tool-wafer handling system with a robot.

Isolation pod with vaccum or inert atmosphere

Standard mechanical interface (SMIF)

Wafer Cassette

Figure 4.13 Wafer transfer microenvironment.

Figure 4.14 Minienvironment system elements.

The WIT–minienvironment strategy has the advantage of allowing the upgrade of existing fabrication processes. Another benefit is that yield losses from contamination are lowered. This critical benefit can be delivered with less facility construction and, therefore, at lower operating costs. WIT allows aisle air cleanliness at a higher level, which reduces construction and operating costs. With the wafer isolated, there are fewer restrictions on operator clothing, procedures, and the like.

Temperature, humidity, and smog

In addition to control of particulates, the air parameters of temperature, humidity, and smog must be specified and controlled in a clean room. Temperature control is necessary for operator comfort and process control. Many of the wet processes of etching and cleaning take place in non-temperature-controlled baths and rely on the room temperature for control. This control is important. A rule of thumb is that chemical reactions, such as an etch rate, change by a factor of 2 with a 10°C change in temperature. A typical temperature range is 72°F, ± 2°F.

The relative humidity is also a critical process parameter, especially in patterning areas. In this area, thin films of a polymer are put on the wafer to act as an etch stencil. If the humidity is too high, the wafers collect moisture, preventing the polymer from sticking. The situation is the same as applying paint on a wet surface. On the other side, low humidity can foster the buildup of static charge on the wafer surface. The charge causes the wafer to attract particles out of the air. Relative humidity is controlled at between 15 and 50 percent.

Smog is another airborne contaminant in a clean room. Again, the problem is most critical in the patterning areas. A step in the patterning process is similar to photographic film developing, a chemical process. Ozone, a major component of smog, interferes with the development process and must be controlled. Ozone is filtered out of the air by installing carbon filters in the incoming air ducts.

Clean-Room Construction

Selection of a clean air strategy is the first step in the design of a clean room. Every clean room is a tradeoff of cleanliness versus cost. Whatever the final design, every clean room is built upon basic principles. The overall design principle is to build a sealed room that is supplied with clean air, is built with materials that are noncontaminating, and includes systems to prevent accidental contamination from the outside or from operators.

Construction materials

The inside of a clean room is constructed entirely of materials that are nonshedding. This includes wall coverings, process station materials, and floor coverings. All piping holes are sealed, and even the light fixtures must have solid covers. Additionally, the design should minimize flat surfaces that can collect dust. Stainless-steel materials are favored for process stations and work surfaces.

Clean-room design

Both the design of a clean room and its operation must be set up to keep dirt and contamination from getting into the room from the outside. Figure 4.15 shows a layout for a typical class 10 VLSI processing area. Nine techniques are used to keep out and control dirt. They are:

1. Adhesive floor mats
2. Gowning area
3. Air pressure
4. Air showers
5. Service bays
6. Double-door pass-throughs
7. Static control

Figure 4.15 Fab area with gowning area, air showers, and service aisles.

8. Shoe cleaners

9. Glove cleaners

Adhesive floor mats. At the entrance to every clean room is a floor mat with an adhesive surface. The adhesive pulls off and holds dirt adhering to the bottoms of shoes. In some clean rooms, the entire floor has a surface treated to hold dirt. Most mats have layers of the adhesive sheets. When the top sheet becomes dirty it is removed to expose the next clean one.

Gowning area. A major part of a clean room is the gowning area or anteroom. This area is a buffer between the clean room and the plant. It quite often is supplied with filtered air from ceiling HEPA filters. In this area, the operator's clean-room apparel is stored in lockers. It is also the area where the clean-room personnel change into their clean-room garments. The management of this area varies with the degree of cleanliness required in the clean room. Quite often it is managed to the same stringent requirements of the clean room itself. Often the gowning area is divided into two sections by a bench. The operators don the garments on one side and put on their shoe coverings on the bench. The purpose is to keep the area between the bench and the clean room at a higher cleanliness level.

A good clean-room procedure is to ensure that the doors between the factory and the clean room are never opened at the same time. This procedure ensures that the clean room is never exposed directly to the dirtier factory areas. Clean-room management also includes lists of materials and garments that can and cannot come into the gowning room. Some areas will provide hallway lockers for coats, etc.

Air pressure. A key design element is the air pressure balances between the clean room, the gowning room, and the factory. The well-designed facility will have the air of these three sections balanced such that the highest pressure is in the clean room, the second highest in the gowning area, and the lowest in the factory hallways. The higher pressure in the clean room prevents airborne particles from entering when doors are opened.

Air showers. The final design element protecting the clean room from outside contamination is the air shower located between the governing room and the clean room. Clean-room personnel enter the air shower where high-velocity air jets blow off particles on the outside of the garments. An air shower will have an interlocking system to prevent both doors from being opened at the same time.

Service bays. A clean room is really a series of rooms (Fig. 4.15) within the factory, each contributing to the maintenance of the clean room. In the center is the processing clean room. Surrounding it is a bay area that is maintained at some designated class number that is generally higher than the clean room. Class 1000 and class 10,000 are typical for service bays. In the bay are the process chemical pipes, electrical power lines, and clean-room materials. Critical-process machines are backed up to the wall dividing the clean room and the bay. This arrangement allows technicians to service the equipment from the back without entering the clean room.

Double-door pass-throughs. The bay also serves as a semiclean area for the storage of materials and supplies. They are put into the clean room through double-door, pass-through units that protect the cleanliness of the clean room. Pass-through units may be simple double-door boxes or have a supply of positive-pressure filtered air with interlocking devices to prevent both doors from being opened at the same time. Often the pass-throughs are fitted with HEPA filters. All materials and equipment brought into the clean room should be cleaned prior to entry.

Static control. One of the consequences of moving to higher-density circuits with submicron feature sizes is the vulnerability of the devices to smaller particles of contamination. One source of particulate contamination is the static charges that build up on the wafers, the storage boxes, work surfaces, and equipment. Each of these items can carry static charges as high as 50,000 V (volts) that attract aerosols out of the air and from personnel garments. The attracted particles end up contaminating the wafers. Statically held particulates are very difficult to remove with standard brush-and-wet cleaning techniques.

Most static charge is produced by *triboelectric* charging. This occurs when two materials initially in contact are separated. One surface becomes positively charged as it loses electrons, and the other becomes negatively charged as it gains electrons. The triboelectric series table in Fig. 4.16 shows the charging potential for some materials found in a clean room.[9]

Static also represents a device operational problem. It occurs in devices with thin dielectric layers, as in MOS gate regions. An electric static discharge (ESD) of up to 10 A (amperes) is possible. This level of ESD can physically destroy an MOS device or circuit. ESD is a particular worry in device-packaging areas. This problem requires that sensitive devices, such as large-array memories, be handled and shipped in holders of antistatic materials.

Photolithography image sources (photomasks and reticles) are particularly sensitive to ESD. A discharge can vaporize and destroy their chrome patterns.

Some equipment problems are caused by static, especially in robots, wafer handlers, and measuring equipment. Wafers usually come to

Air
Human Skin
Glass
Human Hair
Nylon
Wool   Increasingly Positive
Lead
Aluminum
Paper
Steel
Cotton
Steel
Hard Rubber
Nickel-copper
Brass-silver
Synthetic Rubber   Increasingly Negative
Polyurethane
Polypropylene
PVC
Silicon
Teflon

Figure 4.16 Triboelectric series. (*Hybrid Circuit Technology Handbook, Noyes Publications*)

the equipment in carriers made of PFA-type materials. This carrier material is chosen for its chemical resistance, but it is not conductive. Therefore, when a static charge builds up on the wafers, it cannot be dissipated to the carriers. When the carrier comes close to a piece of metal on the equipment, the wafer charge will discharge to the equipment. The electromagnetic interference (EMI) thus produced will interfere with machine operation.

Static is controlled by prevention (of charge buildup) and discharge techniques (Fig. 4.17). Prevention techniques include use of antistatic materials in garments and in-process storage boxes. In some areas, a topical antistatic solution may be applied to the walls to prevent the buildup of static charge. These solutions work by leaving a neutralizing residue on the surface. Generally, they are not used in critical stations because of the possible contaminating effect of the residue.

Discharge techniques include the use of ionizers and grounded static-discharge straps. Ionizers are placed just underneath the HEPA filters where they function to neutralize any charge buildup in the filtered air. Ionizers are also placed on nitrogen blow-off guns for the same effect. Some stations will have a portable ionizer blowing ionized air directly on the wafers being processed. Static discharge is also accomplished by grounding operators with wrist straps, having grounded mats at critical stations, and grounding work surfaces. Comprehensive static control programs, including prevention and discharge methods, personnel

Figure 4.17 Static-charge reduction techniques.

training, and third party monitoring, are a feature in most advanced fabrication-processing plants.

Shoe cleaners. In any contamination-controlled area, the dirtiest region is the floor. Removal of dirt from the sides of shoes and shoe coverings is accomplished by shoe cleaners stationed at the clean-room door. Typical machines have rotating brushes to remove the dirt, an internal vacuum to capture the loosened dirt, and bags to hold the dirt for later removal from the area.[10]

Glove cleaners. Maintaining clean gloves in the fabrication area is also a problem. Instructing operators to discard gloves when they become contaminated or dirty is one strategy. However, some contamination is hard to see and the decision to discard becomes a judgment that can vary from operator to operator. Another approach is to discard the gloves after every shift. This can get very expensive, so some fabrication areas use glove cleaners that clean and dry the gloves in an enclosure.[11]

Personnel-Generated Contamination

Clean-room personnel are one of the biggest sources of contamination. A clean-room operator, even after showering and sitting still, can give off between 100,000 and 1,000,000 particles per minute.[12] This number increases dramatically when a person is in motion. At 2 miles an hour, a human being gives off up to 5 million particles per minute. The particles come from flakes of dead hair and normal skin flaking. Additional particle sources are hair sprays, cosmetics, facial hair, and exposed clothing. The table in Fig. 4.18 lists the percentage increase in particulate generation over the background contamination level for different operator activities.[13]

Normal clothing can add more millions of particles to the area even under a clean-room garment. In clean rooms with very high cleanliness levels, operators will be directed to wear street clothing that is made of tight-weave, nonshedding materials. Garments made of wools and cottons are to be avoided, as are ones with high collars.

A human breath also contains high levels of contaminants. Every exhale puts numerous water droplets and particles into the air. The breath of smokers carries millions of particles for a long time after a cigarette is finished. Body fluids such as saliva contain sodium, a killer to many semiconductor devices. While healthy human beings are sources of many contaminants, sick individuals are even worse. Specifically, skin rashes and respiratory infections are additional

Normal Breath	No Particles
Smoker's Breath after Smoking	500 %
Sneezing	2000%
Sitting Quietly	20 %
Rubbing Hands On Face	200%
Walking	200%
Stamping Foot on Floor	5000%

Figure 4.18 Activity-caused increase in particles. (*Hybrid Microcircuit Technology Handbook, Noyes Publications*)

sources of contaminants. Some fabrication areas reassign personnel with certain health problems.

Given the scope of the problem, the only feasible way to render humans acceptable in a clean room is to cover them up. The style and material of clothing selected for clean-room personnel depends on the level of cleanliness required. For a typical class 10 area, the clothing material will be nonshedding and may contain conductive fibers to draw off static charge. Materials commonly used are polyester and GORE-TEX™. The tradeoff is the greater filtering ability of the material at the expense of operator comfort. The reclean versus discard issue applies to clean-room suits as well as gloves. Most ULSI fabrications have found that reusable gowns, even with the cost of cleaning, provide an overall lower cost than disposables.[14]

Every part of the body is covered up (see Fig. 4.17). The head will have an inner cap that keeps the hair in place. This is covered by an outer shell that is designed to fit close to the face and has snaps or a tail for securing the headgear under the body-covering smock. Covering the face will be a mask. Masks vary from surgical types to full ski mask style designs. In some clean rooms, both an inner and outer face mask are required. The eyes, which are a major source of fluid particles, are covered by glasses (usually safety glasses) with side shields. In contamination-critical areas, the operators might wear a covering that totally encloses the head and face. Modeled after a space suit, the head gear attaches to a belt filter, blower, and pump system. Fresh air is supplied by the pump while the filter ensures that no breath-generated contaminants are discharged into the room.[15]

Body covers are oversuits (called "bunny suits") that have closures for the legs, arms, and neck. Well-designed suits will have covers over the zippers and no outside pockets. The feet are covered with shoe coverings, some with attached leggings that come up the leg. In static-sensitive areas, straps are available to drain off static charge. Hands are covered with at least one pair of gloves. Favored are medical-type PVC gloves that permit good tactile feeling. Glove materials for chemical handling include orange latex (acid protection), green nitrile (solvent protection), and silver multilayered PVA solvent (special solvents).[16] In some areas, an inner pair of cotton gloves is permitted for comfort. Gloves should be pulled up over the sleeves to prevent contamination from traveling down the arm and into the clean room.

Skin flaking can be further controlled with the use of special lotions that moisten the skin. Any lotions used must be sodium- and chlorine-free.

In general, the order of gowning is from the head down. The theory is that dirt stirred up at each level is covered up by the next lower garment. Gloves are put on last. The garments and procedures needed to control contamination from clean-room workers are well known. However, the primary level of defense is the dedication and training of the operators. It is easy for an area to become lax in maintaining clean-room discipline and to suffer high levels of contamination.

Process Water

During the course of fabrication processing, a wafer will be chemically etched and cleaned many times. Each of the etching or cleaning steps is followed by a water rinse. Throughout the entire process, the wafer may spend a total of several hours in water rinse systems. Given the vulnerability of semiconductor devices to contamination, it is imperative that all process water be treated to meet very specific cleanliness requirements.

Water from a city system contains unacceptable amounts of the following contaminants:

1. Dissolved minerals

2. Particulates

3. Bacteria

4. Organics

5. Dissolved oxygen

6. Silica

The dissolved minerals come from salts in normal water. In the water, the salts separate into ions. For example, salt ($NaCl$) breaks up into Na^+ and Cl^- ions. Each is a contaminant in semiconductor devices and circuits. They are removed from the water by reverse osmosis (RO) and ion-exchange systems.

Resistivity Ohms-cm 25°C	Dissolved Solids (ppm)
18,000,000	0.0277
15,000,000	0.0333
10,000,000	0.0500
1,000,000	0.500
100,000	5.00
10,000	50.00

Figure 4.19 Resistivity of water versus concentration of dissolved solids.

The process of removing the electrically active ions changes the water from a conductive medium to a resistive one. This fact is used to improve the quality of deionized (DI) water. Deionized water has a resistivity of 18,000,000 $\Omega \cdot$ cm at 25°C. It is called 18-megohm (MΩ) water. Figure 4.19 shows the effect on the resistivity of water when various amounts of dissolved minerals are present.

The resistivity of all process water is monitored at many points in the fabrication area. The goal and specification is 18 MΩ in VLSI areas, although some fabrication areas will run with 15-MΩ water levels. Solid particles (particulates) are removed from the water by sand filtration, earth filtration, and/or membranes to submicron levels. Bacteria and fungi find water a favorable host. They are removed by sterilizers that use ultraviolet radiation to kill the bacteria and filters to remove them from the stream of water.

Organic contaminants (plant and fecal materials) are removed with carbon-bed filtration. Dissolved oxygen and carbon dioxide are removed with forced draft decarbonators and vacuum degasifiers.[17] Water specifications for DRAM products are shown in Fig. 4.20.

Contaminant	64Kb	1Mb	16Mb
Resistivity (MΩcm)	>17	>18	>18.1
TOC (micro-g/L)	<200	<50	<10
Bacteria (#/liter)	<250	<50	<1
Particles (#/ml)	<50	<50	<1
Critical size (micron)	0.2	0.2	0.05
Dissolved oxygen (micro-g/l)	<200	<100	<20
Sodium (micro-g/l)	<1	<1	<0.01
Chlorine (micro-g/l)	<5	<1	<0.01
Manganese (micro-g/l)	N/A	<1	<0.005

Figure 4.20 DRAM water specs. (*Semiconductor International, July 1994, p. 178*)

The cost of cleaning process water to acceptable levels is a major operating expense of a fabrication area. In most fabrications, the process stations are fitted with water meters that monitor the used water. If the water falls within a certain range, it is recycled in the water system for cleanup. Excessively dirty water is treated as dictated by regulations and discharged from the plant. A typical fabrication system is shown in Fig. 4.21. Water stored in the system is blanketed with nitrogen to prevent the absorption of carbon dioxide. Carbon dioxide in the water interferes with resistivity measurements, causing false readings.

Process Chemicals

The acids, bases, and solvents used to etch and clean wafers and equipment have to be of the highest purity for use in a fabrication area. The contaminants of concern are metallics, particulates, and other chemicals. Unlike water, process chemicals are purchased and used as they come into the plant. Industrial chemicals are rated by grade. They are commercial, reagent, electronic, and semiconductor grade. The first two are generally too dirty for semiconductor use. Electronic grade and semiconductor grade are cleaner chemicals, but levels of cleanliness vary from manufacturer to manufacturer.

Trade organizations such as SEMI are establishing cleanliness specifications for the industry. However, most semiconductor plants purchase their chemicals to in-house generated specifications. Of primary

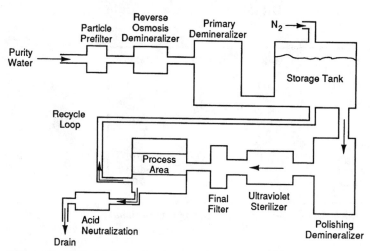

Figure 4.21 Typical deionized water system.

concern are the metallic mobile ionic contaminants. These are usually limited to levels of 1 part per million (ppm) or less. Some suppliers are making available chemicals with MIC levels of only 1 part per billion (ppb). Particulate filtering levels are specified at 0.2 μm or lower.

Chemical purity is indicated by its assay number. The assay number indicates the percentage of the chemical in the container. For example, an assay of 99.9 percent on a bottle of sulfuric acid means that the bottle contains 99.9 percent sulfuric acid and 0.01 percent of other substances.

The delivery of a clean chemical to the processing area involves more than just making a clean chemical. Care must be taken to clean the inside of the containers, using containers that do not dissolve, using containers with particulate-free labels, and placing the clean bottles in bags before shipping. Chemical bottle use is found only in older and lower technology fabrications.

Most firms purchase clean process chemicals in bulk quantities. They are decanted into smaller vessels, distributed to the process stations from a central location by pipes or distributed directly to the process station from a miniunit. Bulk chemical distribution systems (BCDS) offer cleaner chemicals at a lower cost. Special care must be maintained to ensure that piping and transfer vessels are cleaned regularly to prevent contamination. A particular worry with BCDS and secondary vessel filling is cross contamination. Vessels should be reserved for one type of chemical only.

The requirement for cleaner chemicals, tighter process control, and lower costs is being met by several techniques and various types of equipment. One is the point-of-use (POU) chemical mixer (another version of a BCDS). Several mixers, connected to a wet bench or automatic machine, mix chemicals and deliver them directly to the process vessel or unit. Another is the chemical reprocessor. These units are located on the drain side of a wet-process station. Discharged chemicals are refiltered and (in some cases) recharged for reuse. Critical recirculating etch baths are connected to filters to keep a clean supply of chemicals at the wafer. A newer process is point-of-use chemical generation (POUCG). Chemicals such as NH_4OH, HF, and H_2O_2 are made at the process station by combining the proper gases with DI water. Elimination of the contamination associated with chemical packaging and transferring gives this method the potential of producing chemicals in the parts per trillion (ppt) range.[18]

Process Gases

In additional to the many wet (liquid) chemical processes, a semiconductor wafer is processed with many gases. The gases are both the air-separation gases, oxygen, nitrogen, and hydrogen, and specialty gases such as arsine and carbon tetrafluoride.

Like the wet chemicals, they have to be delivered clean to the process stations and tools. Gas quality is measured in four categories:

1. Percentage of purity

2. Water vapor content

3. Particulates

4. Metallic ions

Extremely high purity is required for all process gases. Of particular concern are the gases used in oxidation, sputtering, plasma etch, chemical vapor deposition (CVD), reactive ion etch, ion implantation, and diffusion. All of these processes involve chemical reactions that are driven by an energy source. If the gases are contaminated with other gases, the anticipated reaction can be significantly altered or the result on the wafer changed. For example, chlorine contamination in a tank of argon used for sputtering can end up in the sputtered film with disastrous results for the device. Gas purity is specified by the assay number, with typical values ranging from 99.99 to 99.999999 percent,[19] depending on the gas and the use in the process. Purity is expressed as the number of 9's to the right of the decimal point. The highest purity is referred to as "six 9's pure."

The challenge in fabrication areas is maintaining gas purity from the manufacturer to the process station. Starting at the source, the gas passes through a piping system and a gas panel containing valves and flowmeters until it reaches the connection to the tool. Leaks in any part of the system are disastrous. Outside air (especially oxygen) can enter into a chemical reaction, changing the composition of the process gas and the desired reaction in the process. Contamination of the gas can come from outgassing of the system materials. A typical system features stainless-steel piping and valves, with some polymer components such as connectors and seals. For ultraclean systems, the stainless-steel surfaces have electropolished and/or double-vacuum melt inside surfaces to reduce outgassing.[20] Another technique, called *oxygen passivation* (OP), is the growing of an iron oxide film to further reduce outgassing. Polymer components should be avoided. Gas panel design should eliminate dead spaces that become repositories for contamination. Also critical is the use of clean welding processes to eliminate gases absorbed into the piping from the welding gases.

The control of water vapor is also critical. Water vapor is a gas and can enter into unwanted reactions just like other contaminating gases. In fabrication areas, processing silicon wafers with water vapor present is a particular problem. Silicon oxidizes easily wherever free oxygen or water is available. Control of unwanted water vapor is necessary to prevent accidental oxidation of silicon surfaces. Water-vapor limits are 3 to 5 ppm.

The presence of particulates and/or metallics in a gas has the same effect on the processing as in wet chemicals. Consequently, gases are filtered to the 0.2-μm level and metallic ions are controlled to parts per million or lower.

The air-separation gases are stored on the site in the liquid state. In this state they are very cold, a situation that freezes some contaminants in the bottom of the tanks. Specialty gases are purchased in high-pressure cylinders. Since many of the specialty gases are toxic or flammable, they are stored in special cabinets outside the plant.

Quartz.   Wafers spend a lot of process time in quartz holders such as wafer boats, furnace tubes, and transfer holders. Quartz can be a significant source of contamination, both from outgassing and particulates. Evan high-purity quartz contains heavy metals that can outgas into diffusion and oxidation tube gas streams, especially during high-temperature processes. Particulates come from the abrasion of the wafers in wafer boats and the scrapping of the boats against the furnace tubes. (Solutions to this problem are discussed in Chap. 11.) Both electric and flame-fused quartz processes are used to produce quartz surfaces acceptable for semiconductor use.[21]

Equipment

Successful contamination control is dependent on knowing the sources of contamination. A 1985 analysis (Fig 4.22) indicated that process equipment was the largest source of particulates. By the 1990s, equipment-induced particulates rose to 75 to 90 percent of all particle sources.[22] This does not mean that the equipment is getting dirtier. Advances in particulate control in air, chemicals, and from personnel has shifted the focus to equipment. Defect generation is part of equipment specifications. Generally, the number of particles added to the wafer per product pass (ppp) through the tool is specified. The term *particles per wafer pass* (pwp) is also used.

Reduction of particle generation starts with design and material selection. Other factors are the transport of the particles to the wafer and deposition mechanisms, such as static. Most equipment is assembled in a clean room with the same class number as the customer's fabrication area. The clustering of several process tools with one clean microenvironment for loading and unloading minimizes the contami-

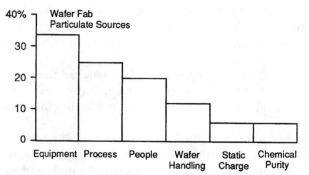

Figure 4.22 Sources of particulate contamination. This analysis, shown at SEMI Forecast by Dr. C. Rinn Cleavelin, Texas Instruments, revealed equipment-generated particles as the top enemy in 1985. (*Courtesy of Semiconductor Equipment and Materials Institute.*)

nation generation associated with multiple loading stations. (Cluster tools are discussed in Chap. 15.) There is interest in having in-situ particle monitors in the process chambers. Challenges to this level of automation and detection include correlation in-chamber levels with on-wafer levels and the lifetime of detectors in the hostile and corrosive environments of most process chambers.

Clean-Room Materials and Supplies

In addition to the process chemicals, it takes a host of other materials and supplies to process the wafers. Each of these must meet cleanliness requirements. Logs, forms, and notebooks used will be either of nonshedding coated paper or of a polymer plastic. Pencils are not allowed and pens are the nonretracting type.

Wafer storage boxes are made of specific non-particle-generating materials as are carts and tubing materials. Cart wheels and tools are used without greases or lubricants. In many areas, mechanics' tools and tool boxes are cleaned and left inside the clean room.

Clean-Room Maintenance

Regular maintenance of the clean room is essential. The cleaning personnel must wear the same garments as the operators. Clean-room cleaners and applicators, including mops, must be carefully specified. Normal household cleaners are far too dirty for use in clean rooms. Special care must be taken when using vacuum cleaners. Special clean-room vacuums with HEPA filtered exhausts are available. Many clean rooms will have built-in vacuum systems to minimize dirt generation during cleaning.

The wipe-down of process stations is done with special low-particulate wipes and sponges. The wiping procedure is critical. Wall surfaces should be wiped from top to bottom and deck surfaces from back to front. Cleaning chemicals in spray bottles should be sprayed into wipes, not onto the surfaces. This simple procedure minimizes unwanted overspray onto the wafers and equipment. Clean-room cleaning has itself become a supporting technology to semiconductor processing. Many fabrication plants use outside certification services[23] to identify and document cleanliness levels, practices, and control procedures.

Figure 4.23 summarizes the specifications for clean-room cleanliness. (The detection of the various forms of contamination is detailed in Chaps. 8 and 14.)

Parameter	ULSI Target Specifications
Air	
Particles	Class 1 to 0.01
	0.3- 0.1
Humidity	15-50%
Temperature	68-72 Degrees Farenheit
Photochemical smog	2 pphm (parts per hundred million)
Deionized Water	
Resistivity	17-18.1 meg-ohms
Bacteria	<250 - <1 #/liter
Particles	<50 - <1 per ml
Metals	<1 - <0.1 microg/l
Chemicals	
Particles	50-200/ ml max
Trace metals	10-59 ppb
Moisture	approx 0.10 wt %
Gases	
Purity	> 99.9 %
Water vapor	<20 ppbv
Filtration	< 1 per cu ft
Oxygen	< 20 ppbv
Static charge	< 25 - 50 volts

Figure 4.23 Typical ULSI clean-room specs.

Key Concepts and Terms

Aerosols	Minienvironments
Air shower	Mobile ionic contaminants
Anteroom	Particulates
Bacteria	Process gases
Clean-room class number	Process wet chemicals
Clean-room design	Service bay
Contamination sources	Temperature, humidity, smog
DI water	Total clean-room strategy
Gowning requirements	Tunnel strategy
HEPA filter	Vertical laminar flow
Micron	VLF work station strategy
Microenvironments	

Review Questions

1. State the class number required for VLSI processing and ULSI processing.

2. List three techniques used to keep contamination out of a clean room.

3. Draw a cross section of a "total" clean room, showing the filters and air patterns.

4. What does VLF mean?

5. What is the DI water specification?

6. Name three contaminants found in normal water.

7. What are the two ways that static is harmful in a fabrication area?

8. Explain the problem associated with high humidity in a fabrication area.

9. What is a mobile ionic contaminant and why is it unwanted in a fabrication area?

10. How does a service bay contribute to the cleanliness of a clean room?

11. Describe a minienvironment.

12. Describe and sketch an SMIF/microenvironment unit.

References

1. "New Challenges for robotics and automation," *Solid State Technology,* April 1994, p. 67.
2. R. Iscoff, "Wafer Cleaning: Wet Methods Still Lead the Pack," *Semiconductor International,* July 1993, p. 60.
3. Ibid.
4. Clean Room and Work Station Requirements, Federal Standard 209E, Sec. 1–5, September 11, 1992, Office of Technical Services, Dept. of Commerce, Washington, D.C.
5. A. Bonora, "Minienvironments and Their Place in the Fab of the Future," *Solid State Technology,* September 1993, p. 47.
6. Operator Training Course, Class-10 Technologies, Inc., San Jose, Calif., 1983, p. 13.
7. "SI Honors Top Fabs of 1994," *Semiconductor International,* April 1994, p. 78.
8. B. Newboe, "Minienvironments: Better Cleanrooms for Less," *Semiconductor International,* March 1993, p. 54.
9. J. Licari and L. Enlow, *Hybrid Microcircuit Technology Handbook,* Noyes Publications, Park Ridge, NJ, 1988, p. 281.
10. Dryden Engineering Product Description, Santa Clara, CA, 1995.
11. Ibid.
12. *The New American Revolution,* Araclean Services, La Grange, IL., 1984, p. 9.
13. J. Licari and L. Enlow, *Hybrid Microcircuit Technology Handbook,* Noyes Publications, Park Ridge, NJ, 1988, p. 280.
14. R. Iscoff, "Cleanroom Apparel: A Question of Tradeoffs," *Semiconductor International,* March 1994, p. 65.
15. Dryden Engineering Product Description, Santa Clara, CA, 1995.
16. R. Iscoff, "Cleanroom Apparel: A Question of Tradeoffs," *Semiconductor International,* March 1994, p. 66.
17. R. Governal, "Ultrapure Water: A Battle Every Step of the Way," *Semiconductor International,* July 1994, p. 177.
18. L. Peters, "Point-of-use Generation: The Ultimate Solution for Chemical Purity," *Semiconductor International,* January 1994, p. 62.
19. P. Carr, et al., "RTP Characterization Using In-situ Gas Analysis," *Semiconductor International,* November 1993, p. 75.
20. H. Kobayashi, "How Gas Panels Affect Contamination," *Semiconductor International,* September 1994, p. 86.
21. M. Hill, et al., "Quartzglass Components and Heavy-metal Contamination," *Solid State Technology,* March 1994, p. 49.
22. A. Busnaina, "Solving Process Tool Contamination Problems," *Semiconductor International,* September 1993, p. 73.
23. Dryden Engineering Service Description, Santa Clara, CA, 1995.

5

Overview of
Wafer Fabrication

Overview

This chapter will present an overview of the wafer-fabrication process. The four basic planar operations performed on the wafer are explained along with the process sequences used to create the circuit components on the chip surface. Circuit design is traced from the functional diagram to the production of a photomask. Finally, wafer and chip features and terminology are detailed.

Objectives

Upon completion of this chapter you should be able to:

1. Identify and explain the four basic wafer operations.

2. Identify the parts of a wafer.

3. Draw a flow diagram of the circuit-design process.

4. Explain the definition and use of a composite drawing and mask set.

5. Draw cross sections showing the doping sequence of basic operations.

6. Draw cross sections showing the metallization sequence of basic operations.

7. Draw cross sections showing the passivation sequence of basic operations.

8. Identify the "parts" of an integrated circuit chip.

Figure 5.1 Wafer-fabrication stage.

Figure 5.2 Wafer terminology.

Goal of Wafer Fabrication

The four stages of semiconductor manufacturing are materials preparation, crystal growth and wafer preparation, wafer fabrication, and packaging (see Chap. 1). The first two stages have been explored in Chap. 3. In this chapter the fundamentals of wafer fabrication are explained.

Wafer fabrication is the series of processes used to create the semiconductor devices in and on the wafer surface. The polished starting wafers come into fabrication with blank surfaces and exit with the surface covered with hundreds of completed chips (Fig. 5.1).

Wafer Terminology

A completed wafer is illustrated in Fig. 5.2. The regions of a wafer surface are:

1. Chip, die, circuit, microchip, or bar. All of these terms are used to identify the identical patterns covering the majority of the wafer surface.

2. Scribe lines, saw lines, streets, and avenues. These areas are small avenues of space between the chips used to separate the chips from each other. Generally the scribe lines are blank, but some companies place alignment targets (see photomasking) in them.

3. Engineering die, test die. These chips are different from the regular device or circuit die. They contain special devices and circuit elements that can be electrically tested during fabrication processing.

4. Edge die. The edges of the wafer contain partial die patterns. The partial die will not function. The number and area occupied by the edge die is a function of the chip size and the wafer diameter. One of the driving forces behind larger wafer diameters is to minimize the area occupied by the edge die.

5. Wafer crystal planes. The cutaway section illustrates the crystal structure of the wafer under the circuit layers. The diagram shows that the chip edges are oriented to the wafer crystal structure.

6. Wafer flats. The depicted wafer has a major and minor flat, indicating that it is a P-type <100> oriented wafer (see Chap 3. for the flat code).

Basic Wafer-Fabrication Operations

Wafer-fabrication areas around the world produce billions of chips with thousands of different functions and designs. However, there are a limited number of transistor types. Most are variations on either a bipolar or MOS structure (see Chap. 16). Additionally, the choices for wafer materials are few in number (as explained in Chap. 2). Even so, the variations in manufacturing processes are almost infinite, with each company exercising its own version of a standard process. But within all this process diversity there are only four (Fig. 5.3) basic operations performed on a wafer in the fabrication process. They are:

1. Layering
2. Patterning
3. Doping
4. Heat treatments

- Layering
- Patterning
- Doping
- Heat Treatments

Figure 5.3 Basic wafer-fabrication operations.

Layering

Basic semiconductor operations are analogous to the basic operations of metal bending, welding, casting, painting, and so on, that are performed in automobile factories. While these operations are common to all auto factories they are used in different ways to produce a large variety of vehicles.

Layering is the operation used to add thin layers to the wafer surface. An examination of the simple MOS transistor structure in Fig. 5.4 shows a number of layers that have been added to the wafer surface. These layers are either insulators, semiconductors, or conductors. They are of different materials and are grown or deposited by a variety of techniques.

Layers are added to the surface by two major techniques, growing and deposition (Fig. 5.5). Oxidation is a technique of growing a silicon dioxide layer on a silicon wafer. Common deposition techniques are chemical vapor deposition (CVD), evaporation, and sputtering. Figure 5.6 lists common layer materials and layering processes. (The details of each are explained in the process chapters. The role of the different layers in the structures is explained in Chap. 16.)

Patterning

Patterning is the series of steps that results in the removal of selected portions of the added surface layers (Fig. 5.7). After removal, a *pattern* of the layer is left on the wafer surface. The material removed

Deposited Passivation Layer

Deposited Metal Layer

Grown Oxide Layers

n   n

P

Figure 5.4 Cross section of completed metal gate MOS transistor with grown and deposited layers.

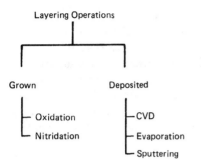

Figure 5.5 Layering operations.

Layers	Thermal Oxidation	Chemical Vapor Deposition	Evaporation	Sputtering
Insulators	Silicon Dioxide	Silicon Dioxide Silicon Nitrides		Silicon Dioxide Silicon Monoxide
Semiconductors		Epitaxial Silicon Poly Silicon		
Conductors			Aluminum Aluminum/Silicon Aluminum/Copper Nichrome Gold	Tungston Titanium Molybdenum Aluminum Aluminum/Silicon Aluminum/Copper

Figure 5.6 Table of layers, processes, and materials.

Figure 5.7 Patterning.

may be in the form of a hole in the layer or just a remaining island of the material.

The patterning process is known by the names photomasking, masking, photolithography, and microlithography. It is the patterning operation that creates the surface parts of the devices that make up a circuit. The goal of the operation is to create in or on the wafer surface the *parts* of the device or circuit in the exact dimensions (feature size) required by the circuit design and to locate them in their proper location on the wafer surface.

Patterning is the most critical of the four basic operations. This operation sets the critical dimensions of the devices. Errors in the patterning process can cause distorted or misplaced patterns that result in changes in the electrical functioning of the device or circuit. Misplacement of the pattern can have the same bad results. Another problem is defects. Patterning is a high-tech version of photography, but performed at incredibly small dimensions. Contamination in the process steps can introduce defects. This contamination problem is magnified by the fact that patterning operations are performed on the wafer from 5 to 20-plus times in the course of the wafer-fabrication process.

Doping

Doping is the process that puts specific amounts of dopants in the wafer surface through openings in the surface layers (Fig. 5.8). The two techniques are thermal diffusion and ion implantation, which are detailed in Chap. 11.

Thermal diffusion is a chemical process that takes place when the wafer is heated to the vicinity of 1000°C and exposed to vapors of the proper dopant. Ion implantation is a physical process in which the dopant atoms are ionized, accelerated to a high speed, and "shot" into the wafer surface.

The purpose of the doping operation is to create either N-type or P-type pockets in the wafer surface (Fig. 5.9). These pockets form the N-P junctions required for operation of the transistors, diodes, capacitors, and resistors of the circuit.

Thermal Diffusion

Ion Source

Figure 5.8 Doping.

Ion Implantation

Oxide Layer
Wafer

Doped N-or
P-Type Region

Figure 5.9 Formation of doped N- or P-type region in wafer surface.

Heat treatments

Heat treatments are the operations in which the wafer is simply heated and cooled to achieve specific results. In the heat treatment operations, no additional material is added or removed from the wafer.

An important heat treatment takes place after ion implantation. The implantation of the dopant atoms causes a disruption of the wafer crystal structure which is repaired by a heat treatment, called *anneal,* at about 1000°C. Another takes place after the conducting stripes of metal are formed on the wafer. These stripes carry the electrical current between the devices in the circuit. To ensure good electrical conduction, the metal is "alloyed" to the wafer surface by a heat treatment, which takes place at 450°C. Figure 5.10 shows a list of the basic operations and the principle process options used to achieve each.

Manufacturing Semiconductor Devices and Circuits

The manufacture of a semiconductor circuit is a multistep process requiring up to hundreds of individual steps. The four basic operations are used in specific sequences to build the parts of the devices in and on the wafer surface. In this section, the building of a simple device called a MOS metal gate transistor is illustrated to explain the manufacturing sequence. (The functions of the individual parts of this type of transistor and the operation of the transistor are explained in Chap. 14.)

Circuit design

Circuit fabrication begins with a definition of a specific function and circuit design to accomplish the function. The circuit designer starts with a block functional diagram of the circuit such as the logic diagram in Fig. 5.11. Next, the designer translates the functional dia-

Basic Operation	Process	Options
Layering	Oxidation	Atmospheric
		High Pressure
		Rapid Thermal Oxidation (RTO)
	Chemical Vapor Deposition (CVD)	Atmospheric Pressure
		Low Pressure (LPCVD)
		Plasma Enhanced (PECVD)
		Vapor Phase Epitaxy (VPE)
		Metaloranic CVD (MOCVD)
	Molecular Beam Epitaxy (MBE)	
	Physical Vapor Deposition (PVD)	Vacuum Evaporation
		Sputtering
Patterning	Resist	Positive
		Negative
	Exposure Systems	Contact
		Proximity
		Scanning Projection
		Stepper
	Exposure Sources	High Pressure Mercury
		X-Rays
		E-Beams
	Imaging Processes	Single Layer Resist
		Multilayer Resist
		Antireflecting layers
		Off-Axis Illumination
		Annular Ring Illumination
		Planarization
		Contrast Enhancement
	Etch	Wet Chemistry-Liquid/vapor
		Dry (Plasma)
		Lift-Off
		Ion Milling
		Reactive Ion Etch (RIE)
Doping	Diffusion	Open Tube-Horizontal/Vertical
		Closed Tube
		Rapid Thermal Processing (RTP)
	Ion Implantation	Medium/High Current
		Low/High Voltage (energy)
Heating	Thermal	Hot Plates
		Convection
		RTP
	Radiation	Infrared (IR)

Figure 5.10 Summary of wafer-fab operations/processes.

Figure 5.11 Example functional logic design of a simple circuit.

gram to a schematic diagram (Fig. 5.12). This diagram shows the number and location of the various circuit components. Each component is represented by a symbol. Accompanying the schematic diagram are the electrical parameters (voltage, current, resistance, etc.) required to make the circuit work.

These first two steps are common to any electrical-circuit design. The third step, circuit layout, is unique to semiconductor circuits. The semiconductor components are physical entities in and on the wafer surface and their dimensions (along with the doping levels) determine the electrical parameters of each. Layout starts with the translation of each of the circuit components into their actual physical dimensions relative to their final dimensions in and on the wafer surface.

Layout is a complicated process performed with sophisticated computer-aided design (CAD) systems. First, each component design is set in the computer. Next, the components are assembled into the circuit layout patterns. The computer draws a composite picture of the circuit surface showing all of the sublayer patterns. This drawing is called a *composite* (Fig. 5.13). The composite drawing is analogous to the blueprint of a multistory office building as viewed from the top and showing all of the floors.

Since buildings and semiconductor circuits are built one layer at a time, it is necessary to separate the composite drawing into the individual layer drawings. Figure 5.13 illustrates the composite and individual layer patterns for a metal gate MOS transistor. After separation, each drawing is digitized. Digitizing is the translation of the layer drawings to a digital database. Finally the layer patterns are drawn on a computerized X-Y plotting table.

Figure 5.12 Example circuit schematic diagram with component symbols.

Reticle and masks

The information from the digitizing step is used to produce a reticle. A *reticle* is a "hard copy" of the individual drawing recreated in a thin layer of chrome deposited on a glass or quartz plate (Fig. 5.14*a*). The reticle may be used directly in the patterning process or may be used to produce a photomask. A *photomask* is also a glass plate with a thin chrome layer on the surface. After production, it is covered with many copies of the circuit pattern (Fig. 5.14*b*). It is used to pattern a whole wafer surface in one pattern transfer. (Reticle and mask-making processes are explained in detail in Chap. 11.)

The process sequence from circuit design to wafer patterning is shown in Fig. 5.15. Reticles and masks are produced in a separate department or are purchased from outside vendors. They supply the fabrication area with a separate set of reticles or masks, called a *mask set,* for each type of circuit.

Example fabrication process

The actual construction of a circuit proceeds in steps starting with a polished wafer. The cross-section sequence in Fig. 5.16 shows the basic operations required to form a MOS metal-gate transistor structure.

Figure 5.13 Composite and layer drawings for 5-mask silicon gate transistor.

Step 1: Layering. The building starts with an oxidation of the wafer surface to protect it and serve as a doping barrier. This silicon dioxide layer is called the *field oxide*.

Step 2: Patterning. The patterning process leaves two holes in the field oxide that define the source and drain areas of the transistor.

Step 3: Doping. Next, the wafer goes to a doping process where an N-type dopant is put through the openings to create two N-type pockets in the wafer surface.

Step 4: Patterning. Here another patterning operation is used to remove the field oxide in the gate region between the source and drain.

Step 5: Layering. The exposed silicon in the gate region is reoxidized with a thin layer known as the gate oxide. At the same time, an oxide grows in the source and drain holes.

(a)

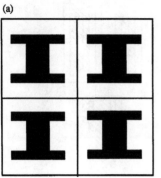

(b)

Figure 5.14 (a) Chrome on glass reticle. (b) Photomask of same pattern.

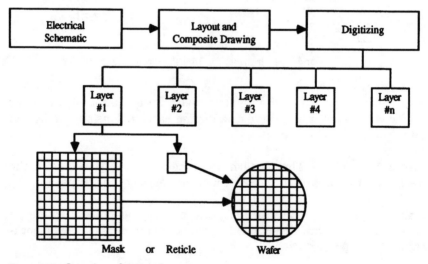

Figure 5.15 Overview of patterning sequence.

Cross Section	Step	Operation	Name/Purpose
		Starting wafer	
	1	Layering	Field Oxide
	2	Patterning	Source/drain holes
	3	Doping Layering	N-type doping and reoxidation of source/drain
	4	Patterning	Gate region is formed
	5	Layering	Gate oxide is grown
	6	Patterning	Contact holes are patterned into source/drain regions
	7	Layering	Conducting metal layer is deposited
	8	Patterning	Metal layer is patterned
	9	Heat Treatment	Metal is alloyed to layer
	10	Layering	Protective passivation layer is deposited
	11	Patterning	Passivation layer is removed over metal pads

Figure 5.16   Formation of metal gate MOS transistor.

Step 6: Patterning. In this patterning step, two holes are patterned in the reoxidized source and drain regions. These holes are called *contact holes* and will allow direct contact of the metallization layer with the source and drain regions.

Step 7: Layering. Next in the sequence is the deposition of a conductive metal layer. This layer blankets the entire surface of the wafer.

Step 8: Patterning. After deposition, the wafer goes back to the patterning area where portions of the metallization layer are removed from the chip area and the scribe lines. The remaining portions connect all the parts of the surface components to each other in the exact pattern required by the circuit design.

Step 9: Heat Treatment. Following the metal-patterning step, the wafer goes through a heating process in a nitrogen gas atmosphere. The purpose of the step is to "alloy" the metal to the exposed source and drain regions and the gate region to ensure good electrical contact.

Step 10: Layering. The final layer of this device is a protective layer known variously as a *scratch* or *passivation layer.* Its purpose is to protect the components on the chip surface during the testing and packaging processes and during use.

Step 11: Patterning. The last step in the sequence is a patterning process that removes portions of the scratch protection layer over the metallization terminal pads on the periphery of the chip. This step is known as the *pad mask.*

The 11-step process illustrates how the four basic fabrication operations are used to build a particular transistor structure. The other components (diodes, resistors, and capacitors) required for the circuit are formed along with the transistors. For example, in this sequence resistor patterns are put on the wafer at the same time as the source–drain pattern for the transistor. Thus the subsequent doping operation creates the source and drain *and* the resistors. Other transistor types, such as bipolar and silicon gate MOS, are formed by the same basic four operations, but using different materials and in different sequences.

Fabrication process sequences

The process illustrated shows three sequences that show up in all process flows. The doping sequence (steps 1, 2, and 3) is used to create doped regions in the wafer surface. The metallization sequence (steps 6, 7, 8, and 9) is used to provide the metal interconnections of the circuit. The passivation sequence (steps 10 and 11) is used in most process flows to provide the top layer protection for the chip.

Figure 5.17 shows this process flow and a basic bipolar flow with the sequences marked. Figure 5.18 shows the same metal gate MOS process flow but separated into the basic operations.

Chip Terminology

Figure 5.19 is a photomicrograph of a bipolar medium-scale integration (MSI) integrated circuit. This level of integration was chosen so that the surface details could be seen. The components of higher-density circuits are so small that they cannot be distinguished on a photomicrograph of the entire chip. The chip features are:

1. A bipolar transistor
2. The circuit designation number
3. Bonding pads for connecting the chip into a package
4. A piece of contamination on a bonding pad
5. Metallization lines
6. Scribe (separation) line

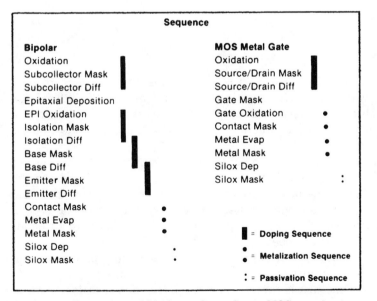

Figure 5.17 Comparison of bipolar and metal gate MOS process steps with sequences indicated.

Figure 5.18 Metal gate MOS process flow.

7. Unconnected component
8. Mask alignment marks
9. Resistor

Wafer Sort

Following the wafer-fabrication process comes a very important testing step, wafer sort. This test is the report card on the fabrication process. During the test, each chip is electrically tested for electrical performance and circuit functioning. Wafer sort is also known as *die sort* or *electrical sort*.

For the test, the wafer is mounted on a vacuum chuck and aligned to thin electrical probes that contact each of the bonding pads on the chip (Fig. 5.20). The probes are connected to power supplies that test the circuit and record the results. The number, sequence, and type of tests are directed by a computer program. Wafer probers are automated so that after the probes are aligned to the first chip (manually or with an automatic vision system), subsequent testing proceeds without operator assistance.

The goal of the test is three-fold. First is the identification of working chips before they go into the packaging operation. Second is characterization of the electrical parameters of the device or circuit.

Figure 5.19 Chip terminology.

Figure 5.20 Wafer sort.

Individual chips may function to specifications, but engineers need to track the individual device or circuit parameter distributions produced by the fabrication area. The third goal of wafer sort is not really an on-chip test. It is to determine the yield of working chips, which will give the fabrication personnel feedback on overall performance. The working and nonworking chips are identified by a drop of ink on the *nonworking* chips or by a computer map of the wafer indicating the locations of the working and nonworking chips.

Wafer sort is one of the principal yield calculations in the chip production process. The cost of wafer probing has become a significant part of chip production, estimated at 30 percent or higher of the manufacturing cost for larger chips.[1] Larger and more dense chips require longer times to probe and increased power supplies; wafer-handling mechanics and computer systems have to be more sophisticated to perform the tests and track results. The vision systems allowing automatic probing must also evolve in sophistication (and cost) with expanding die size. Cutting the time it takes to test chips is therefore a goal. Chip designers are being asked to include test modes for memory arrays, and test designers are exploring ways to streamline test sequences, using stripped-down tests once the chip is fully characterized, scan tests of the circuit, and parallel testing of different circuit parts. The details of wafer sort and parameters yield are addressed in Chap. 6.

Packaging

Wafer fabrication and wafer sort complete stage three of the chip production process. Stage four is packaging where the chips are incorporated into a protective package (Fig. 5.21). The wafers are transferred to a packaging area on the same site or to a remote location. Many semiconductor producers package their chips in offshore facilities. (Chapter 18 details the packaging process.)

Wafer Preparation	Fabrication	Electrical Test (Die Sort)	Assembly
Sand to Silicon Crystal Growth Wafer Prep	Individual Circuits Fabricated In and On Wafer	Each Circuit is Electrically Tested	Good Die are Assembled into Packages and Tested

Figure 5.21 Integrated circuit manufacturing sequence.

Summary

The semiconductor manufacturing process is long and complicated. It includes many variations, depending on the type of product, level of integration, feature size, and the like. Separating the entire manufacturing sequence into its four stages and identifying the four basic operations in the wafer-fabrication process provides a method of understanding the complex reality of chip manufacturing. Several simple processes are used to illustrate the basic wafer-fabrication operations. The actual processes used are described in the process chapters (Chaps. 7 through 13), and actual processes for manufactured circuits are explained in detail in Chaps. 16 and 17. Industry drivers and trends are explained in Chap. 15.

Key Concepts and Terms

Chip, die, device, circuit, bar   Engineering test die

Chip terminology   Heat treatment operation

Circuit-design steps   Layering operation

Circuit layout   Metallization sequence

Composite drawing   Wafer sort

Review Questions

1. Name one layering technique.

2. Which basic operations use the ion implant technique?

3. List the four basic wafer-fabrication operations.

4. Draw and label cross sections of a chip illustrating a doping sequence.

5. Describe a composite drawing.

6. Which basic operation uses a photomask?

7. What parameter is tested at wafer sort (wafer thickness, defect density, circuit function)?

8. What step in the circuit-design process uses a CAD system?

9. List the steps and basic operations in the metallization sequence.

10. What is the role of the holes created in the contact mask step?

Reference

1. R. Iscoff, "VLSI Testing: The Stakes Get Higher," *Semiconductor International,* September 1993, p. 58.

6

Process Yields

Overview

High process yields are essential to the production of reliable chips at a profit. In this chapter, the major yield measurement points are identified along with the major influences on each of them. Typical yields for the different yield points and for different circuits are presented.

Objectives

Upon the completion of this chapter you should be able to:

1. Name the three major yield measurement points in the process.
2. Explain the effect of wafer diameter, die size, die density, number of edge die, and defect density on the wafer sort yield.
3. From a list of individual process step yields, calculate the accumulative fabrication yield.
4. Be able to explain and calculate an overall process yield.
5. Explain the four major influences on fabrication yield.
6. Sketch a yield versus time curve for different process and circuit maturities.
7. Explain the relationship between high process yields and device reliability.

Yield Measurement

Maintaining and improving production yields is the life blood of semiconductor manufacturing. To the casual observer it would seem that the industry is fixated on their production yields. That observation is accurate. It is accurate because of the demanding nature of the process and the sheer number of processes required to produce a packaged chip. These two factors result in a production process that typically ships only 20 to 60 percent of the chips it commits into the wafer-fabrication line.

These yields seem extraordinarily low to most manufacturing engineers. Yet when one considers the goal of producing millions of micron-size patterns in layers that are equally thin, at very stringent cleanliness levels, all within the confines of a $\frac{5}{8}$-in^2 chip, it is a testament to the industry that any chips are produced at all.

Another factor that helps keep yields depressed is the nonrepairable nature of most production mistakes. While defective automobile parts can be replaced, no such option is available in semiconductor manufacturing. Defective chips or wafers generally cannot be recovered. In some cases, chips that fail performance tests can be downgraded for a less demanding use. Scrapped wafers may find a new life as control wafers or "dummies" (see oxidation) in some processes.

Added to these process factors is the volume nature of the business. High capital costs and a higher than average percentage of engineering personnel translate to a high-overhead situation. This high overhead coupled with competition that keeps downward pressure on selling prices requires that most chip producers run a high-volume, high-yield process.

Given all of the factors, the preoccupation with yield is understandable. Most suppliers of equipment and materials tout the yield improvements possible with their products. Likewise, process engineering groups have as their prime responsibility the maintenance and improvement of process yields. Yield percentages are calculated at every process station. However, three major yield points are calculated and monitored to measure and control the process (Fig. 6.1).

Accumulative Wafer-Fabrication Yield

The first major yield measurement point is at the completion of wafer fabrication. This yield is called by a variety of names, including fab yield, line yield, accumulative fab yield, or "cum" yield.

The cum yield is expressed as a percentage and calculated two different ways. One is simply to divide the number of wafers exiting the fabri-

Major Yield Measurement Points

Manufacturing Stage	Measured
Wafer Fabrication- Yield =	$\dfrac{\text{\# Wafers Out}}{\text{\# Wafers Started}}$
Wafer Sort- Yield =	$\dfrac{\text{\# Functioning (Good) Die}}{\text{\# Die On Wafer(s)}}$
Packaging- Yield =	$\dfrac{\text{\# Packages Die Passing Final Test}}{\text{\# Good Die Started Into Packaging}}$

Figure 6.1 Major yield measurement points.

cation area by the number that were started. This simplistic calculation is seldom used because most fabrication lines process a number of different circuits simultaneously. Different circuits have different feature sizes and density factors. Most often the fabrication line will be running a number of different process flows to produce various products, each with a different number of steps and its own degree of difficulty. In these situations a cum yield is calculated for each product type.

Also, a wafer-fabrication line carries a large inventory of in-process wafers, and the cycle times stretch out to 4 to 6 weeks. It is not unusual for one or more of the product types to get bogged down somewhere in the process. The number of wafers exiting the process seldom relate directly to the number started. Using a simplistic counting of wafers in and wafers out would not reflect the true yield of each of the individual circuit types.

The first step in calculating the cum yield is to count the number of wafers that leave each of the individual processes (called *station yields*) and then divide that sum by the number that entered the process.

$$\text{Station yield} = \frac{\text{Number of wafers leaving station}}{\text{Number of wafers entering station}}$$

The station yields are in turn multiplied together to calculate the overall cum yield.

$$\text{Cum yield} = Y(\text{station 1}) \times Y(\text{station 2}) \times \cdots \times Y(\text{station } n)$$

Figure 6.2 lists an 11-step process such as the one illustrated in Chap. 5. Typical station yields are listed in column 3 and the accumulated yield in column 5. For a single product, the cum yield calculated from

STEP	Wafers In	Yield*	Wafers Out	Accumulated Yield
1. Field Oxidation	1000	99.5	995	99.5
2. Source/Drain Mask	995	99.0	985	98.5
3. Source/Drain Doping	985	99.3	978	97.8
4. Gate Region Mask	978	99.0	968	96.8
5. Gate Oxidation	968	99.5	964	96.4
6. Contact Hole Mask	964	94.0	906	90.6
7. Deposit Metal Layer	906	99.2	899	89.9
8. Metal Layer Mask	899	97.5	876	87.6
9. Alloy Metal Layer	876	100	876	87.6
10. Passivation Layer Deposition	876	99.5	872	87.2
11. Passivation Layer Mask	872	98.5	859	85.9

*Yield values are typical for the particular steps.

Figure 6.2 Accumulated (wafer fab) yield calculation.

the station yields is the same as the yield calculated by dividing the number of wafers *out* by the number of wafers *in*. The accumulated yield equals the simple cum yield calculation for this individual circuit.

Typical wafer-fabrication cum yields vary from 50 to 95 percent depending on a number of factors detailed in the following text. The calculated cum yield is used for production planning and by engineering and management as a measure of the process effectiveness.

Wafer-Fabrication Yield Limiters

Wafer-fabrication yield is limited by five dominant factors. They are:

1. Number of process steps
2. Wafer breakage and warping
3. Process variation
4. Process defects
5. Mask defects

Number of process steps

Note in the calculation in Fig. 6.2 that each individual process operation yield must be in the 90 percent range to produce the 85.9

percent cum yield. Illustrated is a fairly simple 11-step process. Most VLSI and ULSI circuits require 50 to 100 major process operations, and each operation requires several steps, which in turn involve a number of substeps. It is easy to appreciate the continual pressures on fabrication areas to maintain high cum yields. The first cum yield limiter is the number of process steps. The more complicated the circuit, with a high number of steps, the lower the expected cum yield.

More process steps also increase the probability that one of the other four yield limiters will affect the wafer during the process. This factor is a tyranny of numbers. For example, in order to achieve a 75 percent accumulated fabrication yield with a 50-step process, each of the individual steps would have to be 99.4 percent! A further tyranny of this type of calculation is that the cum yield can never exceed the lowest individual step yield. If one process step can only achieve a 50 percent yield, the overall cum yield can never be higher than 50 percent.

More manufacturing pressure is present in the fact that for each major process operation there are a number of steps and substeps. In the illustrated 11-step process, the first operation is oxidation. A simple oxidation requires three steps: cleaning, oxidation, and evaluation. Each of the steps requires substeps. (Figure 6.3 lists the six substeps for a typical oxidation cleaning step.) Each of the substeps represents an opportunity to contaminate the wafer, break it, or make another mistake.

Substep	Number of Wafer Handlings
1. Wafers are removed from carrier and placed in cleaning boat.	2
2. Wafers are cleaned, rinsed, and dried.	1
3. Wafers are removed from cleaning boat, inspected, and placed on oxidation boat.	2
4. Boat is removed from furnace.	0
5. Wafers are removed from boat and placed back in carrier.	1
6. Test wafers are removed from carrier and measured.	2
TOTAL NUMBER OF HANDLINGS	8

Figure 6.3 Substeps of oxidation process.

A figure of merit for a profitable merchant semiconductor manufacturer is a minimum of a 75 percent cum yield, with automated lines achieving up to a 90 percent (or better) yield.

Wafer breakage and warping

During the course of the fabrication process, the wafers are handled many times by a combination of manual and automatic techniques. Each handling presents an opportunity to break the relatively fragile wafers. Recall that a typical 6-in-diameter wafer is only about 25 mils (0.025 in) thick. Careful wafer handling is a required operator skill and automatic handlers must be maintained to minimize breakage.

The many heat treatments add to the susceptibility of the wafers to breaking. Strains are induced in the crystalline material that make them vulnerable to breaking in subsequent steps. In manual processes, partial wafers may be processed. However, with automatic processing machines only full-diameter wafers can be processed. Therefore, any breakage, however small, is a cause for rejecting the wafer.

Silicon wafers are relatively easy to handle with good practices and automatic equipment has reduced wafer breakage to a low level. Gallium arsenide wafers, however, are not that resilient and breakage is a major wafer-yield limiter. In gallium arsenide fabrication lines, where the circuits and devices are very high performance and command a high selling price, partial wafers may be processed, especially through the manual processes.

Along with minimizing breakage, the wafer surfaces must remain flat throughout the processing. This is especially true on fabrication lines that use patterning techniques which project the pattern onto the wafer surface. If the surface is warped or wavy, the projected image will become distorted and have out-of-specification image dimensions. Warping comes about from rapid heating and/or cooling of the wafers in tube furnaces. (The solution for this problem is addressed in Chap. 7.)

Process variation

As the wafer comes through the fabrication process it receives a number of doping, layering, and patterning processes, each of which must meet incredibly stringent physical and cleanliness requirements. On a daily basis, even the most sophisticated processes will vary, inducing some variation in the result on the wafer surface. Occasionally, a process will exceed its process limits and produce out-of-specification wafers. A primary benefit of automated processing is the minimization of these process variations.

A goal of process engineering and process control programs is not only to keep each process operating within its control specifications, but to maintain a constant distribution of the process parameters. Most processes exhibit a parameter distribution described by mathematicians as a *normal distribution*. (This distribution is explained in Chap. 15.) It is sometimes called a *central theorem distribution,* which means that most of the data points fall close to the average and there are fewer data points at the extremes. Sometimes a process will produce data points that fall within the specified limits but with most of the points near one of the extremes. Though considered nominally within specification, the distribution is altered and chances are that the final circuit performance will also be altered, possibly out of specification. The challenge in the fabrication area is to keep the individual process distributions constant.

Throughout the process there are a number of inspections and tests (see Chap. 14) designed to detect unwanted variations as well as frequent calibration of the equipment parameters to process specifications. Some of these tests are performed by production personnel and some by quality control organizations. All of the tests and process specifications allow for some variation.

Process defects

Process defects are defined as isolated regions (or spots) of contamination or irregularities on the wafer surface. These defects are often called *spot defects*. Within a circuit, *one* very small defect can render the circuit inoperable. These are called *killer defects* (see Fig. 6.11). Unfortunately, these small isolated defects are not always detectable within the fabrication process. They become evident at wafer sort as rejected chips.

The major sources of these defects are the various liquids, gases, room air, personnel, process machines, and water used in the fabrication area. Particulates and other small contaminants become lodged in or on the wafer surface. Many of these defects occur in the patterning process. Recall that the patterning process requires using a thin, fragile layer of photoresist to protect the wafer surface during the etch steps. Any holes or tears in the photoresist layer from particulates will end up as tiny etched holes in the wafer surface layer. These holes are called pinholes and are a major concern of photomasking engineers. Consequently, the wafers are inspected often, usually after each major step, for contamination. Wafers that exceed the established allowable density are rejected.

Mask defects

A photomask or reticle is the source of the pattern that is transferred to the wafer surface in the patterning process. Defects on the mask or reticle end up on the wafer as defects or pattern distortions. There are three common defects that originate from masks or reticles. The first is contamination from dirt or stains on the clear part of the mask or reticle. In optical lithography, dirt can block the light and print onto the wafer as though it was an opaque part of the pattern. Second are cracks in the quartz. They too can block the patterning light or scatter the light, causing unwanted images and/or distorted images. Third are pattern distortions that may have occurred in the mask- or reticle-making process. These include pinholes or chrome spots, pattern extensions or missing parts, breaks in the pattern, or bridges between adjacent patterns (see Fig. 6.4).

Control of mask-generated defects is more critical for devices and circuits with smaller feature sizes, higher densities, and larger die sizes.

Wafer-Sort Yield Factors

After fabrication, the wafers go to the wafer-sort tester. During the test, each chip will be tested electrically for device specifications and functionality. Up to several hundred individual electrical tests may be performed on each circuit. While these tests measure the electrical performance of the device(s), they indirectly measure the precision and cleanliness of the fabrication processes. Because of natural process variations and undetected defects, the wafer may have passed all the in-process checks and still have many chips that do not function.

Since wafer sort is a comprehensive test, many factors influence the yield. They are:

1. Wafer diameter
2. Die size (area)
3. Number of processing steps
4. Circuit density
5. Defect density
6. Crystal defect density
7. Process cycle time

Figure 6.4 Mask defects: (a) spot; (b) hole; (c) inclusion; (d) protrusion; (e) break; (f) bridge. (*Source: Solid State Technology, July 1993, page 95*).

Wafer diameter and edge die

The semiconductor industry went to round wafers with the introduction of silicon. The first wafers were less than 1 in in diameter. Since that time there has been a regular progression to larger-diameter wafers, with 150 mm (6 in) being the standard of VLSI fabrication lines in the late 1980s and 200-mm wafers being developed for production use in the 1990s. Projections call for 300- and 400-mm diameter wafers early in the 21st century.[1]

The move to larger-diameter wafers has been driven by production efficiency, increasing die sizes, and the influence on the wafer-sort yield. Production efficiency is easily understood when one considers that there is an incremental increase in the cost of processing a larger-diameter wafer while the number of available whole chips on the wafer can increase substantially as illustrated in Fig. 6.5.

The effect of increasing the wafer diameter also has positive effects on the wafer-sort yield. Fig. 6.6 shows two wafers of the same diameter but with different die sizes. Note that the smaller-diameter wafer has a very large proportion of its surface covered with partial die, die that cannot function. The larger-diameter wafer with its greater number and percentage of whole die will, if all other factors are equal, have a higher wafer-sort yield.

Wafer diameter and die size

Another driving force for larger-diameter wafers is the trend to larger die sizes. As shown in Fig. 6.6. increasing the die size *without* increasing the wafer diameter also results in a wafer surface with a smaller percentage of whole die.

Maintaining a decent wafer-sort yield as the die size increases requires increasing the wafer diameter. Figure 6.7 lists the number of

Figure 6.5 Effect of larger wafer diameter on percentage of partial die.

Figure 6.6 Effect of processing larger die on larger wafers.

Die Size	Wafer Diameter (mm)			
(mils)	100	125	150	200
200x200	283	452	660	1194
300x300	126	201	293	531
400x400	71	113	165	298
500x500	45	72	106	191

Figure 6.7 Die size versus number of die on wafer.

various size chips that will fit on different size wafers. The bottom line is that larger-diameter wafers are more cost-effective.

Wafer diameter and crystal defects

In Chap. 3 the concept of a crystal dislocation was introduced. A crystal dislocation is a point defect *in the wafer* that comes from a local discontinuity of the crystal structure. Dislocations exist throughout the crystal structure and, like contamination and process defect density, affect the wafer-sort yield.

Dislocations also are generated during the fabrication process. They generate (or nucleate) at sites where there are chips and abrasions of the edge of the wafer. These chips and abrasions come from poor handling techniques and automatic handling equipment. The abraded area causes a crystal dislocation. Unfortunately, the dislocation is propagated into the center of the wafer (Fig. 6.8) during subsequent heat treatments, such as oxidations and diffusions. The length

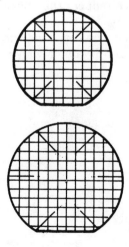

Figure 6.8 Effect of dislocations on wafer-sort yield for different wafer diameters.

of the dislocation line into the interior of the wafer is a function of the thermal history of the wafer. Consequently, wafers receiving more process steps and/or more heating steps will have more and longer dislocation lines. One obvious solution to the problem is larger-diameter wafers. This solution leaves a larger number of unaffected die in the center of the wafer.

Wafer diameter and process variations

The process variations discussed under "Wafer-Fabrication Yield Limiters" affect the wafer-sort yield. In the fabrication area, process variations are detected by sampling inspection and measurement techniques. The nature of inspection sampling is that not all of the variations and defects are detected so that wafers are passed on with some number of problems. These problems show up at wafer sort as failed devices.

Process variations occur at a higher rate around the edge of the wafer. In the high-temperature processes performed in tube furnaces, there is always some temperature nonuniformity across the wafers. The change in temperature results in uniformity differences on the wafer. Variations occur more at the wafer's outer edges where heating and cooling occur at a faster rate. Another contributor to this wafer edge phenomenon is contamination and physical abuse of the wafer layers that emanate from handling and touching the wafers on their edges. Also, in the patterning process there can be feature size uniformity problems in the mask-driven processes (full-mask projection, proximity, and contact exposure). The nature of the light systems is such that the center will be of higher uniformity than the outside edges. In the reticle-driven masking processes (steppers) there is a smaller area of exposure (one or several die), which reduces the image variations across the wafer.

All of these problems result in a lower wafer-sort yield around the edge of the wafer, as illustrated in Fig. 6.9. Larger-diameter wafers

Figure 6.9 Typical location of functioning die after wafer sort.

• **Nonfunctioning Die**

help to maintain wafer-sort yields by having a larger area of unaffected die in the center of the wafer.

Die area and defect density

The die size also affects wafer-sort yield relative to the defect density on the wafer surface. The relationship is illustrated in Fig. 6.10. In Fig. 6.10a, a wafer is shown with five defects and no die pattern. This situation illustrates a background defect density produced by all the fabrication area factors, regardless of the die size, device type, process control requirements, and so on. The wafers in Figs. 6.10b and c illustrate the effect of this background defect density on the wafer-sort yield for two different die sizes. The larger the die size for a given defect density, the lower the yield.

Circuit density and defect density

The defects on the wafer surface result in die failures by causing a malfunction of some part of the die. Some of the defects are located in nonsensitive parts of the die and do not cause a failure. However, the trend is to higher levels of circuit integration, which has come about because of smaller feature size and higher-density die components. The result of these trends is a higher probability that any given defect will be in an active part of the circuit, thus lowering the wafer-sort yield as illustrated in Fig. 6.11.

Number of process steps

The number of process steps was earlier shown to be a limiter of the cum yield. The more steps there are, the more opportunities there are to break or otherwise damage a wafer. This effect also influences the wafer-sort yield. As the number of process steps increases, the background defect density increases unless procedures are implemented to

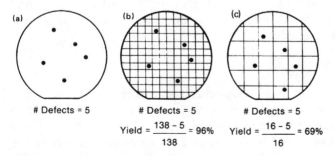

(a) # Defects = 5

(b) # Defects = 5 $\text{Yield} = \dfrac{138 - 5}{138} = 96\%$

(c) # Defects = 5 $\text{Yield} = \dfrac{16 - 5}{16} = 69\%$

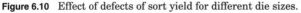

Figure 6.10 Effect of defects of sort yield for different die sizes.

Failed   **Passed**

Figure 6.11 "Killer" defects and nonfatal defects.

lower it. A higher background defect density affects more chips, thus lowering the wafer-sort yield.

Feature size and defect size

Smaller feature sizes make maintaining an acceptable sort yield difficult because of two major factors. First, the smaller images are more difficult to print (see "Mask Defects" and Chap. 8). Second, the smaller images are vulnerable to ever-smaller defect sizes as well as the overall defect density. The 10 to 1 ratio rule of minimum feature size to allowable defect size has been discussed. One assessment is that at a defect density of 1 defect per cm^2 a circuit with a 0.35-μm feature size will have a wafer-sort yield 10 percent less than a 0.5-μm circuit processed under the same conditions.[2]

Process cycle time

The time that the wafers are actually being processed can be measured in days. But because of queuing at the process stations and temporary slowdowns due to process problems, the wafer often stays in the fabrication area for several weeks. The longer the wafer stays in-process, the more opportunity there is for contamination, which in turn will lower wafer-sort yield.

The move to just-in-time manufacturing (see Chap. 15) is one attempt to increase yields and decrease the manufacturing costs associated with high in-line inventories.

Wafer-sort yield formulas

The ability to understand and predict wafer-sort yields with some accuracy is essential to the operation of a profitable and reliable chip supplier. Over the years, a number of models have been developed that relate process, defect densities, and chip size parameters to the wafer-sort yield. Four yield model formulas are shown in Fig. 6.12.

(a) Exponential $Y = \dfrac{1}{e^{AD}}$

(c) Murphy $Y = \left(\dfrac{1 - e^{DA}}{DA}\right)^2$

(b) Seeds $Y = \dfrac{1}{(1 + DA)^n} \left(\dfrac{r - \sqrt{A}}{\sqrt{A}}\right)^2$

(d) "New" $Y = \left(\dfrac{N_{tr}\, d_d\, D_1}{\lambda^{p-2}}\right)$

where Y = percent functioning die
A = die area
D = defect density
n = number of masking steps
r = radius of wafer

N_{tr} = number of transistors
d_d = number of minimum feature size "squares"
D_1 = defect characterization parameter
p = defect characterization parameter
λ = minimum feature size

Figure 6.12 Wafer-sort yield models.

Each relates different parameters to the wafer-sort yield. As the chips get larger, the number of process steps increases, and the feature size decreases, the sensitivity to smaller defect sizes increases and thus more background defects become killer defects.

Exponential model. The exponential relationship (Fig. 6.12a) is the simplest and one of the first yield models[3] developed. It is applicable for wafers that contain over 300 die and MSI circuits of lower densities. The model does not have a factor for edge die as does the Seeds model. The exponential model is a valuable learning tool because it clearly shows the primary relationships among die area, defect density, and wafer-sort yield. To understand the relationship in this mathematical model, remember that e is a constant with a value of 2.718. The exponential model shows that any increase in die area or defect density will increase the value of the divisor on the right-hand side of the equation and drive the wafer-sort yield down.

Seeds model. The Seeds model adds two additional parameters to the job of predicting wafer-sort yield, the number of process steps (n) and the wafer radius–die size relationship r (r and a). In most yield models, the factor for processing steps (n) is actually the number of patterning steps. Experience has proved that the patterning steps contribute the greatest number of point defects and therefore have a direct bearing on sort yield.

Murphy model. Perhaps the most widely used yield model is the Murphy relationship (Fig. 6.12c). While not including a factor for the

Figure 6.13 Murphy's yield model, showing die yield as a function of die size defect density.

number of processing steps, it has proved a good predictor for LSI-, VLSI-, and ULSI-level circuits. Figure 6.13 shows a graphical solution of the Murphy model, where the relationships of sort yield, defect density, and die size are clearly shown.

High-density, small-feature-size model. A newer model (Fig. 6.12d) factors in the minimum feature size (λ), the area occupied by a transistor (d_a), and the design density (lower values of d_a). This model is complicated and several of the parameters had to be fine-tuned to the experimental data. (The reader is directed to reference 4 for a better understanding of the model.) But it illustrates the complexity of relating wafer-sort yield to the many sensitive factors in high-density, small-feature-size chips.[4] Complex circuits rarely have comparable designs and processes vary from company to company as does the basic background defect density. These factors make the development of an accurate universal yield model difficult. Most chip companies have developed their own models that reflect their particular manufacturing processes and product designs.

The defect density used in all the models is not the same as a defect density determined by optical inspection of the wafer surface. The defect density that shows up in the yield models is all-inclusive; it includes contaminants and surface and crystal defects. Further, it predicts only the defects that destroy die: the "killer defects." Defects that fall in noncritical areas of the chip are not part of the models nor are situations where two or more defects fall in the same sensitive area.

It is also important to keep in mind that the yield numbers predicted by the formulas are those expected from a process that is basically under control. In reality the wafer-sort yield will vary from wafer to wafer because of the normal process variations in the fabrication process. A typical wafer-sort yield plot is shown in Fig. 6.14. Note that wafer 13 falls far below the normal range of sort yields. In a situation like this, the process engineer would look for some catastrophic process failure such as an out-of-spec layer thickness or a doping layer that is too deep or too shallow.

Assembly and Final Test Yields

After wafer sort, the wafers go to the assembly (packaging) process. There they are cut into die and packaged into a protective enclosure. During this series of steps, there are a number of visual inspections and quality checks of the assembly process.

At the conclusion of the packaging process, the packaged die go through a series of physical, environmental, and electrical tests, known collectively as the *final tests*. (The details of the processes, inspections, and final tests are described in Chap. 18.) After the final tests, the third major yield is calculated, which is the ratio of die passing the final tests compared with the number of die that passed the wafer-sort test.

Overall Process Yields

The overall process yield is the mathematical product of the three major yield points (Fig. 6.15). This number, usually expressed as a percent,

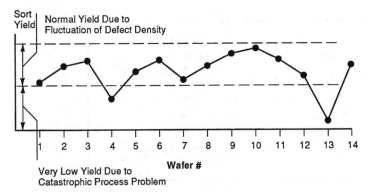

Figure 6.14   Plot of wafer-sort yields.

Overall Yield Formula

(Wafer Fab Yield)	(Wafer Sort Yield)	(Packaging Yield)	= Overall Yield

$$\frac{\text{\# Wafers Out}}{\text{\# Wafers Started}} \text{ X } \frac{\text{\# Functioning (Good) Die}}{\text{\# Die On Wafer(s)}} \text{ X } \frac{\text{\# Packages Passing Final Test}}{\text{\# Die Started Into Packaging}} = \text{Overall Yield}$$

Figure 6.15   Overall yield formula.

gives the percentage of shipped die as compared with the number of whole die on the starting wafer. It is an inclusive measurement of the success of the entire process.

Overall yields vary with several major factors. In Fig. 6.16 is a list of typical process yields and their calculated overall yield. In the first two columns are major process factors that influence the individual and overall yields.

First is the integration level of the particular circuit. The more highly integrated the circuit, the lower the expected yield in all categories. Higher integration levels assume a corresponding decrease in feature size. Column 2 lists the maturity of the manufacturing process. Process yields almost always follow an S curve pattern (Fig. 6.17) through the lifetime of the product in manufacturing. In the beginning, the yield rises rather slowly as the initial bugs are worked out of the process. This is followed by a period when the yields rise rapidly, eventually leveling off because of the limits imposed by the process maturity, including die size, integration level, circuit density, and defect density. If the product goes into a period of steadily declining volume, the causes may be older equipment and less engineering attention.

As the table in Fig. 6.16 shows, overall yields can vary from very low (maybe even zero for new or poorly designed products) to the 90

Integration Level	Product Maturity	Fab Yield %	Sort Yield %	Assembly Final Test %	Overall Yield %
ULSI	Mature	95	85	97	78
ULSI	Mid	88	65	92	53
ULSI	Introduction	65	35	70	16
LSI	Mature	98	95	98	91
Discrete	Mature	99	97	98	94

Figure 6.16   Typical yields for various products.

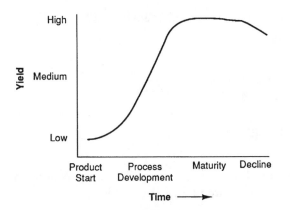

Figure 6.17 Yield changes with process maturity levels.

percent range for simpler or more mature products. Semiconductor producers consider their yield performance very proprietary since profit and production control are a direct function of the process yields.

An examination of the yield values in the table reveals that wafer sort is the lowest of the three yield points. This fact illustrates why yield-improvement programs are directed at the many factors that influence the wafer-sort yield. Improvement of wafer-sort yields has the biggest impact on productivity. The advent of larger and more complex chips (such as megabit memory chips) has shifted productivity improvements to include other factors including the cost of ownership of equipment (see Chap. 15). Successful competition in the megachip era will require wafer-sort yields in the 85–90 percent range.[5]

Key Concepts and Terms

Assembly and final test yields	Typical overall yields
Defect density	Wafer-fabrication yield factors
Murphy's yield formula	Wafer-sort yield factors
Overall yield factors	Wafer-sort yield formulas
Three major yield points	

Review Questions

1. Name the three yield measurement points in the manufacturing process.

2. Of the three yield points, which is typically the lowest?

3. Indicate if the wafer-sort yield would go up or down with the following changes:

   a. Change to larger-diameter wafers

   b. Change to smaller die size

   c. More process steps

   d. Decrease in defect density

   e. Increase in die density

4. If 1000 wafers are started into the process and 875 are passed to wafer sort, what is the accumulated fabrication yield?

5. Calculate the overall yield for a process that has an 82 percent fabrication yield, a 47 percent wafer-sort yield, and a 92 percent assembly and final test yield.

6. Which situation would you expect to have the highest overall yield, a 22-step LSI process or a 33-step VLSI process?

References

1. APT Presentation, "Overall Road Map Technology Characteristics," Industry Strategy Symposium sponsored by The Semiconductor Equipment and Materials Institute, January 1995.

2. Bill Walker, "Motorola VP Defines Sub-Micron Manufacturing Challenges," *Semiconductor International,* Cahners Publishing, October 1994, p. 21.

3. S. M. Sze, *VLSI Technology,* McGraw-Hill Publishing Company, New York, 1983, p. 605.

4. W. Maly, H. T. Heineken, F. Agricola, "A Simple New Yield Model," *Semiconductor International,* Cahners Publishing, July 1994, p. 148.

5. Bill George, Susan Billat in, "Process Control: Covering all of the Bases," *Semiconductor International,* Cahners Publishing, September 1993, p. 80.

7

Oxidation

Overview

The ability of a silicon surface to form a silicon dioxide passivation layer is one of the key factors in silicon technology. In this chapter, the uses, formation, and processes of silicon dioxide growth are explained. Detailed is the all-important tube furnace, which is a mainstay of oxidation, diffusion, heat treatment, and chemical vapor deposition processes. Basic wafer-cleaning techniques (used before most processes) are described.

Objectives

Upon completion of this chapter, you should be able to:

1. List the three principal uses of a silicon dioxide layer in silicon devices.
2. Describe the mechanism of thermal oxidation.
3. Sketch and identify the principal sections of a tube furnace.
4. List the two oxidants used in thermal oxidation.
5. Sketch a diagram of a dryox oxidation system.
6. Draw a flow diagram of a typical oxidation process.
7. Explain the relationship of process time, pressure, and temperature on the thickness of a thermally grown silicon dioxide layer.
8. Describe the principles and uses of rapid thermal, high-pressure, and anodic oxidation.

Silicon Dioxide Layer Uses

The oxidation process is one of the primary layering operations in silicon technology (Fig. 5.1). When a silicon surface is exposed to oxygen, it is converted to silicon dioxide, whose molecule is composed of one silicon atom and two oxygen atoms (SiO_2). We encounter silicon dioxide daily. For example, it is the chemical composition of ordinary window glass. Its semiconductor version, however, is purer and formed in a specific way.

Although silicon is a semiconducting material, silicon dioxide is a dielectric material. This combination, a dielectric layer formed on a semiconductor, along with other properties of silicon dioxide, makes it one of the most commonly used layers in silicon devices. Silicon dioxide layers find use in devices to passify the silicon surface, to act as doping barriers and surface dielectrics, and to serve as dielectric parts of devices.

Surface passivation

In Chap. 4, the extreme sensitivity of semiconductor devices to contamination was examined. While a major focus of a semiconductor facility is the control and elimination of contamination, the techniques are not always 100 percent effective. Silicon dioxide layers play an important role in protecting semiconductor devices from contamination.

Silicon dioxide performs this role in two ways. First is the physical protection of the surface and underlying devices. Silicon dioxide layers formed on the surface are very dense (nonporous) and very hard. Thus a silicon dioxide layer (Fig. 7.1) acts as a contamination barrier by preventing dirt in the processing environment from getting to the sensitive wafer surface. The hardness of the layer protects the wafer surface from scratches and abuse endured by the wafer in the fabrication processes.

The second way silicon dioxide protects devices is chemical in nature. Regardless of the cleanliness of the processing environment, some electrically active contaminants (mobile ionic contaminants) end up in or on the wafer surface. During the oxidation process the top layer of silicon is converted to silicon dioxide. Contaminants on the

**Surface
Passivation**

Figure 7.1 Surface passivation with silicon dioxide layers.

surface end up in the new layer of oxide, away from the electrically active surface. Other contaminants are drawn up into the silicon dioxide film where they are less harmful to the devices. In the early days of MOS device processing, it was common to oxidize the wafers and then remove the oxide before further processing to rid the surface of unwanted mobile ionic contamination.

Doping barrier

In Chap. 5 doping was identified as one of the four basic fabrication operations. Doping requires creating holes in a surface layer through which specific dopants are introduced into the exposed wafer surface through diffusion or ion implantation. In silicon technology, the surface layer is most often silicon dioxide (Fig. 7.2). The silicon dioxide left on the wafer acts to block the dopant from reaching the silicon surface. All of the dopants used in silicon technology have a very slow rate of movement in silicon dioxide as compared to silicon. While the dopants penetrate to the required depth in the exposed silicon, they only penetrate a short distance into the silicon dioxide surface. It only takes a relatively thin silicon dioxide layer to block the dopants from reaching the silicon surface.

Another factor favoring the use of silicon dioxide is that it possesses a coefficient of thermal expansion similar to that of silicon. In high-temperature processes such as oxidation and diffusion doping, the wafer expands and contracts as it is heated and cooled. The silicon dioxide expands and contracts at close to the same rate as silicon, which means that the wafer will not warp during the heating and cooling.

Surface dielectric

Silicon dioxide is classified as a dielectric. This means that under normal circumstances it does not conduct electricity. When dielectrics are used in electrical circuits or devices they are referred to as *insulators.* Acting as an insulator is an important role of silicon dioxide layers. Figure 7.3 shows a cross section of a wafer with a conductive layer of metal on top of a layer of silicon dioxide. The oxide prevents shorting

Figure 7.2 Silicon dioxide layer as dopant barrier.

Figure 7.3   Dielectric use of silicon dioxide layer.

of the metal layer to the underlying metal just as the insulation on an electric cord prevents the wires from shorting. In this capacity the oxide must be continuous, that is, having no holes or voids.

The oxide must also be thick enough to prevent a phenomenon known as *induction*. Induction can occur when the separating layer of oxide is thin enough to allow an electrical charge in a metal layer to cause a buildup of charge in the wafer surface. The surface charge can cause shorting and other unwanted electrical effects. A thick enough layer will prevent an induced charge in the wafer surface. An oxide layer thick enough to prevent induction from the metal layers is called the *field oxide*.

Device dielectric

At the other end of induction phenomena is MOS technology. In a MOS transistor, a thin layer of silicon dioxide is grown in the gate region (Fig. 7.4). The oxide functions as a dielectric whose thickness is chosen *specifically* to allow induction of a charge in the gate region under the oxide. The gate is the part that controls the flow of current through the device (see Chap. 16). The dominance of MOS technology for ultra-large-scale integrated (ULSI) circuits has made the formation of gate regions a prime focus of process development and concern. Thermally grown oxides are also used as the dielectric layer in *capacitors* formed between the silicon wafer and a surface conduction layer (Fig. 7.5).

Figure 7.4   Silicon dioxide as field oxide and in MOS gate.

Figure 7.5   Silicon dioxide layer in solid-state capacitor.

Silicon dioxide dielectric layers are also used in structures with two or more metallization layers. In this application, the silicon dioxide layers are deposited with *chemical vapor deposition* (CVD) techniques rather than thermal oxidation (see Chap. 12).

Thermal Oxidation Mechanism

The silicon dioxide layers used in silicon-based devices vary in thickness. At the thin end of the scale are advanced MOS gate oxides, in the 35 to 80 Å range.[1] At the thick end are field oxides. Figure 7.6 lists the thickness ranges for the major uses.

Most of the silicon dioxide layers are grown on the silicon surface by thermal oxidation. Thermal oxide growth is a simple chemical reaction as shown in Fig. 7.7. This reaction takes place even at room temperature. However, an elevated temperature is required to achieve quality oxides in reasonable process times for practical use in circuits and devices. Oxidation temperatures are between 900 and 1200°C.

Although the formula in Fig. 7.7 shows the reaction of silicon with oxygen, it does not illustrate the growth mechanism of the oxide. To understand the growth mechanism, consider a wafer placed in a heated chamber and exposed to oxygen gas (Fig. 7.8a). Initially, the oxygen atoms combine readily with the silicon atoms. This stage is called *linear* because the oxide grows in equal amounts for each unit of time (Fig. 7.8b). After approximately 500 angstroms (Å) of oxide is grown, a limit is imposed on the linear growth rate. [An angstrom is one ten-thousandth of a micron (μm); in other words, there are 10,000 Å in 1 μm.]

Silicon Dioxide Thickness, Å	Application
60–100	Tunneling Gates
150–500	Gate Oxides, Capacitor Dielectrics
200–500	LOCOS Pad Oxide
2,000–5,000	Masking Oxides, Surface Passivation
3,000–10,000	Field Oxides

Figure 7.6 Silicon dioxide thicknesses chart.

$$\text{Si (solid)} + \text{O}_2 \text{ (gas)} \xrightarrow{\triangle} \text{SiO}_2 \text{(solid)}$$

Figure 7.7 Reaction of silicon and oxygen to form silicon dioxide.

(a)

(b)

Figure 7.8 Silicon dioxide growth stages: (a) initial, (b) linear, and (c) parabolic.

(c)

For the oxide layer to keep growing, the oxygen and silicon atoms must come in contact. However, the initially grown layer of silicon dioxide separates the oxygen in the chamber from the silicon atoms of the wafer surface. For oxide growth to continue, either the silicon in the wafer must migrate through the already grown oxide layer to the oxygen in the vapor or the oxygen must migrate to the wafer surface. In the thermal growth of silicon dioxide, the oxygen migrates (the technical term is *diffuses*) through the existing oxide layer to the silicon wafer surface. Thus the layer of silicon dioxide consumes silicon atoms from the wafer surface. Thus, the oxide layer grows *into* the silicon surface.

With each succeeding new growth layer, the diffusing oxygen must move further to reach the wafer. The effect is a slowing of the oxide growth *rate* with time. This stage of oxidation is called the *parabolic* stage. When graphed, the mathematical relationship of the oxide thickness, growth rate, and time takes the shape of a parabola. Other terms used for this second stage of growth are *transport-limited reaction* and *diffusion-limited reaction,* which mean that the growth rate is limited by the transportation and diffusion, respectively, of the oxygen through the oxide layer already grown. The linear and parabolic stages of growth are illustrated in Fig. 7.9. The formula in Fig. 7.10 expresses the fundamental parabolic relationship for oxide layers above approximately 1200 Å.

Thus a growing oxide goes through two stages, the linear stage and the parabolic stage. The change from linear to parabolic is dependent on the oxidizing temperature and other factors (see Influences on the oxidation rate). In general oxides less than 1000 Å (0.1 μm) are controlled by the linear mechanism. This is the range of most MOS gate oxides. A more detailed explanation of the growth mechanism of silicon dioxide is presented in *Silicon Processing* by Wolf and Tauber.[2]

$$X = B/A \; t$$
Linear oxidation of silicon

$$X = \sqrt{B \; t}$$
Parabolic oxidation of silicon

X = oxide thickness
B = parabolic rate constant
B/A = linear rate constant
t = oxidation time

Figure 7.9 Linear and parabolic growth of silicon dioxide.

$$R = \frac{X^2}{t}$$

where R = silicon dioxide growth rate
X = oxide thickness
t = oxidation time

Figure 7.10 Parabolic relationship of SiO_2 growth parameters.

The major implication of this parabolic relationship is that thicker oxides require much more time to grow than thinner oxides. For example, growth of a 2000-Å (0.20-μm) film at 1200°C in dry oxygen requires 6 min (Fig. 7.11).[3] To double the oxide thickness to 4000 Å requires some 220 min, over 36 times as long. This longer oxidation time presents a problem for semiconductor processing. When pure dry oxygen is used as the oxidizing gas, the growth of thick oxide layers requires even longer oxidation times, especially at the lower temperatures. Generally, process engineers want to have the shortest process times possible as is consistent with quality control. The 220 min in the example given is excessive, since only one oxidation would be possible in one shift of operation.

One way to achieve faster oxidations is to use water vapor (H_2O) instead of oxygen as the oxidizing gas (oxidant). The growth of silicon dioxide in water vapor proceeds by the reaction shown in Fig. 7.12. In the vapor state the water is in the form H-OH−. It is composed of one atom of hydrogen (H) and a molecule of oxygen and hydrogen with a negative charge (OH−). This molecule is called the *hydroxyl ion*. The hydroxyl ion diffuses through the oxide layers already on the wafer faster than oxygen. The net effect is a faster oxidation of the silicon, as shown in the growth curves in Fig. 7.11.

(a)

(b)

Figure 7.11 Silicon dioxide thickness versus time and temperature in (a) dry oxygen and (b) steam.

$$\text{Si (solid)} + H\,O_2(\text{gas}) \xrightarrow[\triangle]{} SiO_2(\text{solid}) + 2H_2 \text{ (gas)} \uparrow$$

Figure 7.12 Reaction of silicon and water vapor to form silicon dioxide and hydrogen gas.

Water vapor at the oxidation temperatures is in the form of steam, and the process is called either *steam oxidation, wet oxidation,* or *pyrogenic steam.* The term wet oxidation comes from the time when liquid water was the primary water vapor source. An oxygen-only oxidation process is called *dry oxidation.* If oxygen only is used, it must be free of any water vapor (dry) or the oxide growth would be that of water vapor.

Notice in the reaction of water vapor and silicon that there are two hydrogen molecules ($2H_2$) on the right side of the equation. Initially these hydrogen molecules are trapped in the solid silicon dioxide layer, making the layer less dense than an oxide grown in dry oxygen. However, after a heating of the oxide in an inert atmosphere, such as nitrogen (see "Oxidation Processes"), the two oxides become similar in structure and properties.

Influences on the oxidation rate

The original oxide thickness vs time curves were determined on <111>-oriented, undoped wafers.[4] MOS devices are fabricated in <100>-oriented, doped wafers. Wafer orientation and doping type and level both influence the oxidation rate for a particular temperature and oxidant environment. Other factors influencing the oxidation growth are impurities intentionally included in the oxide (such as HCl) and oxidation of polysilicon layers.

Wafer orientation. The orientation of the wafer has an effect on the oxidation growth rate. <111> planes have more silicon atoms than <100> planes. The larger number of atoms allows for a faster oxide growth on <111>-oriented wafers than for <100>-oriented wafers. Figure 7.13 shows the growth rates for the two orientations in steam. This difference is seen more in the linear growth stage and at lower temperatures.

Wafer dopant redistribution. The silicon surface being oxidized always has dopants. A production silicon wafer starts into the line doped as either an N type or P type. Later on in the process, wafers have dopant(s) in the surface from diffusion or ion implant operations. The

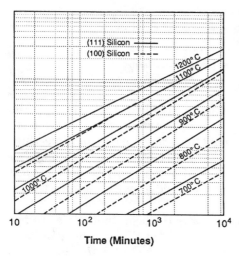

Figure 7.13 Oxidation of <111> and <100> silicon in steam.

Time (Minutes)

dopant elements used and their concentration both have effects on the oxidation growth rate. For example, oxides grown over a highly doped phosphorus layer are less dense than those grown over the other silicon dopants. These phosphorus-doped oxides also etch faster and present an etching challenge in the patterning operation due to resist lifting and rapid undercutting.

Another effect on oxidation growth rate is the distribution of the dopant atoms in the silicon after the oxidation is completed.[5] Recall that during thermal oxidation the oxide layer grows *down* into the wafer. A question is "What happens to the dopant atoms that were in the layer of silicon converted to silicon dioxide?" The answer depends on the conductivity type of the dopant. The N-type dopants of phosphorus, arsenic, and antimony have a higher solubility in silicon than in silicon dioxide. When the advancing oxide layer reaches them, they move down into the wafer. The silicon–silicon dioxide interface acts like a snowplow pushing ahead an ever-greater pile of snow. The effect is that there is a higher concentration (called *pile-up*) of N-type dopants at the silicon dioxide–silicon interface than was originally in the wafer.

When the dopant is the P-type boron, the opposite effect happens. The boron is drawn up into the silicon dioxide layer, causing the silicon at the interface to be depleted of the original boron atoms. Both of these effects, pile-up and depletion, have significant impact on electrical performance of devices. The exact effects of pile-up and depletion on the dopant concentration profile are illustrated in Chap. 17.

Doping concentration effects on the oxidation rate vary with the dopant type and concentration level. In general, heavily doped regions oxidize faster than lightly doped regions. Heavily doped phosphorus regions can oxidize at two to five times the undoped oxidation rate.[6] However, doping-induced effects are more pronounced in the linear stage (thin oxides) of oxidation.

Oxide impurities.   Certain impurities, particularly chlorine from hydrochloric acid (HCl), are included in the oxidizing atmosphere for inclusion in the growing oxide (see "Oxidation Processes"). These impurities have an influence on the growth rate. In the case of HCl, the growth rate can increase from 1 to 5 percent.[7]

Oxidation of polysilicon.   Polysilicon conductors and gates are a feature of most MOS devices and circuits. The device and circuit processes require oxidation of the polysilicon. As compared to single-crystal silicon, the oxidation rate of polysilicon can be faster, slower, or similar. A number of factors related to the formation of the polysilicon structure will influence subsequent oxidation. These include the polysilicon deposition method, deposition temperature, deposition pressure, the type and concentration of doping, and the grain structure of the polysilicon.[8]

Differential oxidation rates and oxide steps.   After the initial oxidation step in a device, or circuit fabrication process, the wafer surface exhibits a variety of conditions. Some areas have the field oxide, some are doped, some are polysilicon regions, and so on. Each of these areas has a different oxidation rate and will increase in oxide thickness depending on the surface condition. This oxidation thickness difference is called *differential oxidation*. For example, oxidation of a MOS wafer after a polysilicon gate has been formed next to lightly doped source or drain areas (Fig. 7.14a) results in a thicker oxide growth on the gate because silicon dioxide grows faster on polysilicon.

Differential oxidation rates cause the formation of steps in the wafer surface. Figure 7.14b shows a step created by the oxidation of an exposed area next to a relatively thick field oxide. The oxide will grow faster in the exposed area since additional oxide growth in the field oxide is limited by the parabolic rate limitation. In the exposed area, the faster-growing oxide will use up more silicon than that under the field oxide. This step is shown in Fig. 7.14b.

Thermal Oxidation Methods

The oxide formation reaction formulas include a triangle under the reaction direction arrows. These triangles indicate that the reaction

Before reoxidation   (a)

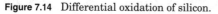

After reoxidation   (b)

Figure 7.14   Differential oxidation of silicon.

requires energy to proceed. In silicon technology that energy is usually supplied by heating the wafers and is called *thermal oxidation.* Silicon dioxide layers are grown either at atmospheric pressure or at high pressure. An atmospheric pressure oxidation takes place in a system without intentional pressure control—the pressure is simply that of the atmosphere for the location. There are two atmospheric techniques, tube furnaces and rapid thermal systems (Fig. 7.15).

Horizontal Tube Furnaces

The mainstay equipment for the thermal oxidation of silicon is the horizontal tube furnace. Horizontal tube furnaces have been used in

Thermal Oxidation		
• Atmospheric Pressure	Tube furnace	Dry Oxygen Wet Oxygen Bubbler Flash System Dry Oxidation
	Rapid Thermal	Dry Oxygen
• High Pressure	Tube Furnace	Dry or Wet Oxygen
Chemical Oxidation		
• Anodic Oxidation	Electrolytic Cell	Chemical

Figure 7.15   Oxidation methods.

the industry since the early 1960s for oxidation, diffusion, heat treating, and various deposition processes. They were first developed for diffusion processes in germanium technology and to this day are often called simply *diffusion furnaces*. The more correct generic term is a *tube furnace*.

A cross section of a basic single horizontal tube furnace is shown in Fig. 7.16. It consists of a long ceramic tube made of mulite (a ceramic), with three to five coils of copper tubing on the inside surface. Each of the coiled tubes defines a zone and is connected to a separate power supply operated by a proportional band controller. Inside the furnace tube is a quartz reaction tube that serves as the reaction chamber for the oxidation (or other processes). The reaction tube may itself be inside a ceramic liner called a *muffle*. The muffle acts as a heat sink, fostering a more even heat distribution along the quartz tube.

Thermocouples are positioned against the quartz tube and send temperature information to the proportional band controllers. The controllers proportion power to the coils, which in turn heat the reaction tube by radiation and conduction. Radiant heating comes from the energy given off by the coils and impinging on the tube. Conduction takes place where the coils touch the tube. These controllers are very sophisticated and can control temperatures in the

Figure 7.16 Cross section of single horizontal tube furnace with three heating zones.

center zone (flat zone) to ± 0.5°. For a process that operates at 1000°C, this variation is only ± 0.05 percent. For the oxidation, the wafers are placed on a holder and positioned in the flat zone. The oxidant gas is passed into the tube where the oxidation takes place. The details and the options used in actual practice are discussed in the following sections.

Horizontal tube furnace systems

A production tube furnace is an integrated system of various sections (Fig. 7.17). They are

1. Source cabinet
2. Furnace section:

   Temperature control system
   Reaction chamber(s)
3. Wafer load station:

   Process automation (elevators, etc.)
4. Wafer-cleaning station

Reaction chamber.   The basic tube furnace operation was previously described. An important part of the system is the reaction chamber. This part protects the wafers from outside contamination and serves as a heat sink to even out the temperature inside the tube. Tube furnace reaction tubes are round with a gas inlet end and a wafer load

Figure 7.17 Tube furnace.

end. The gas inlet end, also known as the source end, tapers down to a ground fitting to provide a leak-free connection to the gas sources. The load end also has a ground fitting to receive either an end cap that keeps dirt out of the furnace or a wafer-transfer unit (see "Wafer Loading").

The traditional reaction chamber material is high-purity quartz. Quartz is a highly purified glass favored for its inherent stability at high temperatures and its basic cleanliness. Drawbacks to quartz are its fragility and the presence of some metallic ions. Quartz also allows sodium ions from the heating coils to pass into the reaction chamber. Another drawback is the tendency of quartz to break up and sag at temperatures above 1200°C. The breakup is called *devitrification* and results in small flakes of the quartz tube surface falling onto the wafers. Sagging impedes the easy placement of the wafer holders in and out of the tube.

Quartz tubes are manufactured by two methods: electric-fused and flame-fused. As the names imply, two different heat sources are used to fuse the quartz. Both materials have about the same levels of trace metals, but in one evaluation flame-fused tubes produced devices with better characteristics.[9]

Quartz tubes require periodic cleaning. One method is in a tank of hydrofluoric (HF) acid or a solution of HF and water. This cleaning process takes place out of the fabrication area, necessitating the removal and cooling of the tube. The cooling and heating hasten the devitrification process. The HF cleans by removing a thin layer of the quartz. This continual etching eventually weakens the tube wall, limiting its lifetime.

Some firms do an in situ tube cleaning with a portable plasma generator. The plasma generator is positioned inside the tube where it activates etching gases passed into the tube. Both nitrogen trifluoride (NH_3) and the etching gases etch away contaminants. A second in situ cleaning method uses etching gases in the tube that are activated by a plasma field created in the tube. This type of cleaning is more applicable to tube furnaces used for chemical vapor deposition (CVD) processes where the buildup of reaction-created particles is greater than in oxidation processes.

An alternative material to quartz process tubes and wafer holders is silicon carbide. Silicon carbide is structurally stronger and does not break down with repeated heating and cooling. This resistance to temperature cycling also makes the material a better metallic ion barrier over a longer period of time. During processes where oxygen is in the tube, the inside surface grows a thin layer of silicon dioxide. When this oxide is removed in HF, there is no attack of the tube material, which also contributes to the extended lifetime. The widespread use of silicon carbide tubes and wafer holders has been slowed by their high cost and weight.

Temperature control system. The temperature control system consists of thermocouples touching the reaction tube that are connected to proportional band controllers that feed the power to the heating coils. Proportional band controllers maintain even temperatures in the tube by feeding in or turning off the current to the coils in proportion to the deviation of the tube temperature from the set point. The closer the tube is to the set-point temperature, the smaller the amount of power that is fed to the coils. This system allows fast recovery of the tube to a cold load without overshoot. *Overshoot* is the raising of the tube temperature too high above the desired process temperature as a result of applying too much power to the coils (Fig. 7.18).

Setup of the temperature control system usually is by long thermocouples sheathed in quartz tubes. The thermocouples are placed inside the tube and moved in increments through the furnace zones and the temperature is recorded. Adjustments are made to the controllers until the desired temperatures in the processing section of the tube (flat zone) are achieved.

Advanced systems have thermocouples positioned against the outside of the tube wall. These feed a microprocessor that, in turn, gives the information to the controllers. This system is called *autoprofiling.*[10] Good process control requires that the temperature profile inside the furnace be checked periodically with thermocouples inserted in the tube and the temperature measured by independent recorders.

The processing of larger-diameter wafers has brought with it the concern of wafer warping. Wafers that are heated or cooled rapidly

Figure 7.18 Temperature levels during oxidation.

will warp to the point of being useless. The degree of warping increases with higher process temperatures, that is, those above 1150°C.

Two methods are employed to minimize warping of wafers in tube furnaces. One is called *ramping* (or temperature ramping). Ramping is the procedure of maintaining the furnace at a temperature several hundred degrees below the process temperature. The wafers are slowly inserted into the furnace at this lower temperature and, after a short stabilization period, the controllers automatically take the furnace up to the process temperature. At the end of the process cycles, the furnace is cooled to the lower temperature before the wafers are removed. During the ramping process the controllers must maintain the temperature control in the flat zone.

Tube furnaces are maintained at close to the process temperature 24 hours a day due to the devitrification of quartzware and the length of time it takes to stabilize the flat zone. In the interest of economy, some fabrication areas will keep the furnaces at the lower temperature. This is called an *idle* condition.

The second antiwarping procedure is the slow loading of the wafer boat into the tube. At loading rates of about 1 in/min, warping is minimized. For large-diameter wafers and large batch sizes, both methods are used.

Another requirement of the heating system is a fast recovery time after the wafers are loaded in the tube. A full load of 200 six-inch-diameter wafers can drop the tube temperature as much as 50°C.[11] The heating system works to bring the flat zone to temperature as fast as possible without introducing warping conditions. Figure 7.18 illustrates a typical temperature-time recovery curve for a five-zone tube furnace.

Furnace section. A production-level tube furnace will contain three or four tubes (reaction chambers) and a separate temperature control system for each of the tubes. The tubes are arranged vertically above each other in a stack. They are referred to as three- or four-stack furnaces or a furnace bank. The load end of the tube sticks into an exhaust chamber of the furnace enclosure. The exhaust draws away the spent and heated gases as they exit the tube. This section of the furnace is called a *scavenger* after its role in *scavenging* the gases from the tubes. The scavenger is connected to the facility's exhaust system, which contains a scrubber to remove toxic gases from the withdrawn gases.

The tubes within a bank may be all used for the same purpose, such as oxidations, or be designated for different operations, such as oxidation, diffusion, alloy, or CVD. The exact designation depends on the process requirements and the volume of wafers being processed. However, each of the tubes is dedicated to a particular use. To change the use requires a change of tubeware and contamination monitoring to ensure that contaminants from a previous use are not present in the individual tube section.

Source cabinet. Each individual tube process requires a number of gases to accomplish the desired chemical reaction. In the case of oxidation, the gas oxidants of oxygen or water vapor have been detailed. In addition, an oxidation process will use a nitrogen gas flow. In fact, all of the tube processes require a source of nitrogen. The nitrogen is used during the loading and unloading stage of the process. When a tube is in the idle condition, nitrogen is kept constantly flowing through the tube. The flow serves to keep dirt out of the system and maintain the preestablished flat zone.

Each of the tube processes requires that the gases be delivered to the tube in a specified sequence, at a specified pressure, at a specific flow rate, and for a specific time. The equipment used to regulate the gases is located in a cabinet attached to the furnace section of the system and is known as the *source cabinet*. There is a separate unit called a *gas control panel* or a *gas flow controller* connected to each tube. The gas control panel consists of solenoids, pressure gauges, mass flow controllers or flow meters, filters, and timers. In its simplest version, the gas flow controller consists of manually operated valves and timers. In production systems, the sequencing and timing of the various gases into the tube is controlled by a microprocessor. Required gases are plumbed to the panel. During operation, a timer opens a solenoid to admit the required gas to the panel. Its pressure is controlled by a pressure gauge. Flow amount into the tube is controlled by a flow meter or mass flow controller.

A *mass flow meter* is preferred in place of a flow meter because of its inherent superior control. Stoichiometric considerations require that the same amount of material, as measured by mass, be delivered into the reaction chamber. Standard flow meters measure the volume of material, but equal volumes of the same material may contain different amounts of material due to pressure and temperature differences. Semiconductor processes use a thermal type of mass flow meter. The system consists of a heated gas passage tube with two temperature sensors. When no gas is flowing, the temperature sensors are at the same temperature. With the introduction of a gas flow the downstream sensor will indicate a higher temperature. The difference between the two sensors is related to the amount of heat mass (not volume) that has moved downstream. The meter has a feedback mechanism to control the gas flow such that a steady amount of material flows through the meter. The source section also contains the microprocessor-controlled valves that allow the gas into the reaction chamber in the right sequence for the right amount of time. A general schematic of a mass flow meter is shown in Fig. 7.19. Mass flow meters can be set for a specific value, and on-board sensors will measure and control the output with a feedback control system.[12] The piping material used in gas flow controllers is stainless steel to maintain high levels of cleanliness and to minimize chemical reactions between the gas and tube material.

Often the gas flow controller is called a *jungle*. This term came about when the gas controllers were built in-house and had the look of a "jungle" of tubing and valves. Gases are supplied to the gas flow controller either through piping from the liquid gas supplies in the pad section of the facility or by smaller lecture bottles of gas located at the process tool.

Some processes require a chemical that is difficult to deliver in gas form. In this situation a bubbler and liquid source is used. A *bubbler* consists of a quartz vial designed to admit a gas into a liquid. As the gas bubbles through the liquid and mixes with the source vapors at the top of the bubbler, it picks up the source chemicals and carries them into the tube. Bubblers are used in oxidation, diffusion, and CVD processes. An oxidation bubbler is shown in Fig. 7.20. Diffusion and CVD bubblers generally have a cylindrical shape.

Oxidant sources

Dry oxygen. When oxygen is used as the oxidant, it is supplied from the facility source or from tanks of compressed oxygen located in or near the source cabinet. It is imperative that the gas be dry, that is, not contaminated with water vapor. The presence of water vapor in the oxygen would increase the oxidation rate and cause the oxide layer to be out of specification. Dry oxygen oxidation is the preferred method for growing the very thin (<1000 Å) gate oxides required for MOS devices.

Water vapor sources. There are several methods used to supply water vapor (steam) into the oxidation tube. The choice of method depends on the level of thickness and cleanliness control required of the oxide layer in the device.

Bubblers. The traditional method of creating a steam vapor in the tube is by a bubbler (see "Source Cabinet" section). For oxidation, the bubbler liquid is deionized (DI) water (Fig. 7.20) heated close to the boiling point (98 to 99°C), which creates a water vapor in the space above the

Figure 7.19 (a) Operating principle of mass flow meter; (b) cutaway.

Figure 7.20 Bubbler water vapor source.

liquid. As the carrier gas is bubbled through the water and passes through the vapor, it becomes saturated with water. Under the influence of the elevated temperature inside the tube, the water vapor becomes steam and causes the oxidation of the silicon surfaces.

Drawbacks to bubbler systems are changes in the amount of water vapor entering the tube as the water level in the bubbler changes and fluctuations in the water temperature. With bubblers there is always concern about contamination of the tube and oxide layer from dirty water or dirty flasks. This contamination potential is heightened by the need to open the system periodically to replenish the water.

Flash systems. A *flash system* is also a glass or quartz system (Fig. 7.21). The system is connected to a continuous source of deionized water through a narrow tube designed to allow small drops of the water into the flask. The drops fall onto a heated quartz surface maintained at a high enough temperature to "flash" the drops into steam. A carrier gas also fed into the system takes the steam created into the oxidation tube. Flash systems suffer the same control problems as bubblers but are somewhat cleaner in that they do not have to be replenished with water.

Dry oxidation (dryox). Bubblers and flash systems are adequate for thick and noncritical oxide layers, but new levels of thickness control

Figure 7.21 Flash system water vapor source.

and cleanliness came with the introduction of MOS devices. The heart of a MOS transistor is the gate structure, and the critical layer in the gate is a thin, thermally grown oxide. Liquid-water-steam systems are unreliable for growing thin, clean gate oxides. The answer was found in dry oxidation (or dry steam) process (Fig. 7.22).

In the dry oxidation system, gaseous oxygen and hydrogen are introduced directly into the oxidation tube. Inside the tube, the two gases mix and, under the influence of the high temperature, form steam. The result is a wet oxidation in steam. Dryox systems offer improved control and cleanliness over liquid systems. First, gases can be purchased in a very clean and dry state. Second, the amounts going into the tube can be very precisely controlled by the mass flow controllers. Dryox is the preferred general oxidation method for production for all advanced devices.

A drawback to dryox systems is the explosive property of hydrogen. At oxidation temperatures, hydrogen is very explosive. Precautions used to reduce the explosion potential include separate oxygen and hydrogen lines to the tube and flowing excess oxygen into the tube. The excess oxygen ensures that every hydrogen molecule (H_2) will combine with an oxygen atom to form the nonexplosive water molecule, H_2O. Other precautions used are hydrogen alarms and a hot filament in the source cabinet and in the scavenger end of the furnace to immediately burn off any free hydrogen before it can explode.

Chlorine-added oxidation. The thinner MOS gate oxides require very clean layers. Improvements in cleanliness and device performance are achieved when chlorine is incorporated into the oxide.[13] The chlorine tends to reduce mobile ionic charges in the oxide layer, reduce structural defects in the oxide and silicon surface, and reduce charges at the oxide-silicon interface. The chlorine comes from the inclusion of anhydrous chlorine (Cl_2), anhydrous hydrogen chloride (HCl), trichloroethylene (TCE), or trichloroethane (TCA) in the dry oxygen gas stream. The gas sources, chlorine, hydrogen chloride, and oxygen, are metered into the tube from separate flow meters in gas flow controller. The liquid sources, TCE and TCA, are carried into the tube as vapors from liquid bubblers. For safety and ease of delivery, TCA is the preferred source of chlorine. The oxidation-chlorine cycle may take place in one step or be preceded or followed by a dry oxidation cycle.

O_2 $\longrightarrow$

H_2 $\longrightarrow$

Figure 7.22 "Dryox" (dry steam) water vapor source.

Wafer-cleaning station

When the wafers come to a tube furnace, they first go to the attached wafer-cleaning station. The need for stringent cleanliness control has been stressed throughout this text. It is especially important before the tube operations because of the heating of the wafers. Contamination left on the wafer surfaces can pit the surface, diffuse into the surface, or interfere with the quality of the layer being grown.

The cleaning station may be a VLF "wet" bench that has built-in tanks for cleaning chemicals, units for rinsing the wafers in deionized water, and drying units. The station may be, or include, an automatic cleaning machine. (Wafer cleaning is detailed in the section on "Oxidation Processes.")

Wafer-load station

After cleaning, the wafers are passed to the in-line load station. Here the wafers are inspected for cleanliness and loaded into holders for insertion into the tube. The station is located under a ceiling-mounted HEPA filter or in an VLF hood. The loading station and the cleaning station are located next to each other in such a manner that the wafers stay in a clean environment during the transfer.

A quick surface inspection of the wafers is normally done with the aid of an ultraviolet light. These high-intensity light sources allow the operator to see small particles and stains not visible to the naked eye. Sometimes a microscope inspection of the surface is performed.

Manual wafer loading. Wafers are processed through the cleaning steps in quartz, Teflon, or Teflon-derivative wafer holders, also called *boats* or *cassettes*. They are transferred to quartz or silicon carbide holders for the furnace process. The boats come in various designs

Figure 7.23 Wafer boat styles. (a) Flat or slab; (b) slotted flat; (c) cradle; (d) slotted cradle boats on paddle.

(Fig. 7.23) depending on the degree of control required of the operation and the loading density. The most productive designs have the wafers standing up in machined slots, positioned crosswise to the gas flow. In tube operations where uniformity is critical, the wafers may be loaded parallel to the gas flow.

The wafers are manually removed from the chemical cleaning boats with vacuum wands or limited-grasp tweezers (Fig. 7.24). Vacuum wands are attached to a vacuum source and are designed to allow grasping of the wafers from the backside. This arrangement minimizes damage and contamination of the sensitive front side of the wafer.

A faster boat-loading method is by dump transfer. In this method the cleaning boats and the tube boats are the same size and designed to fit together with a pin-and-slot arrangement. The operator positions the empty tube boat over the loaded cleaning boat and, holding them tightly together, rotates them until the empty boat is under the full one. During the rotation the wafers are "dumped" into the empty boat. The loaded boat is placed inside a transfer tube called an *elephant*. The elephant has a ground joint that mates with the joint on the load end of the tube. It also has a hole in the back end to admit a quartz rod that hooks into the back end of the wafer boat. The operator pushes the boat into the flat zone of the tube furnace (Fig. 7.25).

Automatic wafer loading. Production efficiency and VLSI cleanliness requirements have led to the development of a variety of automatic

(a)

(b)

Storage Box

Process Boat

(c)

Figure 7.24 Manual wafer handling devices. (a) Vacuum pickup; (b) limited grasp tweezer; (c) flip transfer boats.

Transfer Tube
(Elephant)   Process Tube

Push Rod   Slotted
Paddle   Cradle
                  Boats

Figure 7.25 Transfer tube loading of wafers.

boat- and tube-loading mechanisms. Automatic boat loaders come in
two styles: dump and pick-and-place. The dump style is an automated
version of manual dump transfer. Pick-and-place machines (sometimes called *robots*) pick each of the wafers out of one cassette and
place it into an empty one. Some versions of pick-and-place machines
pick up the entire load of wafers and transfer them in one operation
to the empty cassette. A challenge to any wafer boat-loading system is
the correct placement of test wafers within the load of device wafers
as well as dopant source wafers (see "Concept of Diffusion" in Chap.
11). These wafers must be *picked* from other boats. An additional
challenge is the loading of two wafers back to back in the same slot.
This procedure is used to increase the productivity of the operation.

The loaded cassettes are put into the furnace by a number of techniques. One version uses an elephant but automates the push-pull
action with a push rod connected to a motor. Another version has a
number of cassettes placed by the operator on a platform called a
paddle. The paddle moves the cassettes into the tube under the control of a motor (Fig. 7.25).

Boat loading can be a source of severe wafer contamination from
particles scraped from the inside of the tube as the cassettes are
shoved down the tube. One technique for solving this problem is to
use cassettes or paddles with small wheels. The rolling wheels kick
up fewer particles than does scraping. Another solution is a cantilever
system, which features a rigid rod with the cassettes positioned on
the end. The rod moves the cassettes down the center of the tube
without ever touching its sides. Some systems leave the cassettes suspended in the center of the tube while others give the cassettes a soft
landing on the tube bottom.

Automated push-pull machines are necessary to achieve the control
required for slow entry and exit of the wafers to prevent warping. A
typical push-pull rate is 1 in/min. A standard four-stack furnace system requires an automatic push-puller for each of the tubes. Although
these systems load and unload automatically, they require an operator to place the loaded boats on the paddle.

An overriding goal of process equipment design is "hands-off" operation. That need is addressed in tube furnace systems built on the elevator design.[13] The concept employs a storage buffer of loaded tube cassettes whose tube destination is known by a computer. When it is time for the cassettes to be loaded in the tube, a robot selects the proper cassettes and moves to the correct tube position and loads them into the tube. The advantages of these systems are the computer production control and the need for only one loading mechanism per furnace stack. For a properly designed system, robotic activity is less contaminating to the wafers than human operators.

Oxidation process automation

Tube furnaces have evolved from operator-controlled manual systems to various levels of automation. A modern production tube furnace is automated from wafer load to wafer unload in a one-button operation. A goal is to automate the process from wafer cleaning to wafer unload. The operator places the cassettes of cleaned wafers on the load mechanism and pushes the start button. The computer takes over and directs the tube furnace through the predetermined process, which is called a *recipe*. The times, temperatures, gas sequences, and push-pull rates are programmed into an onboard computer or a host computer. Full automation of a tube operation involves delivery of the proper wafers to the station with robots or automatic equipment taking the wafers entirely through the process and passing them on to the next process. The various components of a furnace process may be organized in a *cluster* arrangement (see Chap 15).

Vertical Tube Furnaces

Horizontal tube furnaces have enjoyed great popularity over the years. A lot of process development has taken place in their four decades of use. However, the needs for greater contamination control and the move to ever-increasing wafer diameters as well as the need for more productive processing has pushed horizontal tube furnaces to a limit. While contamination control has led to the development of cantilever systems, they have had to get larger and stronger as the wafers have grown in diameter and weight. Also, the larger wafer diameters require larger tube diameters, which puts additional pressures on the maintenance of extended temperature flat zones. Few firms still use horizontal tube furnaces for 200-mm and larger wafers.[14]

The larger wafer loads have caused a lengthening of the furnaces and their associated load stations. In clean-room terms, the "footprint" of the equipment is increasing. The problem is that larger foot-prints require larger and more expensive clean rooms.

There are also process problems associated with larger-diameter horizontal tubes. One is keeping the gas streams in a laminar flow pattern in the tube. *Laminar gas flow* is uniform with no separation of the gases into layers and without the turbulence that causes uneven reactions within the tube.

These considerations have resulted in the development of vertical tube furnaces.[15] Several designs are available and are becoming the furnace configuration of choice. In this configuration, the tube is held in a vertical position (Fig. 7.26) with loading taking place from the top or bottom. Tube materials and heating systems are the same as for horizontal systems.

The wafers are loaded in standard cassettes and lowered or raised into the flat zone. This action is accomplished without the particulates generated by the cassettes scraping the sides of the tubes. In this configuration, the wafers are in the most dense loading for a tube furnace. An added plus for vertical tube furnaces is the ease of rotating the wafers in the tube, which produces a more uniform temperature across the wafer.

Process uniformity is also enhanced by a more uniform (laminar) gas flow in a vertical tube. In a horizontal system, gravity tends to

Vertical
Process Tube

Controls

Loading
Station

Figure 7.26 Vertical tube furnace.

separate mixed gases as they flow down the tube. In a vertical system, the gas moves parallel to gravity, thus minimizing the gas separation problem. The boat rotation minimizes gas turbulence. Vertical furnaces are capable of producing process variations that are 60 percent of those in horizontal furnaces.[16]

Particle generation associated with boat scraping in horizontal systems is virtually eliminated in vertical systems. Also, the smaller area required for loading results in a cleaner system. Particle densities can reach into the $0.01/cm^2$ range.[17]

Perhaps one of the most appealing cost aspects of vertical tube furnaces is the small footprint. The system is smaller than a conventional four-stack system. Vertical systems offer the possibility of locating the furnaces outside the clean room with only a load station door opening into the clean room. In this arrangement, the clean-room footprint of the furnace is practically zero and maintenance can take place from the service chase. Another possible arrangement of vertical furnaces is in an island or cluster configuration. The furnaces are arranged around a central robot that alternately loads several furnaces. A simpler design translates into a more reliable furnace with lower maintenance costs and longer periods of uptime.

Rapid Thermal Processing (RTP)

Each advance in the density of circuits, reduction of feature size, and increase in wafer diameter strains the applicability of traditional processes. Ion implantation has replaced thermal diffusion due to its inherent doping control. However, ion implantation requires a follow-on heating operation, called *annealing,* to cure out crystal damage induced by the implant process. The annealing step has been traditionally done in a tube furnace. Although the heating anneals out the crystal damage, it also causes the dopant atoms to spread out in the wafer, an undesirable result. This problem led to the investigation of alternate energy sources to achieve the annealing without the spreading of the dopants. The investigations led to the development of rapid thermal process (RTP) technology.

RTP technology is based on the principle of radiation heating (Fig. 7.27). The wafer is automatically placed in a chamber fitted with gas inlets and exhaust outlets. Inside, a heat source above (and possibly below) the wafer provides the rapid heating. Heat sources include graphite heaters, microwave units, and plasma arc and tungsten halogen lamps.[18] Tungsten halogen lamps are the most popular.[19] The radiation from the heat source couples into the wafer surface and brings it up to the process temperature at rates of 75–125°C/s. The

Lamp Type System

Figure 7.27   RTP design (Source: *Semiconductor International, May 1993*).

Figure 7.28   Example RTP time/temperature curve. (Source: *Semiconductor International, May 1993*).

same temperature would take minutes to reach in a conventional tube furnace. A typical time-temperature cycle is shown in Fig. 7.28. Likewise, cooling takes place in seconds. With radiation heating, because of its very short heating times, the body of the wafer never comes up to temperature. For the ion implant annealing step, this means that the crystal damage is annealed while the implanted atoms stay in their original location.

Use of RTP reduces the *thermal budget* required for a process. Every time a wafer is heated to near-diffusion temperatures, the doped regions in the wafer continue to spread down and sideways (see Chap. 11). Every time a wafer is heated and cooled, more crystal dislocations form (see Chap. 3). By minimizing the total time and thermal budget, RTP allows more dense designs and fewer failures from dislocations. Another advantage is single-wafer processing. The move to larger-diameter wafers has introduced uniformity requirements that, in many processes, is best met by a single-wafer process tool.

RTP technology is a natural choice for the growth of thin oxides used in MOS gates. The trend to smaller feature sizes on the wafer surface has brought along with it a decrease in the thickness of layers added to the wafer. Layers undergoing dramatic reduction in thickness are thermally grown gate oxides. Advanced production devices are requiring gate oxides less than 100 Å thick. Oxides this thin are sometimes hard to control in conventional tube furnaces due to the problem of quickly supplying and removing the oxygen from the system. RTP systems can offer the needed control by their ability to heat and cool the wafer temperature very rapidly. RTP systems used for oxidation, called *rapid thermal oxidation* (RTO) systems, are similar to the annealing systems but have an oxygen atmosphere instead of an inert gas. A typical time/temperature/thickness relationship for RTO is shown in Fig. 7.29.

Other processes using RTP technology include wet oxide growth, localized oxide growth, source/drain activation after ion implant, LPCVD polysilicon, amorphous silicon, tungsten, salicide contacts, LPCVD nitride, and LPCVD oxide.[20]

During the 1980s, RTP was touted as a mainstream process for advanced devices and circuits. The hope was that RTP could become the replacement for virtually every heat-based process. In reality, development has somewhat lagged that hope. A primary hurdle is temperature uniformity. Temperature uniformity control across a wafer is different in a radiation chamber then in a furnace tube. In an RTP system, the wafer never comes to thermal stability. The problem is particularly acute at the wafer edges. Another problem comes from the number and variety of layers already on an in-process wafer. These different layers each absorb the heating radiation in a different way, resulting in temperature variations across the wafer, which contribute to temperature nonuniformity. This phenomenon is called *emissivity* and is a property of the particular material used and wavelength of the heating radiation source. Temperature nonuniformity creates nonuniform process results in and on the wafer surface and, if the temperature differential is high enough, crystal slip at the wafer's edge.

Solutions to the temperature uniformity problem include lamp placement, such as top and bottom lamps, and control of individual lamps in the system. Some systems have a heated annular ring to

Figure 7.29 Oxidation of silicon by RTO (*Source: Ghandi, VLSI Fabrication Principles*).

keep the edge of the wafer within the required temperature range. Process temperatures are usually measured by thermocouples. However, thermocouples require back contact with the heated wafer (which is impractical in a single-wafer system), and they have a response time that is longer than some RTP heating cycles. Optical pyrometers, which gauge temperatures by measuring characteristic energies given off by the heated object, are preferred. However, they too are prone to errors, especially on wafers with a number of layers. The difficulty is relating the emission given off by the wafer to the actual temperature on the surface. Solutions to this problem include a backside seal layer of silicon nitride to minimize backside emissivity variations[21] and open-loop lamp control. *Open-loop control* is based on converting lamp control to direct current (dc) to avoid voltage variations to the lamps. Other approaches involve elaborate sampling of and/or filtering of the radiation coming off the wafer to more closely relate the measurement to the wafer-surface temperature. It has also been suggested that measurement of the wafer expansion, which is directly due to temperature increases, may be a more reliable and direct measurement technique.[22]

The benefits of RTP, especially in a high-production, single-wafer processing environment, still outweigh the disadvantages and RTP development will continue.

High-Pressure Oxidation

The thermal budget problem also was an impetus for the development of high-pressure oxidation. Two high temperature oxidation problems are the growth of dislocations in the bulk of the wafer and the growth of hydrogen-induced dislocations along the edge of openings in layers on the wafer surface.[23] In the first case, the dislocations cause various device performance problems. In the latter case, surface dislocations cause electrical leakage along the surface or the degradation of silicon layers grown on the wafer for bipolar circuits.

The growth of dislocations is a function of the temperature the wafer is processed at and the time it spends at that temperature. A solution to this problem is to perform thermal oxidation processes at a lower temperature. This solution by itself causes the production problem of longer oxidation times. The solution that addresses both problems is high-pressure oxidation (Fig. 7.30). These systems are configured like conventional horizontal tube furnaces but with one

Figure 7.30 High-pressure oxidation.

major exception: the tube is sealed and the oxidant is pumped into the tube at pressures of 10 to 25 atm (10 to 25 times the pressure of the atmosphere). The containment of the high pressure requires encasing the quartz tube in a stainless steel jacket to prevent it from cracking.

At these pressures, the oxidation proceeds at a faster rate than in atmospheric systems. A rule of thumb is that a 1-atm increase in pressure allows a 30°C drop in the temperature. In a high-pressure system, that increase relates to a drop of 300 to 750°C in temperature. This reduction is sufficient to minimize the growth of dislocations in and on the wafers.

Another option using high-pressure systems is to maintain the regular process temperature and reduce the time of the oxidation. Other considerations concerning high-pressure systems focus on the safe operation of the system and possible contamination from the additional pumps and piping needed to create the high pressures inside the tube.

Very thin MOS gate oxide growth is a candidate for high-pressure oxidation. The thin oxide must have structural integrity (no holes, etc.) and have a dielectric strength high enough to prevent charge induction in the gate region. Gate oxides grown in high-pressure processes have higher dielectric strength than similar oxides grown at atmospheric pressure.[24] High-pressure oxidation is also a solution for the *bird's beak* problem that occurs during local oxidation of silicon (LOCOS) (see Chap. 16). An unwanted bird's beak of oxide grows into the active region of a MOS device, as shown in Fig. 7.31. High-pressure oxidation can minimize the bird's beak encroachment into the device area and minimize field oxide thinning during LOCOS processing.[25]

In addition to oxidation, high-pressure systems are finding some use in CVD epitaxial depositions and for flowing glass layers on the wafer surface.[26] Both of these processes are of higher quality when performed at lower temperatures.

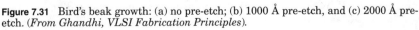

Figure 7.31 Bird's beak growth: (a) no pre-etch; (b) 1000 Å pre-etch, and (c) 2000 Å pre-etch. (*From Ghandhi, VLSI Fabrication Principles*).

Incoming Wafers ——▶ Log and Preclean
                         ▼
                Etch
                 ▼
       Load on Oxidation Boat
                 ▼
       Calibrate Furnace
               ▼
       Load Boat-Oxidize
              ┃   Cycle 1-2-3
              ▼
      Unload Oxidation Boat
               ▼
      Evaluate Oxide and Log
              ┗━━━▶ Transfer

Figure 7.32   Oxidation process flow.

Preoxidation Wafer Cleaning

The processing of the wafers through the oxidation process is divided into several distinct steps, as shown in the flow diagram in Fig. 7.32. After the wafers are logged into the station, they receive a thorough cleaning. Clean wafers are essential at all stages of the fabrication process but are especially necessary before any of the operations performed at high temperature. Up to 20 percent of all process steps relate to wafer cleaning.[27] The cleaning techniques described are used throughout the wafer-fabrication process. (Where appropriate, the reader will be referred to the techniques in this chapter.)

The story of semiconductor process development is in many respects the story of cleaning technology keeping pace with the increasing need for contamination-free wafers. Wafer surfaces have four general types of contamination. Each represents a different problem on the wafer and each is removed by different processes. The four types are

1. Particulates

2. Organic residues

3. Inorganic residues

4. Unwanted oxide layers

Particulate removal

Particulates on the wafer surface vary from very large ones (50-μm size) to very tiny ones less than a micron in size. The larger ones can be cleaned off by conventional chemical baths and rinses. The smaller particulates are more difficult to remove because they are often held to the surface by strong intermolecular forces. A particulate cleaning process is most often a series of steps designed to remove both the large and small particles.

The simplest particulate removal process is to blow off the wafer surface using a spray of filtered high-pressure nitrogen from a hand-held gun located in the cleaning stations. In fabrication areas where small particles are a problem, the nitrogen guns are fitted with ionizers that strip static charges from the nitrogen stream and neutralize the wafer surface.

Nitrogen blow-off guns are effective in removing most large particles. Since the guns are hand-held, the operators must use them in a manner that does not contaminate other wafers in the station or the station itself. Blow-off guns are not generally used in class 1 or 10 clean rooms.

Wafer scrubbers

The stringent wafer cleanliness requirements for epitaxial growth led to the development of mechanical wafer surface scrubbers. These machines are used wherever particulate removal is critical.

The scrubbers hold the wafer on a rotating vacuum chuck (Fig. 7.33). A rotating brush is brought in near contact with the rotating wafer while a stream of deionized water is directed onto the wafer surface. Often a detergent will be added to the water to increase the cleaning effectiveness and prevent static buildup. The combination of the brush and wafer rotations creates a high-energy cleaning action at the wafer surface. The liquid is forced into the small space between the wafer surface and the brush ends where it achieves a high velocity, which aids the cleaning action. Caution must be exercised to keep the brushes and cleaning liquid lines clean to prevent secondary con-

D.I. Water and Detergent Spray

Fiber Roller

Wafer Rotation

Mechanical Scrubber

Figure 7.33 Mechanical scrubber.

tamination. Also, the brush height above the wafer must be maintained to prevent scratching the wafer surface.

Scrubbers are designed as standalone units with automatic loading capabilities or are built into other pieces of equipment to clean the wafers automatically before processing.

High-pressure water cleaning

The removal of statically attached particles first became a necessity for the cleaning of glass and chrome photomasks. A high-pressure water spray process was developed, which uses a small stream of water pressurized 2000 to 4000 psi. The stream is swept across the mask or wafer surface, dislodging both large and small particles. Often a small amount of surfactant will be added to the water stream to act as a destatic agent.

Organic residue removal

Organic residues are compounds that contain carbon, such as oils in fingerprints. These residues can be removed in solvent baths such as acetone, alcohol, or TCE. In general, solvent cleaning of wafers is avoided whenever possible due to the difficulty of drying the solvent completely off the wafer surface. Also, solvents always contain some impurities that may themselves represent a contamination source.

When solvents are used, there is a series of baths. A typical process may be one bath of TCE (or other solvent) followed by an isopropyl alcohol rinse, followed by an acetone rinse. Each of the succeeding rinses dissolves the previous solvent. Acetone is the most volatile of the solvents used and is the easiest to remove from the surface. In some processes, the acetone step is followed by a rinsing with deionized water and drying. The use of acetone is discouraged because of its high flammability.

Inorganic residue removal

Inorganic residues are those that do not contain carbon. Examples are inorganic acids, such as hydrochloric or hydrofluoric acid, that may be introduced from other steps in wafer processing. The inorganic residues are cleaned from the wafers in a variety of cleaning solutions described in the following section.

Chemical cleaning solutions

A wide range of cleaning processes exists in the semiconductor industry. Each fabrication area has different cleanliness needs and different experiences with different solutions. The solutions described in this section are those in common use, although there are numerous variations and different combinations of solutions from one fabrication area to another. The processes used to clean wafers before doping, deposition, and metallization are also described. (The special case of photoresist removal is addressed in Chap. 8.)

Liquid chemical cleaning processes are generally referred to as *wet processes* or *wet cleaning*. The cleaning takes place in glass, quartz, or polypropylene tanks fitted into the deck of a cleaning station (see Chap. 4). Where heating of the solution is required, the tank may be sitting on a hot plate, be wrapped with heating elements, or have an immersion heater inside. The entire process generally includes one or more chemical cleanings, a wafer rinse step, and a drying step. Often the movement of the wafers through the various baths is controlled by a robot or moving beam.

Sulfuric acid. The most common chemical cleaning solution is hot sulfuric acid with an added oxidant. The sulfuric acid is an effective cleaner in the 90 to 125°C range. At these temperatures, it will remove most inorganic residues and particulates from the surface. Oxidants are added to the sulfuric acid to remove carbon residues. The chemical reaction converts the carbon to carbon dioxide, which leaves the bath as a vapor by the reaction

$$C + O_2 \rightarrow CO_2 \rightarrow (gas)$$

The oxidants normally used are hydrogen peroxide (H_2O_2), ammonium persulfate [$(NH_4)_2S_2O_8$], nitric acid (HNO_3), and ozone (O_2).

Sulfuric acid and hydrogen peroxide. The solution of hydrogen peroxide and sulfuric acid is a common wafer cleaner used at all stages of processing, but especially before the tube processes. It is also used as a photoresist stripper in the patterning operation. Within the industry this solution is known by a number of names, including *Carro's acid* and *piranha etch*. The latter term attests to the aggressiveness and effectiveness of the solution.

Two mixing methods are used. One is the addition of about 30% (by volume) of hydrogen peroxide to a beaker of room-temperature sulfuric acid. In this ratio an exothermic reaction takes place that quickly raises the temperature of the bath into the 110 to 130°C range. This is the temperature range for effective cleaning. As time proceeds, the reaction slows and the bath temperature falls below the effective range. At this point the bath may be recharged with additional hydrogen peroxide or discarded. Recharging the bath eventually results in a lowered cleaning rate due to the conversion of the hydrogen peroxide to water, which dilutes the sulfuric acid.

Another way to use these two chemicals is to heat the sulfuric acid to the effective cleaning temperature range and add small amounts (50 to 100 ml) of hydrogen peroxide before cleaning each batch of wafers. This method, while requiring a heating and temperature control system, maintains the bath at the proper temperature and the water created from the hydrogen peroxide evaporates out of the solution. The use of heated sulfuric acid is preferred for economic and process control reasons. It is also easier to automate this approach to mixing the two chemicals.

Sulfuric acid and ammonium persulfate. A drawback to the use of hydrogen peroxide in cleaning solutions is its reactiveness with itself and other chemicals. Hydrogen peroxide is unstable and is constantly giving off oxygen molecules. Safety demands that hydrogen peroxide bottles have vented caps. Unfortunately, the evolution of the oxygen gradually degrades the effectiveness of the chemical, especially in hot climates or heated storage rooms. Also, hydrogen peroxide is classified as an oxidizer due to its rapid (and often violent) reactions with many other chemicals. This property requires that special precautions be taken to keep it from mixing with other chemicals or splashing onto operators where it can burn the skin.

An oxidant that performs the same role when mixed with heated sulfuric acid but without the drawbacks mentioned is ammonium persulfate (AP),[28] a white crystalline material that contains 6.7% oxygen. The oxygen is released into the bath when it is mixed into the heated sulfuric acid. A typical process would be the addition of 40 g of AP to every 1.5 gal of sulfuric acid in the bath. The AP is added before cleaning every run of wafers. This method maintains the required level of oxidant in the bath. A drawback is the buildup of any contaminants contained in the AP as each succeeding amount is added to the bath. Being a crystal, AP does not have the deterioration problem of hydrogen peroxide or the safety problem. However, it is an operator-controlled process that is less easy to automate.

Ozone. The purpose of oxidant additives is to add oxygen to the solution. Some companies use a flow of gaseous ozone (O_2) directly in the container of sulfuric acid. There is some research indicating that a combination of ozone and DI water provides a cleaning solution for light organic contamination.[29]

RCA clean. In the mid-1960s, Walter Kern, an RCA engineer, developed a two-step process to remove organic and inorganic residues from silicon wafers. The process proved to be very effective and the formulas became known simply as the *RCA cleans.* Whenever an RCA cleaning process is referred to, it means that hydrogen peroxide is used along with some base or acid. The first step, standard-clean-1 (SC-1), uses a solution of water, hydrogen peroxide, and ammonium hydroxide. Solutions vary in composition from 5:1:1 to 7:2:1 and are heated to the 75 to 85°C range. The second step, standard-clean-2 (SC-2) uses a solution of water, hydrogen peroxide, and hydrochloric acid in ratios of 6:1:1 to 8:2:1 and used at the 75 to 85°C temperature range. The original mixtures are shown in Fig. 7.34.

The RCA formulas have proven durable over the years and are still the base process for most prefurnace wafer cleaning. Improvements in chemical purity have kept pace with industry cleaning needs. Depending on the application, the order of the SC-1 and SC-2 steps may be reversed. Where an oxide-free surface is required, an HF step is used, either before, in between, or after the RCA cleans.

There have been many adaptations and changes made to the original cleaning solutions. One problem is the removal of metallic ions from the wafer surface. These ions exist in chemicals and are not dissolved (made soluble) in most cleaning or etching solutions. The addition of a *chelating agent,* such as ethylenediamine-tetra-acetic acid, serves to bind up the ions so they do not redeposit on the wafer.

Oxide removal

The ease of silicon oxidation has been mentioned. The oxidation can take place in air or in the presence of oxygen in the heated chemical cleaning baths. Often the oxide grown in the baths, while thin (100 to 200 Å), is thick enough to block the silicon surface from reacting properly during one of the other process operations. The thin surface oxide

RCA Clean Type	Parts by Volume
Standard Clean (SC-1)	5: DI water
	1: 30% Hydrogen peroxide
	1: 29% Amonium hydroxide
	Process 70 degrees centigrade for 5 min.
Standard Clean 2 (SC-2)	6: DI water
	1: 30% hydrogen peroxide
	1: 37% hydrochloric acid
	Process 70 degrees centigrade, 5-10 min.

Figure 7.34 RCA clean formulas.

can act as an insulator, preventing good electrical contact between the silicon surface and a layer of conducting metal.

The removal of these thin oxides is a requirement in many processes. Silicon surfaces with an oxide are called *hydroscopic*. Surfaces that are oxide free are termed *hydrophobic*. Hydrofluoric acid (HF) is the acid favored for oxide removal. Prior to an initial oxidation, when the surface is only silicon, the wafers are cleaned in a bath of full-strength HF (49%). The HF etches away the oxide but does not etch the silicon.

Later in the processing, when the surface is covered with previously grown oxides, thin oxides in patterned holes are etched away with a water and HF solution. These solutions vary in strength from 100:1 to 10:7 (H_2O to HF). The strength is chosen depending on the amount of oxide already on the wafer, since the water and HF solution will etch both the oxide on the silicon surface in the hole and the oxide covering the rest of the surface. A strength is chosen to ensure the removal of the oxide in the holes while not excessively thinning the other oxide layers.

Spray cleaning.   The standard cleaning technique has been immersion in chemical baths performed in wet benches or automated machines. While wet cleans are projected to carry the industry into the 0.35–0.50-μm era, there are growing concerns. The chemicals are becoming more expensive, immersion in a tank presents a redeposition of contaminants, and smaller and deeper patterns on the wafer surface constrain cleaning efficiency. Alternate cleaning methods have emerged. Spray cleaning, while still based on conventional chemistries, offers several advantages. Chemical costs are decreased because the chemicals used are sprayed directly at the wafer rather than maintained in large reserves in tanks. Less chemical use reduces the cost of treating or removing hazardous chemical waste. Cleaning efficiency is improved because the pressure of the spray assists in cleaning small patterns on surfaces with deep holes. Also, there is less chance of recontamination because the wafer receives only fresh chemicals. Spray methods allow immediate water rinsing after cleaning, eliminating the need to transfer wafers to a separate water rinse station.

Dry cleaning.   The considerations concerning wet immersion methods have spurned interest and development of *vapor* or *gas phase* cleaning. For cleaning, the wafers are exposed to vapors of the cleaning or etching chemical(s). HF and water vapor mixtures have found their way into processes for oxide removal. Systems are available for vapor phase replacement of peroxide-based cleaners.[30]

The ultimate industry dream is all dry cleaning and etching. Dry etching methods (e.g., plasma, see Chap. 9) are well established. Dry cleaning is under development. Ultraviolet (UV) ozone can oxidize and photodissociate contaminants form the wafer surface. Some development has gone into cryogenic cleaning. In one version, high-pressure carbon dioxide (CO_2) is directed at the wafer from a nozzle. As the gas leaves the nozzle, its pressure drops, causing a rapid cooling which in turn forms CO_2 particles, or *snow*. The physical bombardment of the surface is the cleaning mechanism. Argon aerosol is another cryogenic technique.

Water rinsing

The cleaning chemicals can also be contaminants if left on the surface. Therefore, every wet-cleaning sequence is followed by a rinse in deionized water. The water rinse also serves to end the etching action of an oxide etch. Rinsing is done by several different methods.

Overflow rinsers. The need for atomically clean surfaces precludes just dunking the wafers in a tank of water. Thorough rinsing requires a continuous supply of clean water to the wafer surface. One such method is an overflow rinser (Fig. 7.35). Physically the rinser is a box sunk into the cleaning station deck. Deionized water is brought into the bottom of the box and flows through and around the wafers, exiting over a dam into a drain system. The rinsing action of the flowing water is enhanced by a stream of nitrogen bubbles introduced into the rinser from the bottom plate. As the nitrogen *bubbles* up through the water, it aids the mixing of the chemicals on the wafer surface with the water. This type of system is often called a *bubbler*.

A rule of thumb is that adequate rinsing takes a minimum of 5 min (depending on wafer diameter) with a flow rate equivalent to five times the volume of the rinser per minute (the number of turnovers

Figure 7.35 Rinse systems: (a) single overflow; (b) three-stage overflow.

per minute). If the size of the bubbler is 3 L, the flow rate should be a minimum of 15 L/min.

The rinse time is determined by measuring the resistivity of the water as it exits the rinser. The chemicals used in the cleaning processes act as charged molecules in the rinse water and their presence can be inferred from the resistivity of the water. If the water goes into the rinser at an 18-mΩ level, a reading of 15 to 18 mΩ on the exit side indicates that the wafers are cleaned and rinsed. The rinsing step is so critical that generally a minimum of two rinsers are used and the total rinsing time is set at two to five times the minimum determined by resistivity measurements. Often a water resistivity meter is mounted on the outlet to constantly measure the output resistivity. The operators are instructed to leave the wafers in the rinser until the specified resistivity is reached. Some systems will sound an alarm when the correct resistivity is reached.

Cascade rinse systems.   A rinse system based on the overflow principle is the cascade system. A cascade rinse system is two or three overflow rinsers connected to each other. The design is such that the water enters only the end rinser and cascades through the downstream rinsers. The wafers start the rinse process in the downstream rinser and are moved sequentially to the rinser with the direct water supply (Fig. 7.35b). When several boats of wafers are being rinsed simultaneously, this system is more efficient than others in its use of water.

Spray rinsing.   Flowing water removes a water-soluble chemical from the wafer surface by a dilution mechanism. The top layer of the chemical dissolves into the water and is carried away by the flow. This action occurs over and over in a continuous manner. Rinsing is speeded up by a faster water flow rate, which can remove the dissolved chemical more quickly. Actually, it is the number of turnovers that directly influences the rinse rate. This can be understood by considering a very fast water flow rate but in a huge rinse tank. The chemical removed from the wafer surface would be evenly distributed throughout the tank and therefore some would still cling to the surface. The chemical could only be removed from the tank when enough water has flowed in and out of it to carry away the chemical.

One way to have a faster rinse rate is by using a water spray. The spray removes the chemical with a physical force from its own momentum and the many small droplets continually hitting the wafer surface have the effect of an extremely high turnover rinse. Along with more efficient rinsing, a spray rinser uses considerably less water than an overflow rinser. A problem with a spray rinser comes when the resistivity of the exiting water is measured with a resistivity monitor. Carbon dioxide from the air gets trapped in the water spray and the CO_2 molecules act as charged particles and the resistivity meter reads them as contaminants, which they aren't.

Dump rinsers. With respect to rinsing efficiency and water savings, a dump rinser (Fig. 7.36) is an attractive method. The system is like an overflow rinser but with a spray capability. The wafers are placed into the dry rinser and immediately sprayed with deionized water. While they are being sprayed, the cavity of the rinser is rapidly filled with water. As the water overflows the top, a trap door in the bottom swings open and the water is dumped instantly into the drain system. This fill-and-dump action is repeated several times until the wafers are entirely rinsed.

Dump rinsing is also favored because all of the rinsing takes place in one cavity, which saves equipment and space. It is also a system that can be automated, so that the operator only needs to load the wafers in (this can be done automatically) and push a button.

Sonic-assisted cleaning and rinsing. The addition of sonic energy waves to a tank of cleaning chemicals or a rinse system aids and speeds wet processes. Sonic waves are energy waves generated from transducers arranged on the outside of the tank (see Fig 7.37). Two ranges are used. In the 20,000- to 50,000-hertz (1 hertz (Hz) = 1 cycle per second) range they are called *ultrasonic* and in the 850-kilohertz (kHz) range they are called *megasonic* waves.[31] The waves pass through the liquid causing microscopic bubbles to form and collapse rapidly. The phenomenon is called *cavitation*. The energy given off by the cavitation action dislodges particles on the surface. Also, the agitation of the liquid assists the liquid in penetrating small holes and patterns on the wafer.

Figure 7.36 Spray/dump rinser.

Figure 7.37 Ultrasonic/megasonic wafer cleaning/ etching bath.

Spin-rinse dryers (SRDs). After rinsing, the wafers must be dried. This is not a trivial process. Any water that remains on the surface (even atoms) has the potential of interfering with any subsequent operation. Complete drying is accomplished in a centrifugelike piece of equipment. In one version, the wafer boats are put in holders around the inside surface of a drum. In the center of the drum is a pipe with holes that is connected to a source of deionized water and hot nitrogen (Fig. 7.38).

The drying process actually starts with a rinse of the wafers as they rotate around the center pipe that sprays the water. Next, the SRD switches to a high-speed rotation as heated nitrogen comes out of the center pipe. The rotation literally throws the water off the wafer surfaces. The heated nitrogen assists in the removal of small droplets of water that may cling to the wafer.

SRDs are also built for drying single-wafer boats. The boat slips into a rotating holder in the center of a chamber. The water and nitrogen come into the chamber through its side rather than through a pipe in the center. The rinsing and drying take place as the boat spins about its own axis. This type of SRD is called an *axial dryer.* These two machines are used for automatic wafer cleaning and etching. As a wafer cleaner, the

Figure 7.38 Spin rinse dryer styles. (a) Multiboat; (b) single boat axial.

required chemicals are plumbed to the machine and microprocessor-controlled valves direct the right chemicals into the chamber.

Vapor dry. A chemical process for wafer drying relies on the displacement of surface water by heated solvent vapors. The process is called *vapor drying*. The dryer is a tank with a sump of heated solvent [usually isopropyl alcohol (IPA)]. Wafers are suspended in the vapors above the liquid level. When they are withdrawn from the dryer, the solvent vapors evaporate, leaving the wafers dry.

A water jacket surrounding the top of the dryer condenses vapors that run down the sides and replenish the liquid in the sump. Unfortunately this recirculation can leave the solvent and wafers contaminated.

Oxidation Processes

After the wafers are precleaned and etched and loaded on the push-pull mechanism, they are ready for surface oxidation. The actual oxidation proceeds in different gas cycles (Fig. 7.39) within the furnace tube. The first gas cycle occurs as the wafers are being loaded into the tube. Since the wafers are at room temperature and precise oxide thickness is a goal of the operation, the gas metered into the tube during loading is dry nitrogen. The nitrogen is necessary to prevent any oxidation while the wafers are coming up to the required oxidation temperature.

Once the wafers are stabilized at the correct temperature, the flow gas controller switches the gas flow to the selected oxidant. For oxides greater than 1200 Å, the oxidant is usually steam from one of the sources previously discussed. For oxides less than 1200 Å, pure oxygen is usually used because of its greater process control and the cleaner, denser oxide it produces.

The thinner MOS gate oxides require very clean layers. Improvements in cleanliness and device performance are achieved when chlorine is incorporated into the oxide.[32] The chlorine tends to reduce mobile ionic charges in the oxide layer, reduce structural

Cycle	Gas	Purpose
1.	Nitrogen	Temperature Stabilization in an Inert Atmosphere
2.	Oxygen or Water Vapor	Oxide Growth
3.	Nitrogen	Stop Oxidation and Removal of Wafers in an Inert Atmosphere

Figure 7.39 Oxidation process cycles.

defects in the oxide and silicon surface, and reduce charges at the oxide-silicon interface. The chlorine comes from the inclusion of anhydrous chlorine (Cl_2), anhydrous hydrogen chloride (HCl), trichloroethylene (TCE), or trichloroethane (TCA) in the dry oxygen gas stream. The gas sources, chlorine and hydrogen chloride, are metered into the tube along with the oxygen from separate flow meters in gas flow controller. The liquid sources, TCE and TCA, are carried into the tube as vapors from liquid bubblers. For safety and ease of delivery, TCA is the preferred source of chlorine.

The oxidation-chlorine cycle may take place in one step or be preceded or followed by a dry oxidation cycle. After the oxidation cycle, the furnace gas is switched back to dry nitrogen. The nitrogen terminates the oxidation of the silicon by diluting and removing the oxidant used. It also prevents any oxidation during the wafer exit step.

Postoxidation Evaluation

After the wafers are removed from the oxidation boats, they will receive an inspection and evaluation(s). The nature and number of the evaluations depend on the oxide layer and the precision and cleanliness required of the particular circuit being fabricated. (The details of the evaluations performed are explained in Chap. 14.)

A requirement of the oxidation process is a uniform noncontaminated layer of silicon dioxide on the wafer. As the wafers proceed through the wafer-fabrication operations, there is a buildup of both thermally grown oxides and other deposited layers on the wafer surface. These other layers interfere with the determination of the quality of a particular oxide. For this reason each batch of oxidized wafers going into the tube includes a number of test wafers with bare surfaces placed at strategic locations on the wafer boat. Test wafers are necessary for the evaluations that are destructive or require large undisturbed areas of oxide. At the conclusion of the oxidation operation, they are used for the evaluation of the process. Some of the evaluations are performed by the oxidation operator and some are performed off-line in quality control (QC) labs.

Surface inspection

A quick check of the cleanliness of the oxide may be performed by the operator as the wafers are unloaded from the oxidation boat. Each wafer is viewed under a high-intensity ultraviolet (UV) light. Surface particulates, irregularities, and stains are readily apparent in UV light.

Oxide thickness

The thickness of the oxide is of major importance. It is measured on test wafers by a number of techniques (Chap. 14). The techniques are color comparison, fringe counting, interference, ellipsometers, stylus apparatus, and scanning electron microscopes (SEMs).

Oxide and furnace cleanliness

In addition to the physical contaminants of particles and stains, the oxide should have a minimum number of mobile ionic contaminants. These are detected by the sophisticated capacitance-voltage (C/V) technique, which detects the total number of mobile ionic contaminants present in the oxide. It cannot identify the origin of those contaminants, which may come from the tubes, the gases, the wafers, or the cleaning process. Therefore, C/V evaluation is a go–no-go assessment of the wafers and serves as a check of the total furnace operation. In most fabrication lines, C/V analysis is also used to certify the cleanliness of the furnace and its associated parts. An oxide with a low-mobile ionic contamination level certifies that the entire system is clean. When the oxide fails the test, more investigation is necessary to identify the source.

A second oxide-cleanliness-related parameter is dielectric strength. This parameter is a measure of the dielectric (nonconducting) property of the oxide. It is measured by the destructive oxide rupture test.

A third cleanliness factor is the index of refraction of the oxide. Refraction is the property of a transparent substance that causes light to bend as it travels through it. The apparent versus actual location of an object on the bottom of a body of water is an example of refraction. The index of refraction of a pure oxide is 1.46. Variations from this value come about from impurities in the oxide. A constant index of refraction is the starting point for several of the interference thickness-measurement techniques. Variations in the index can lead to erroneous thickness measurements. The index of refraction is measured by interference and ellipseometry techniques (Chap. 14).

Anodic Oxidation

In addition to thermal oxidation, there is one chemical oxidation process used to grow thin oxides on silicon. The method is called *anodic oxidation* and is similar to chemical plating. In this application, the silicon wafer is connected to a positive electrode (the anode) in a bath that also contains a negative electrode. Both are submersed in an electrolyte of KNO_3.[33] When a current is applied between the two electrodes, oxygen is created at the anode and forms silicon diox-

Figure 7.40 Anodic oxidation.

ide on the wafer surface (Fig. 7.40). The oxide formed is generally of lesser quality than a thermally grown oxide, but some renewed interest in the method has been shown for the formation of very thin gate oxides.

The principal use of anodic oxides is in profiling dopant concentrations down into the wafer surface. This is possible because in anodic oxidation the silicon from the wafer surface travels to the top of the oxide layer rather than oxygen traveling to the wafer surface. When an anodic oxide is etched from a silicon surface, it leaves the dopants near the surface undisturbed, unlike a thermally grown oxide. By using a four- or two-point probe to measure the resistivity of the surface and repeating the anodic oxidation and etch steps, the concentration of the dopants at different levels can be determined.

Thermal Nitridation

An important factor in the production of small high-performance MOS transistors is a thin gate oxide. However, in the 100-Å (or less) range, silicon dioxide films tend to be of poor quality and difficult to control. An alternative to silicon dioxide films is a thermally grown silicon nitride (Si_3N_4) film. Si_3N_4 is denser than silicon oxide and has less pinholes in these thin ranges. It also is a good diffusion barrier. Growth control of thin films is enhanced by a flat growth mechanism (after an initial rapid growth). This characteristic is shown (Fig. 7.41) in the growth of silicon nitride formed by the exposure of the silicon surface to ammonia (NH_3) between 950 and 1200°C.[34]

Nitridation kinetics

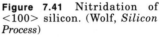

Figure 7.41 Nitridation of <100> silicon. (Wolf, *Silicon Process*)

Some advanced devices use silicon oxynitride (SiO_xN_y) films. They are also called nitrided-oxide or nitroixide films. These are formed from the nitridation of silicon oxide films. Unlike silicon dioxide films, oxynitride films vary in composition depending on the growth process.[35] Another MOS gate structure is a sandwich of oxide-nitrite-oxide (ONO).[36]

Key Concepts and Terms

Dielectric	Oxide
Doping barrier	Oxide growth mechanism
High-pressure oxidation	Oxide layer evaluation
Oxidant sources	Passivation layer
Oxidation	Preoxidation cleaning
Oxidation dopant effects	Rapid thermal processing
Oxidation orientation effects	Silicon dioxide
Oxidation process cycles	Tube furnace

Review Questions

1. Describe the role and movement of oxygen during thermal oxidation.

2. Draw a sketch of a horizontal tube furnace and identify all the sections.

3. Name the water vapor source favored for VLSI oxidation processes.

4. Why are temperature ramping techniques required?

5. What is the advantage of rapid thermal oxidation?

6. Name the two oxidants added to sulfuric acid for wafer cleaning.

7. Make a flow diagram of the oxidation steps, from incoming wafer to evaluation.

8. Name the three factors that determine the oxide thickness.

9. What advantages are offered by high-pressure oxidation?

10. List three uses of silicon dioxide layers in semiconductor devices.

References

1. C. Hu, "MOSFET Scaling in the Next Decade and Beyond," *Semiconductor International,* Cahners Publishing, June 1994, p. 105.
2. S. Wolf and R. Tauber, *Silicon Processing for the VLSI Era.* Lattice Press, Sunset Beach, CA, 1986.
3. P. Gise and R. Blanchard, *Modern Semiconductor Fabrication Technology,* Reston Books, Reston, VA, p. 43.
4. S. M. SZE, *VLSI Technology,* McGraw-Hill Book Company, 1983, p. 137.
5. P. Gise and R. Blanchard, *Modern Semiconductor Fabrication Technology,* Reston Books, Reston, VA, p. 46.
6. S. K. Ghandhi, *VLSI Fabrication Principles,* John Wiley & Sons, Inc., 1994, p. 464.
7. S. M. SZE, *VLSI Technology,* McGraw-Hill Book Company, 1983, p. 147.
8. Ibid., p. 159.
9. M. Hill, D. Helman, and M. Rother, "Quartzglass Components and Heavy Metal Contamination," *Solid State Technology,* PennWall Publishing, March 1994, p. 49.
10. J. Maliakal, D. Fisher, Jr., and A. Waugh, "Trends in Automated Diffusion Furnace Systems for Large Wafers," *Solid State Technology,* December 1984, p. 107.
11. Ibid.
12. C. Murray, "Mass Flow Contollers: Assuring Precise Process Gas Flows," *Semiconductor International,* Cahners Publishing, October 1985, p. 72.
13. P. Burggraaf, "'Hands-Off' Furnace Systems," *Semiconductor International,* September 1987, p. 78.
14. P. Burggraaf, "Verticals: Leading Edge Furnace Technology," *Semiconductor International,* Cahners Publishing, September 1993, p. 46.
15. P. Singer, "Trends in Vertical Diffusion Furnaces," *Semiconductor International,* April 1986, p. 56.
16. P. Burggraaf, "Verticals: Leading Edge Furnace Technology," *Semiconductor International,* Cahners Publishing, September 1993, p. 46.
17. Ibid.
18. P. Singer, "Rapid Thermal Processing: A Progress Report," *Semiconductor International,* Cahners Publishing, May 1993, p. 68.
19. S. Leavitt, "RTP: On the Edge of Acceptance," *Semiconductor International,* March 1987.
20. M. Moslehi, A. Paranjpe, L. Velo, and J. Kuehne, "RTP: Key to Future Semiconductor Fabrication," *Solid State Technology,* Penwall Publishing, May 1994, p. 37.
21. Ibid, p. 38.
22. P. Singer, "Rapid Thermal Processing: A Progress Report," *Semiconductor International,* Cahners Publishing, May 1993, p. 69.
23. D. Toole and P. Crabtree, "Trends in High-Pressure Oxidation," *Microelectronic Manufacturing and Test,* October 1988, p. 1.

24. S. K. Ghandhi, *VLSO Fabrication Principles,* John Wiley and Sons, 1994, p. 466.
25. S. Kim, A. Emami, and S. Deleonibus, "High-Pressure and High-Temperature Furnace Oxidation for Advanced Poly-Buffered LOCOS," *Semiconductor International,* May 1994, p. 64.
26. D. Toole and P. Crabtree, "Trends in High-Pressure Oxidation," *Microelectronic Manufacturing and Test,* October 1988, p. 8.
27. R. Allen, S. O'Brian, L. Loewenstein, M. Bennett, and B. Bohannon, "MMST Wafer Cleaning," *Solid State Technology,* Penwall Publishing, January 1996, p. 61.
28. P. Van Zant, "Ammonium Persulfate as a Stripping and Cleaning Oxidant," *Semiconductor International,* April 1984, p. 109.
29. P. Burggraaf, "Keeping the 'RCA' in Wet Chemistry Cleaning," *Semiconductor International,* June 1994, p. 86.
30. R. Allen, S. O'Brian, L. Loewenstein, M. Bennett, and B. Bohannon, "MMST Wafer Cleaning," *Solid State Technology,* Penwall Publishing, January 1996, p. 62.
31. Wolf and Tauber, *Silicon Processing for the VLSI Era,* p. 519.
32. Ibid., p. 226.
33. P. Schmidt and W. Michel, "Anodic Formation of Oxide Films on Silicon," *Journal of the Electrochemical Society,* April 1957.
34. Wolf and Tauber, *Silicon Processing for the VLSI Era,* p. 210.
35. S. K. Ghandhi, *VLSI Fabrication Principles,* John Wiley and Sons, Inc., New York, NY, 1994, p. 484.
36. P. Singer, "Directions in Dielectrics in CMOS and DRAMs," *Semiconductor International,* Cahners Publishing, April 1994, p. 57.

Photolithography—Surface Preparation to Exposure

Overview

Photolithography processes are used to create the patterns on the wafer surface. This chapter starts with an explanation of a basic ten-step photo process and includes a discussion of photoresist chemistry. The majority of the chapter describes in detail the first four steps, from surface preparation to alignment and exposure. Each of the options available for these steps is explained and evaluated.

Objectives

Upon completion of this chapter, you should be able to:

1. Sketch wafer cross sections showing the basic ten-step photomasking process.

2. Explain the reaction of negative and positive photoresists to light.

3. Describe the correct resist and mask polarities required to produce holes and islands in wafer surface layers.

4. Make a list of the major process options for each of the ten basic steps.

5. Select from the list in objective 4 the processes used to pattern features in micron and submicron sizes.

6. Describe the need for, and process steps used in, double masking, multilayer resist processing, and planarization techniques.

7. Explain the use of antireflective coatings and contrast enhancement in the patterning of "small" feature sizes.

8. List the optical and nonoptical methods used for alignment and exposure.

9. Compare the equipment and advantages of each alignment and exposure method.

Introduction

Patterning is the basic operation that removes specific portions of the top layer(s) on the wafer surface (see Fig. 8.1). *Photolithography* is one of the terms used to identify this basic operation. Other terms used are *photomasking, masking, oxide* or *metal removal* (OR, MR), and *microlithography.*

Photolithography is one of the most critical operations in semiconductor processing. It is the process that sets the surface (horizontal) dimensions on the various parts of the devices and circuits. The goal of the operation is twofold. First, it is to create in and on the wafer surface a pattern whose dimensions are as close to the design requirements as possible. This goal is referred to as the *resolution* of the images on the wafer. The pattern dimensions are referred to as the *feature sizes* or *image sizes* of the circuit.

The second goal is the correct placement (called *alignment* or *registration*) of the circuit pattern on the wafer. The entire circuit pattern must be correctly placed on the wafer surface and the individual parts of the circuit must be in the correct positions relative to each other (Fig. 8.2). Keep in mind that the final pattern is created from several photomasks applied to the wafer in a sequential manner. The registration requirement is similar to the correct alignment of the different floors of a building. It is easy to imagine the effect on the functioning of a stairwell or elevator if the floors did not sit directly on top

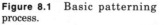

Figure 8.1 Basic patterning process.

#2
GATE
MASK

#1
WELL
MASK

#3
CONTACT
MASK

#4
METAL
MASK

#5
PAD
MASK

Figure 8.2 Five mask set silicon gate transistor.

of each other. In a circuit, the effects of misaligned mask layers can cause the entire circuit to fail.

Control of the dimensions and alignment is difficult because each of the individual process steps is vulnerable to errors. The photomasking process is full of tradeoffs, as is explained in the process step sections. In addition to dimensional control and pattern alignment, defect control during the process steps is critical. Given the number of steps in each patterning operation and the number of mask layers, the masking process is the chief source of wafer defects.

Overview of the Photomasking/Patterning Process

Photolithography is a multistep pattern transfer process similar to photography or stenciling. The required pattern is first formed in reticles or photomasks and transferred into the surface layer(s) of the wafer through the photomasking steps.

The transfer takes place in two steps. First, the pattern on the reticle or mask is transferred into a layer of photoresist (Fig. 8.3). *Photoresist* is a light-sensitive material similar to the coating on regular photographic film. Exposure to light causes changes in its structure and properties. In the example in Fig. 8.3, the photoresist in the region exposed to the light was changed from a soluble condition to an

PROCESS STEP	PURPOSE	
Alignment and Exposure	Precise alignment of mask/reticle to wafer and exposure of photoresist. Negative resist is polymerized.	Mask/reticle Resist Oxide layer Wafer
Development	Removal of unpolymerized resist.	Resist Oxide layer Wafer

Figure 8.3 First pattern transfer—mask/reticle to resist layer.

insoluble one. Resists of this type are called *negatively acting* and the chemical change is called *polymerization.* Removing the soluble portion with chemical solvents (developers) leaves a hole in the resist layer that corresponds to the opaque pattern on the reticle.

The second transfer takes place from the photoresist layer into the wafer surface layer (Fig. 8.4). The transfer occurs when etchants remove the portion of the wafer's top layer that is not covered by the photoresist. The chemistry of photoresists is such that they do not dissolve (or they dissolve slowly) in the chemical etching solutions; they are *etch-resistant,* hence the name *resists* or *photoresists.*

In the example shown in Figs. 8.3 and 8.4, the result is a hole etched in the wafer layer. The hole came about because the pattern in the mask was opaque to the exposing light. A mask whose pattern exists in the opaque regions is called a *clear-field mask* (Fig. 8.5). The pattern could also be coded in the mask in the reverse, in a *dark-field mask.* If the same steps were followed, the result of the process would be an island of material left on the wafer surface (Fig. 8.6).

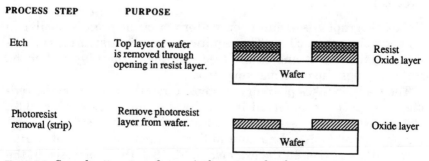

PROCESS STEP	PURPOSE	
Etch	Top layer of wafer is removed through opening in resist layer.	Resist Oxide layer Wafer
Photoresist removal (strip)	Remove photoresist layer from wafer.	Oxide layer Wafer

Figure 8.4 Second pattern transfer—resist layer to surface layer.

Light
Field

Dark   Figure 8.5   Mask-reticle polari-
Field   ties.

Photomasking "Hole"

Photomasking "Island"

Figure 8.6   Photomasking hole and island.

The resist reaction to light just described is a character of negative-acting photoresists. There are also positive-acting photoresists. Within these resists, the light changes the chemical structure from relatively nonsoluble to much more soluble. The term describing this change is *photosolubilization*. Figure 8.7 shows that an island is produced when a light-field mask is used with a positive photoresist.

The result obtained from the photomasking process from different combinations of mask and resist polarities is shown in Fig. 8.8. The choice of mask and resist polarity is a function of the level of dimensional control and defect protection required to make the circuit work. These issues are discussed in the process sections of the chapter.

Ten-Step Patterning Process

Transferring the image from the reticle or mask onto the wafer surface layer is generally a ten-step procedure (Fig. 8.9). Feature size, alignment tolerance, the condition of the wafer surface, and the number of masking layers all influence the level of difficulty. Many photo processes are customized to the particular conditions. However, most are variations or options of the basic ten-step process. The process illustrated is shown with a light-field mask and a negative photoresist. The first image transfer takes place in steps 1 through 7. In

PROCESS STEP	PURPOSE	
Alignment and Exposure	Precise alignment of mask/reticle to wafer and exposure of photoresist. Positive resist is photo-solubilized.	Mask/reticle / Resist / Oxide layer / Wafer
Development	Removal of unpolymerized resist.	Resist / Oxide layer / Wafer
Etch	Removal of surface layer	Wafer
Photo resist removal	Removal of photo-resist	Wafer

Figure 8.7 Image transfer from a light field mask with a positive photoresist to create an island.

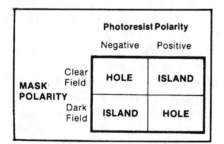

Figure 8.8 Mask and photoresist polarity results.

steps 8, 9, and 10 the image is transferred (second image transfer) into the wafer surface layer.

Understanding this basic process is facilitated by listing the ten steps and drawing the corresponding cross sections for different combinations of resist and mask polarities. It is strongly recommended that the reader master this ten-step process before proceeding to the advanced photolithography processes.

PROCESS STEP	PURPOSE
1. Surface Preparation	Clean and dry wafer surface
2. Photoresist apply	Spin coat a thin layer of photoresist on surface
3. Softbake	Partial evaporation of photoresist solvents by heating
4. Alignment and Exposure	Precise alignment of mask/reticle to wafer and exposure of photoresist. Negative resist is polymerized.
5. Development	Removal of unpolymerized resist.
6. Hard bake	Additional evaporation of solvents
7. Develop inspect	Inspect surface for alignment and defects
8. Etch	Top layer of wafer is removed through opening in resist layer.
9. Photoresist removal (strip)	Remove photoresist layer from wafer.
10. Final inspection	Surface inspection for etch irregularities and other problems

Figure 8.9 Ten-step photomasking process.

Basic Photoresist Chemistry

Photoresists have been used in the printing industry for over a century. In the 1920s they found wide application in the printed circuit board industry. The semiconductor industry adapted this technology to wafer fabrication in the 1950s. Negative photoresists specifically designed for semiconductor use were supplied by the Eastman Kodak Company in the late 1950s. Around the same time, the Shipley Company introduced a line of positive-acting resists. Since then a host of other companies have entered the market with photoresists designed to keep pace with ever-increasing industry demands to print narrower lines with fewer defects and at higher production rates.

The photoresist is the heart of the masking process. The preparation, bake, exposure, etch, and removal processes are fine-tuned to accommodate the particular resist used and the desired results. The selection of a resist and development of a resist process is a detailed and lengthy procedure. Once a resist process is established, it is changed very reluctantly. Rarely will a photoresist engineer risk losing an entire process if only a marginal gain is possible with a new photoresist.

Photoresist composition

Photoresists are manufactured for both general and specific applications. They are tuned to respond to specific wavelengths of light and different exposing sources. They are given specific thermal flow characteristics and formulated to adhere to specific surfaces. These properties come about from the type, quantity, and mixing procedures of the particular chemical components in the resist. There are four basic ingredients in photoresists: polymers, solvents, sensitizers, and additives (Fig. 8.10).

Light-sensitive and energy-sensitive polymers.   The ingredients that contribute the photosensitive properties to the photoresist are special light- and energy-sensitive polymers. Polymers are groups of large, heavy molecules containing carbon, hydrogen, and oxygen that are formed into a repeated pattern. Plastics are a form of polymers.

The most commonly used resists are designed to react to ultraviolet or laser sources. They are called *optical resists*. Others respond to X rays or electron beams.

In a negative resist, the polymers change from unpolymerized to polymerized after exposure to a light or energy source. Physically, the polymers form a cross-linked material that is etch resistant (see Fig. 8.11). In most negative resists, the polymers are the polyisopreme

Component	Function
Polymer	Polymer structure changes from soluble to polymerized (or vice versa) when exposed by the exposure source in an aligner.
Solvent	Thins resist to allow application of thin layers by spinning.
Sensitizers	Controls and/or modifies chemical reaction of resist during exposure.
Additives	Various added chemical to achieve process results, such as dyes.

Figure 8.10 Photoresist components.

Figure 8.11 Negative resist chemistry.

type. Earlier negative photoresists were based on rubber polymers. In 1969, the Hunt Corporation developed the first synthetic polyisopreme polymer structure. Polymerization also occurs when the resist is exposed to heat and/or normal light. To prevent accidental exposure, photomasking areas that process negative resists use yellow filters or yellow lighting.

The basic positive photoresist polymer is the phenol-formaldehyde polymer, also called the phenol-formaldehyde novolak resin (Fig. 8.12). Within the resist, the polymer is relatively insoluble. After exposure to the proper light energy, the resist converts to a more soluble state. This reaction is called *photosolubilization*. The photosolubilized part of the resist can be removed by a solvent in the development process.

Photoresists respond to many forms of energy, which are often referred to by their general category (light, heat radiation, etc.) or by a specific portion of the electromagnetic spectrum [ultraviolet light

Figure 8.12 Phenol-formaldehyde novolak resin structure. (*After: W. S. DeForest, Photoresist: Materials and Processes, McGraw-Hill, New York, 1975.*)

(UV), deep ultraviolet (DUV), I line, etc.] (see the section entitled "Exposure sources"). The exposing energies are described in detail in the section entitled "Alignment and Exposure." There are a number of strategies used to resolve small images (see "Comparison of negative and positive resists"). One is to use a narrower (or single) wavelength-exposing source. The traditional novolak-based positive resist has been fine-tuned for use with I-line exposure sources. However, it does not work very well with DUV sources. Resist manufacturers have developed chemically amplified resists for this exposure source. *Chemically amplified* means the chemical reactions of the polymer to light are increased by chemical additives. X-ray and electron-beam (e-beam) resists are based on polymers that are chemically different from those used for conventional positive and negative resists.

Figure 8.13 contains a list of resist polymers used in production resists. The relative merits of negative and positive resists are listed at the end of this section and the uses for these two types of resists and others are listed at the end of the process sections.

Solvents. The largest ingredient by volume in a photoresist is a solvent. It is the solvent that makes the resist a liquid and allows the resist to be applied to the wafer surface as a thin layer by spinning. Photoresist is analogous to paint, which is composed of the coloring pigment dissolved in an appropriate solvent. It is the solvent that allows the application of the paint onto a surface in a thin layer. For negative photoresist, the solvent is an aromatic type, xylene. In positive resist, the solvent is either ethoxyethyl acetate or 2-methoxyethyl.

Sensitizers. Chemical sensitizers are added to the resists to cause or control certain reactions of the polymer. In negative resists, the untreated polymer responds to a certain range of the ultraviolet spectrum. Sensitizers are added to either broaden the response range or narrow it to a specific wavelength. In negative resists, a compound

Resist	Polymer	Polarity	Sensitivity (Coul/cm^2)	Exposure Radiation
Positive	Novolak (M–Cresol-formaldehyde)	+	$3-5 \times 10^{-5}$	UV
Negative	Poly Isoprene	–		UV
PMMA	Poly–(Methyl Methacrylate)	+	5×10^{-5}	E-Beam
PMIPK	Poly–(Methyl Isopropenyl Ketone	+	1×10^{-5}	E-Beam/ Deep UV
PBS	Poly–(Butene-1-Sulfone)	+	2×10^{-6}	E-Beam
TFECA	Poly–(Trifluoroethyl Chloroacrylate	+	8×10^{-7}	E-Beam
COP (PCA)	Copolymer–(α–Cyano Ethyl Acrylate–α–Amido Ethyl Acrylate)	–	5×10^{-7}	E-Beam X-Ray
PMPS	Poly–(2-Methyl Pentene-1-Sulfone)	+	2×10^{-7}	E-Beam

Figure 8.13   Resist comparison table.

called bis-aryldiazide is added to the polymer to provide the light sensitivity.[1] In positive resists, the sensitizer is o-naphthaquinonediazide.

Additives.   Various additives are mixed with resists to achieve particular results. Some negative resists have dyes intended to absorb and control light rays in the resist film. Positive resists may have chemical *dissolution inhibitor systems*. These are additives that inhibit the dissolution of the nonexposed portions of the resist during the development step.

Photoresist Performance Factors

The selection of a photoresist is a complicated procedure. The primary consideration is the dimensions required on the wafer surface. The resist must first have the capability of producing those dimensions. Beyond that it must also function as an etch barrier during the etching step, a function that requires a certain thickness for mechanical strength. In the role of etch barrier, it must be free of pinholes, which also requires a certain thickness. In addition, it must adhere to the top wafer surface or the etched pattern will be distorted, just as a paint stencil will give a sloppy image if it is not taped tight to the surface. These, along with process latitude and step coverage capabilities, are resist performance factors. In the selection of a resist, the process engineer often must make tradeoff decisions among the various performance factors. The photoresist is one part of a complicated system of chemical processes and equipment that must work together to produce the image results and be productive, that is, yield an acceptable cost of ownership for the whole patterning process.

Resolution Capability

The smallest opening or space that can be produced in a photoresist layer is generally referred to as its *resolution capability*. The smaller the line produced, the better the resolution capability. Resolution capability for a particular resist is related to a particular process, including the exposing source and developing process. Changing the other process parameters will alter the inherent resolution capability of the resist. Generally, smaller line openings are produced with thinner resist film thicknesses. However, a resist layer must be thick enough to function as an etch barrier and be pinhole-free. The selection of a resist thickness is a tradeoff between these two goals.

The capability of a particular resist relative to resolution and thickness is measured by its *aspect ratio* (Fig. 8.14). The aspect ratio is calculated as the ratio of the resist thickness to the image opening. Positive resists have a higher aspect ratio compared to negative resists, which means that for a given image-size opening, the resist layer can be thicker. The ability of positive resist to resolve a smaller opening is due to the smaller size of the polymer. It is a little like the requirement of using a smaller brush to paint a thinner line.

Adhesion capability

In its role as an etch barrier, a photoresist layer must adhere well to the surface layer to faithfully transfer the resist opening into the layer. Lack of adhesion results in distorted images. Resists differ in their ability to adhere to the various surfaces used in chip fabrication. Within the photomasking process, there are a number of steps that are specifically included to promote the natural adhesion of the resist to the wafer surface. Negative resists generally have a higher adhesion capability than positive resists.

Figure 8.14 Aspect ratio.

$$\text{Aspect Ratio} = \frac{W}{T}$$

Exposure speed, sensitivity, and exposure source

The primary action of a photoresist is the change in structure in response to an exposing light or radiation. An important process factor is the speed at which that reaction takes place. The faster the speed, the faster the wafers can be processed through the masking area. Negative resists typically require 5 to 15 seconds of exposure time while positive resists take three to four times longer.

The sensitivity of a resist relates to the amount of energy required to cause the polymerization or photosolubilization to occur. Further, sensitivity relates to the energy associated with specific wavelength of the exposing source. Understanding this property requires a familiarization with the nature of the electromagnetic spectrum (Fig. 8.15). Within nature we identify a number of different types of energy: light, short and long radio waves, X rays, etc. In reality, they are all electromagnetic energy (or radiation) and are differentiated from each other by their wavelengths, with the shorter wavelength radiation having higher energies.

Common positive and negative photoresists respond to energies in the ultraviolet and deep ultraviolet (DUV) portion of the spectrum (Fig. 8.16). Some are designed to respond to particular wavelength peaks (G, H, I lines) within those ranges (see Fig. 8.18). Some resists

Figure 8.15 Electromagnetic spectrum.

Figure 8.16 Ultraviolet and visible spectrum. (*After Elliott.*[1])

are designed to work with x rays or electron beams (e-beams). Resist sensitivity, as a parameter, is measured as the amount of energy required to initiate the basic reaction. The units are milijoules per square centimeter (mJ/cm²).[2] The specific wavelengths the resist reacts to are referred to as the *spectral response characteristic* of the resist. Figure 8.17 is the spectral response characteristic of a typical production resist. The peaks in the spectrum are regions (wavelengths) that carry higher amounts of energy (Fig. 8.18). The different light sources used in masking areas are covered in "Alignment and Exposure."

Figure 8.17   Exposure response curve. (*Source: Shipley Megaposit XP-89131 Photo Resist*)

Figure 8.18   Mercury (Hg) spectrum (*From Silicon Processing for the VLSI Era by Wolfe and Tauber*)

Process latitude

While reading the sections on the individual masking process steps, the reader should keep in mind the fact that the goal of the overall process is a faithful reproduction of the required image size in the wafer layer(s). Every step has an influence on the final image size and each of the steps has inherent process variations. Some resists are more tolerant of these variations, that is, they have a wider process latitude. The wider the process latitude, the higher the probability that the images on the wafer will meet the required dimensional specifications.

Pinholes

Pinholes are microscopically small voids in the resist layer. They are detrimental because they allow etchants to seep through the resist layer and etch small holes in the surface layer. Pinholes come from particulate contamination in the environment, the spin process, and from structural voids in the resist layer.

The thinner the resist layer, the more pinholes. Therefore, thicker films have fewer pinholes, but they also make the resolution of small openings more difficult. These two factors present one of the classic tradeoffs in determining a process resist thickness. One of the principal advantages of positive resists is their higher aspect ratio, which allows a thicker resist film and a lower pinhole count for a given image size.

Particle and contamination levels

Resists, like other process chemicals, must meet stringent standards for particle content, sodium and trace metal contaminants, and water content.

Step coverage

By the time the wafer is ready for the second masking process, the surface has a number of steps. As the wafer proceeds through the fabrication process, the surface gains more layers. For the resist to perform its etch barrier role, it must maintain an adequate thickness over these earlier layer steps. The ability of a resist to cover surface steps with adequate resist is an important parameter.

Thermal flow

During the masking process there are two heating steps. The second one, hard bake, takes place after the image has been developed in the resist layer. The purpose of the hard bake is to increase the adhesion of the resist to the wafer surface. However, the resist, being a plasticlike material, will soften and flow during the hard bake step. The amount of flow has an important effect on the final image size. The resist has to maintain its shape and structure during the bake or the process design must account for dimensional changes due to thermal flow.

The goal of the process engineer is to achieve as high a bake temperature as possible to maximize adhesion. This temperature is limited by the flow characteristics of the resist. In general, the more stable the thermal flow of the resist, the better it is in the process.

Comparison of Positive and Negative Resists

Up to the mid-1970s, negative resist was dominant in the masking process. The advent of VLSI circuits and image sizes in the 2- to 5-μm range strained the resolution capability of negative resists. Positive resists had been around for over 20 years, but their poorer adhesion properties were a drawback and their superior resolution capability and pinhole protection were not needed.

By the 1980s, positive resist was the resist of choice. The transition was not easy. Switching to a positive resist requires changing the polarity of the masks or reticles. Unfortunately, it is not a matter of simply reversing the fields in the mask-making process. Reticle and mask dimensions print differently with the two resists (Fig. 8.19). With negative resist and a light-field mask, the dimension in the resist is smaller than the mask or reticle dimension due to light wrapping (diffraction) around the image. With a positive resist and a dark-field mask, diffraction tends to widen the image. These changes must be considered in making the mask or reticle and designing the other masking processes. In other words, an entirely new process is required to switch resist types.

The images on most of the mask layers are holes. With positive resist, the mask has dark-field polarity, which results in additional pinhole protection for the wafer. Clear-field masks are prone to small cracks in the glass surface (Fig. 8.20). These cracks, called *glass damage,* block the exposing light, creating unwanted holes in the photoresist layer, which in turn etch into the wafer surface as holes. The same is true for dirt particles that locate on the clear area of the mask or reticle. On dark-field masks, most of the surface is covered by chrome, which is hard and less likely to have pinholes. Thus the wafer has fewer unwanted pinholes.

Another problem with negative resists is oxygenation. This is a reaction of the resist to oxygen in the atmosphere, and it can result in

(a) Image size reduction with
light field mask and negative resist

▬ Unpolymerized resist

▨ Polymerized resist

(b) Image size enlargement with
dark field mask and positive photoresist

Figure 8.19 Changes in image size with negative
and positive photoresist.

Exposure

Chrome   Dirt   Glass Crack

(a)

Negative Resist
Oxide
Wafer

(b)

Figure 8.20 (a) Clear-field mask with dirt particle
and glass crack; (b) result in negative resist after
develop.

a thinning of the resist film by as much as 20 percent. Positive resists do not have this property. Cost is always an important consideration. Positive resists are more expensive than negative resists, but for a demanding process the higher cost may be offset by higher yield.

Developing characteristics differ between the two types of resists. Negative resists develop in readily available solvents and feature a high solubility differential between the polymerized and unpolymerized areas. The image dimensions remain relatively constant during the development step. Positive resists have a lower solubility between the polymerized and unpolymerized areas, which requires carefully prepared developer solutions and temperature control of the process. Dissolution inhibitor systems are added specifically to control dimensions during development.

The next-to-last step in the masking process is photoresist removal, which can take place in chemical solutions or in plasma systems. Generally, the removal of positive resists is easier and takes place in chemicals that are more environmentally sound. Positive photoresists are the choice of fabrication areas processing state-of-the-art circuits. A number of fabrication areas producing devices and circuits with image sizes greater than 2 μm still use negative resists. Figure 8.21 shows a comparison of properties of the two resists.

Physical Properties of Photoresists

The performance factors just detailed and the all of the ten basic process steps are related to a number of physical and chemical prop-

Parameter	Negative	Positive
Aspect Ratio (Resolution)		Higher
Adhesion	Better	
Exposure Speed	Faster	
Pinhole Count		Lower
Step Coverage		Better
Cost		Higher
Developers	Organic Solvents	Aqueous
Strippers		
Oxide Steps	Acid	Acid
Metal Steps	Chlorinated Solvent Compounds	Simple Solvents

Figure 8.21 Comparison of negative and positive results.

erties of the resist. These properties are rigorously controlled by resist manufacturers.

Solids content

A photoresist is a liquid that is applied to the wafer by a spinning technique. The thickness of resist left on the wafer is a function of the spin step parameters and several resist properties, two of which are solids content and viscosity.

Recall that the photoresist is a suspension of polymers, sensitizers, and additives in a solvent. Different resists will contain different amounts of these solids. The amount is referred to as the *solids content* of the resist and is expressed as the weight percent in the resist. Solids contents are in the 20 to 40 percent range.[3]

Viscosity

Viscosity is the quantitative measure of liquid flow. High-viscosity liquids, such as tractor oils, flow in a sluggish manner. Low-viscosity liquids, such as water, flow more readily. In both cases, the mechanism of flow is the same. The molecules in the liquid roll over each other as the liquid is being poured. As the molecules roll about, there exists an attraction between them that acts as an internal friction. Viscosity is the measurement of that friction.

Viscosity is measured by several techniques. One is the falling ball viscosimeter. In this instrument, a ball of measured size and weight is dropped into a tube of the liquid and timed as it passes between two marks. The higher the viscosity of the liquid, the longer it takes the ball to transit the marked distance. Another technique is the Ostwalk-Cannon-Fenske method, which is a timed measurement for a given amount of a liquid to pass through a given orifice.

Most photoresist manufacturers measure viscosity with a rotating vane in the resist. The higher the viscosity, the more force is required to move the vane through the liquid at a constant speed. The rotating-vane apparatus illustrates the force-related character of viscosity.

The unit of viscosity is the *centipoise*. It is named after the French scientist Poisseulle, who investigated the viscous flow of liquids. One poise is equal to one dyne second per centimeter. Photoresist is measured in centipoise (one centipoise is one hundredth of a poise). The viscosity unit of centipoise is more correctly named the *absolute viscosity*.

The *kinematic viscosity* is measured by another unit called the centistoke. This value is calculated from the absolute viscosity (centipoise) divided by the density of the resist. Viscosity varies with temperature;

therefore, its specified value is stated at a particular temperature, usually 25°C. Viscosity is a major parameter determining the resist thickness during the spin process. Viscosity is closely related to the solids content. The higher the solids content, the higher the viscosity.

Surface tension

The surface tension of a resist also influences the outcome at spin. Surface tension is a measure of the attractive forces in the surface of the liquid (Fig. 8.22). Liquids with high surface tension flow less readily on a flat surface. It is the surface tension that draws a liquid into a spherical shape on a surface or in a tube.

Index of refraction

The optical properties of the resist play a role in the exposure mechanism. One property is *refraction,* or the bending of light as it passes through a transparent or semitransparent medium. Refraction is the same phenomenon that makes an object in a pool of water appear to be at a different location. This comes about as the light ray is slowed up in the material. The index of refraction is a measurement of the speed of light in a material compared to its speed in air, as shown in Fig. 8.23. It is calculated as the ratio of the reflecting angle to the impinging angle. For photoresists, the index of refraction is close to that of glass, approximately 1.45.

Low Surface Tension
From Low Molecular
Attractive Faces

High Surface Tension
From High Molecular
Attractive Faces

Figure 8.22 Surface tension.

$$N_d = \frac{\text{Sin } \Phi_r}{\text{Sin } \Phi_i}$$

$\Phi = 90°$

Reflective
Substrate

Transparent
Film

(a)   (b)

Figure 8.23 Index of refraction. (a) 90° incident light; (b) angled light is refracted in the transparent film.

Storage and Control of Photoresists

Photoresists are delicate high-technology mixtures. Great care and precision go into their manufacture. Once a photomasking process is developed, its continuing success depends on the day-in, day-out control of the process parameters and a consistent photoresist product. Delivered batch-to-batch consistency is a responsibility of the manufacturer. Maintaining that consistency is the responsibility of the user. Several properties of resists dictate the storage and control conditions required.

Light and heat sensitivity

Both light and heat can activate the sensitive mechanisms in the resist that are intended to be activated in the closely controlled masking process steps. It is imperative that the resist be protected during storage and handling to prevent unwanted reactions that would interfere with the process results. This is the reason that masking areas use yellow light. It is also the reason that resist bottles are brown. The colored glass protects the resist from stray light. Proper transportation and storage of resists require temperature control within the limits specified by the manufacturer.

Viscosity control

Viscosity control is essential for good film thickness control. To maintain the photoresist's viscosity, resist bottles must remain capped prior to use. Opened bottles will allow evaporation of the solvent, which results in a higher solids content, which, in turn, results in a higher viscosity. If photoresist is dispensed from plastic tubing, the material should be tested to ensure that the resist is not leaching plasticizers out of the material. The plasticizers will increase the viscosity of the the kinematic viscosity resist.

Resist is also available in vacuum-sealed pouches. The resist is protected during shipping and storage. During use the pouch continues to collapse as the resist is dispensed, preventing air from reaching the resist surface.

Shelf life

A container of photoresist comes with a recommended shelf life. The problem again has to do with the self-polymerization or photosolubilization of the resist. In time, changes to the polymer will take place, altering the resist performance when it reaches the production line.

Cleanliness

Needless to say, any and all equipment used to dispense photoresist must be maintained in the cleanest condition possible. Besides the effects of particular contamination from the system, the resist tubing must be cleaned regularly due to the possible buildup of dried photoresist. Cleaning agents should be checked for their compatibility with the resist. For example, TCE should not be used with negative resist because it can cause bubbles in the resist.

Photomasking Processes

In the following sections each of the ten basic masking steps is examined. The purpose of the step, technical considerations and challenges, options, and process control methods are described.

Surface preparation

Throughout this text, various analogies will be used to aid the reader in understanding the complicated processes. A good analogy to the photomasking process is painting. Even the amateur painter soon learns that to end up with a smooth film of paint that adheres well, the surface must be dry and clean. The same is true in photomasking technology. Ensuring that the resist will stick to the wafer surface requires surface preparation. This step is performed in three stages: particle removal, dehydration, and priming.

Particle removal.   The wafers coming to the photo masking area almost always come in a clean condition from another process such as oxidation, doping, or CVD. However, during storage, loading, and unloading into carriers, they may pick up some particulate contamination that must be removed from the surface. Depending on the level of contamination and/or the process requirements, several particulate removal techniques may be used (Fig. 8.24). In extreme cases, the wafers may be put through a wet chemical cleaning similar to the preoxidation cleaning processes, including an acid cleaning, water rinsing, and drying. The particular acid used must be compatible with the top layer on the wafer surface. Particle-removal methods are the same as those described in Chap. 7: manual blow-off, mechanical scrubbers, and high-pressure water spray.

Dehydration baking.   It has been mentioned that the wafer surface has to be dry to promote adhesion. A dry surface is called *hydrophobic,* a chemical condition. Liquids form into small droplets on a hydrophobic surface, such as water beads on a newly waxed car. A hydrophobic surface is conducive to good photoresist adhesion (Fig. 8.25). Wafers coming to the masking operation usually have a hydrophobic surface.

Unfortunately, when the wafer is exposed to moisture, either from the air or from postcleaning rinses, the surface condition changes to a *hydrophilic* one. This condition is evidenced by a liquid on the surface

---

- High-pressure nitrogen blowoff
- Wet chemical cleaning
- Rotating brush scrubber
- High-pressure water stream

Figure 8.24   Prespin wafer cleaning methods.

Hydrophilic
Surface

Hydrophobic
Surface

Figure 8.25   Hydrophilic versus hydrophobic surfaces.

$$2\ Si\cdot OH \;\rightleftharpoons\; Si\cdot O\cdot Si \;+\; H_2O$$

spreading out in a wide puddle, such as water on a nonwaxed car surface. A hydrophilic surface is also said to be hydrated. Resist does not adhere very well to hydrated surfaces (Fig. 8.25).

Two important ways to maintain a hydrophobic surface are to keep the room humidity below 50 percent and to coat the wafers with photoresist as quickly as possible after being received from a previous process. Often the wafers will be stored in desiccators purged with dry, filtered nitrogen or in the dry minienvironment of an SMIF box (see Chap. 4). Additional steps may be taken to establish a wafer surface with acceptable adhesion properties. These steps include a dehydration bake and priming with a chemical.

A heating operation may be used to reset the wafer surface to a dehydrated condition. Dehydration bakes take place in three temperature ranges to address three different dehydration mechanisms. In the range of 150 to 200°C (low temperature), surface water is evaporated. At 400°C (medium temperature), water molecules loosely attached to the surface will leave. At temperatures above 750°C (high temperature), the surface is chemically restored to a hydrated condition.

In most masking processes, only low-temperature dehydration bake temperatures are used. This is because this temperature range is easily obtainable with hot plates and chest-type convection and vacuum ovens. Another advantage of low-temperature dehydration is that the wafers do not have to cool down before the spin process. Systems to perform this step can easily be integrated into a spin-bake system, making them dehydration-spin-bake systems. An explanation of these heating systems is in the section entitled "Soft bake."

High-temperature dehydration bakes are rare. One reason is that reaching a temperature of 750°C usually requires the use of a tube furnace, and tube furnaces are large and cannot be integrated into the spin-bake processes. A second reason is the temperature level itself. At 750°C, doped junctions in the wafer can move (which is undesirable) and mobile ionic contaminants on the surface can move into the wafer, causing reliability and performance problems.

Wafer priming.   In addition to dehydration baking, the wafers may go through a chemical priming step to ensure good adhesion of the resist. In painting, primers are a subcoat selected for their ability to adhere to the surface and provide a good surface for the finish paint to stick to. In semiconductor photomasking, the result is similar. The primer chemically ties up molecular water on the wafer surface, thereby increasing its adhesion property.[4]

A number of chemicals provide priming capabilities, but one, hexamethyldisilazane (HMDS), is in universal use. The use of HMDS is described in U.S. Patent 3,549,368 by R. H. Collins and F. T. Deverse (1970) of IBM. The HMDS is mixed with xylene in solutions of from 10 to 100%. The exact mixture is determined by process engineers, based on the particular surfaces and environmental factors in the clean room. Unlike a painting primer, a thickness of only several molecules is sufficient to provide the necessary adhesion promotion.

Immersion priming. The simplest method of applying the primer to the wafer is by dipping the wafers in a beaker of the liquid. This method has the drawbacks of being a manual method and of exposing the wafers to contamination in the HMDS.

Spin priming. In most photomasking areas liquid primers are applied to the wafer while it is on the resist spinner chuck (Fig. 8.26). The dispensing of the HMDS can be manual (from a syringe) or automatic. Automatic spinners (see next section) have a separate system to dispense HMDS onto the wafer surface just prior to the application of the resist. After the spinner dispenses the primer onto the rotating wafer, the chuck is ramped to a higher speed to dry the HMDS layer. A major production advantage of spin priming is that it takes place in-line with the spin step.

Vapor priming. Both immersion and spin priming require the direct contact of the HMDS liquid with the wafer surface. Whenever a liquid is in contact with the wafer, there is the danger of contamination from the liquid. Another consideration is that the HMDS must be dry before the resist is applied. Wet HMDS can dissolve the bottom layer of the resist and interfere with the exposure, development, and etching. Lastly, HMDS is relatively expensive, and in spin priming an excess of HMDS is sprayed on the wafer to ensure adequate coverage. The excess is thrown off the wafer and discarded.

The preceding considerations are overcome by vapor priming techniques. Vapor priming is practiced in three forms. Two are performed at

Figure 8.26   Spin dispense of primer.

(a)   (b)

Figure 8.27 Vapor prime methods. (a) Atmospheric; (b) vacuum bake-vapor prime.

atmospheric pressure and one in a vacuum (Fig. 8.27). One atmospheric system employs a bubbler chamber connected to a desiccator-type chamber. Nitrogen is bubbled through the HMDS and carried into the chamber where it coats the wafers. Another method employs a vapor degreaser with a reserve of liquid HMDS. The HMDS is heated to the vapor point and the wafers are suspended in the vapors for coating.

The third vapor technique is vacuum vapor priming, which uses a sealed flask of HMDS connected to a vacuum oven or single-wafer chamber. The wafers are first heated in the oven in a nitrogen atmosphere. After a temperature of about 150°C is reached, the atmosphere is switched to a vacuum. Once the vacuum level is reached, a valve is opened and HMDS vapors are drawn into the chamber by the low pressure. Within the chamber, the wafers become completely coated as the vapors fill the entire chamber. This method has shown good adhesion longevity even in the presence of high humidity.[5]

Vacuum vapor priming offers the additional advantage of a combined dehydration bake and prime step and a significant reduction in HMDS usage. Vacuum vapor priming practiced in a chest-type oven adds an additional step to the process. Many automatic spinner systems incorporate in-line vacuum vapor primers.

Photoresist spinning

The purpose of the photoresist application step is the establishment of a thin, uniform, defect-free film of photoresist on the wafer surface. These qualities are easy to state, but they require sophisticated equipment and stringent controls to achieve. A typical resist layer varies from 0.5 to 1.5 μm in thickness and has to have a uniformity of plus or minus only 0.01 μm (100 Å). This variation is 1 percent of a 1.0-μm thickness.

The usual methods of applying thin layers of liquids to surfaces are brushing, rolling, and dipping. None of these methods is adequate to achieve the quality resist film necessary for photomasking. The method used is spinning, which was briefly described in the section on priming. Spinners are built in manual, semiautomatic, and automatic designs. The systems differ in the degree of automation and are described in the following text. However, the deposit of the film on the wafer is common to each of the systems.

Spin processes are designed to prevent or minimize the buildup of a bead of resist around the outer edge of the wafer. Called an *edge bead,* the buildup causes image distortions during the exposure and etch processes.

The static spin process. The wafer that is ready for spinning (after priming) is placed on the vacuum chuck. Several cubic centimeters (cm^3) of the photoresist are deposited in the center of the wafer (Fig. 8.28) and allowed to spread out into a puddle. The puddle is allowed to spread until it covers most of the wafer surface. The size of the puddle is a process parameter that depends on the size of the wafer and the type of resist used. The amount of resist deposited in the puddle is critical only in the extremes. Too small an amount will result in incomplete resist coverage, and too much will cause a buildup of a resist rim or result in resist on the back of the wafer (Fig. 8.29).

When the puddle reaches its specified diameter, the chuck is rapidly accelerated to a predetermined speed. During the acceleration, centrifugal forces spread the resist to the wafer edge and throw off excess resist, leaving a thin uniform layer on the wafer. The high-speed spin continues for some time after the resist is spread to allow drying of the resist.

The final thickness of the film is established as the result of the resist viscosity, the spin speed, the surface tension, and the drying characteristics of the resist. In practice, surface tension and the drying characteristics are properties of the resist, and the viscosity–spin speed relationship is determined from curves supplied by the resist

Figure 8.28 Static spin process.

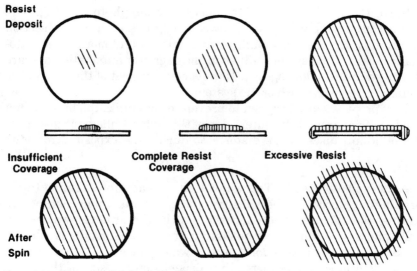

Figure 8.29 Example of resist coverage.

manufacturer or established for the particular spin system used (Fig. 8.30).

Although spin speed is specified to control resist thickness, it is actually the acceleration rate that establishes the final resist thickness. The acceleration characteristic of the spinner must be specified, and it is usually maintained as a constant in the spin process.

Dynamic dispense. The need for uniform resist films on larger-diameter wafers led to the development of the dynamic spin dispensing technique in the 1970s (Fig. 8.31). For this technique the wafer is rotated at a low speed of approximately 500 rpm. While the wafer is rotating, the resist is dispensed onto the surface. The action of the rotation assists in the initial spreading of the resist. Less resist is used and a more uniform layer is achieved. After spreading of the resist, the spinner is accelerated to a high speed to complete the spread and thin the resist into a uniform film.

Moving-arm dispensing. An improvement on the dynamic dispense technique is the addition of a moving-arm resist dispenser (Fig. 8.32). The arm moves in a slow motion from the center of the wafer toward its edge. This action creates more uniform initial and final layers. A moving-arm dispenser also saves resist material, especially for larger-diameter wafers.

Manual spinners. A manual spinner is a simple machine consisting of from one to four vacuum chucks (called *heads*), a motor, a tachometer,

Figure 8.30 Resist thickness versus spin speed. (*Courtesy of KTI Chemicals.*)

Figure 8.31 Dynamic spin dispense.

Figure 8.32 Moving-arm dispense.

and a connection for a vacuum source. Each head is surrounded by a catch cup which serves to collect the excess resist and direct it to a collection vessel. The catch cup also prevents "balls" of resist thrown off the wafers during the acceleration from landing on adjacent wafers. The process starts with the manual removal of particles with a filtered nitrogen gun. The wafer(s) are mounted on the head with tweezers or a vacuum wand and the chuck vacuum is turned on. Next the HMDS is dispensed onto the surface from a syringe or squeeze bottle, and the wafer is spun and dried. Finally, the resist puddle is dispensed from another syringe or squeeze bottle. Most spin processes performed on manual spinners use the static dispense method.

Automatic spinners.   An automatic spinner adds automation to the resist blow-off, resist dispense, and spinning cycles. The nitrogen blow-off is accomplished from a separate tube over the vacuum chuck that is connected to a pressurized nitrogen source (Fig. 8.33). Also in the dispense chamber are a primer resist tube and a resist dispense tube. The resist tube is fed resist from either a nitrogen pressurized vessel or through a diaphragm-type pump. In general, the industry has moved away from pressurized resist dispense systems because of problems that arise from the absorption of nitrogen into the resist. The nitrogen comes out of the resist after the dispense cycle, causing voids in the film. Diaphragm pumps eliminate this problem.

   Automatic resist dispensers have a negative pressure capability that automatically draws the resist back up into the dispenser tube after each dispensing operation. This drawback (or suck back) minimizes the exposed surface of the resist in the tube from drying into a hard ball that can be deposited on the wafer (Fig. 8.34). In fully auto-

Figure 8.33 Automatic spinner diagram.

No Drawback   With Drawback

Resist Ball

Figure 8.34 Automatic dispense
with drawback.

Solvent Evaporation

matic systems, all of the events of the spin process are controlled by
microprocessors. The systems have mechanisms to extract the wafers
from the carriers, place them on the chucks, perform the dispense,
perform a soft bake, and place the wafers back in their carriers. The
standard system configuration is in a line, referred to as a *track*.
Production-level spinners will have from two to four tracks side by
side. A complete system may include a wafer-cleaning module, dehy-
dration-prime station, spinner unit, and soft bake oven along with
wafer load and unload mechanisms.

Backside coating.   In some device processes, it is required that the
oxide on the back of the wafer remain in place through the masking
process. One way this is accomplished is by coating the back of the
wafer with photoresist. The requirement for this backside coating is
simply a thick-enough layer to survive the etch process. Usually the
wafer is backside-coated on a roll-coating machine after the front side
coat (Fig. 8.35). Another backside etch protection system is provided
by adhesive plastic sheets that adhere to the backside of the wafer.

Figure 8.35 Roller coating.

Soft bake

The soft bake step is a heating operation with the purpose of evaporating a portion of the solvents in the photoresist. After the bake, the resist film is still "soft" as opposed to being baked to a varnishlike finish. The solvents are evaporated for two reasons. Remember that the principal role of the solvents is to allow the application of the resist in a thin layer on the wafer. After that role is fulfilled, the presence of the solvent can interfere with the rest of the processing. The first interference occurs during the exposure step. The solvent(s) in the resist can absorb exposing radiation, thus interfering with the proper chemical change in the photosensitive polymers. The second problem is with the resist adhesion. Using the painting analogy, we know that complete drying (evaporation of the solvent) is necessary for good adhesion.

The soft bake parameters of time and temperature are determined by matrix evaluations. A matrix evaluation assesses a number of factors that are varied in a controlled manner to establish the desired results. In photomasking, two major goals are the correct image definition and the adhesion of the resist to the wafer during the etch step. Both of these goals are influenced by the degree of the soft bake. In the extreme, an underbaking will result in incomplete image formation at exposure and excessive lifting (poor adhesion) at the etch step. Overbaking will cause the polymers in the resist to polymerize and not react to exposing radiation.

Temperature and time ranges for the soft bake are provided by the resist manufacturer and are fine-tuned by the masking engineer. Negative photoresists have to be baked in a nitrogen atmosphere, whereas positive resists can be baked in air. A number of different methods are used to accomplish the softbake step, and various pieces of equipment incorporating all three heat transfer methods are used.

The three methods of heat transfer are conduction, convection, and radiation. *Conduction* is the transfer of heat by direct physical contact of the object with a heated surface. Hot plates heat by conduction. In the conduction process, the vibrating atoms of the hotter surface cause the object atoms to vibrate also. As they vibrate and collide, the atoms become hotter.

Some dehydration baking is in convection ovens. Systems using *convection* heating include home forced-air furnaces, hair dryers, air- and nitrogen-fed ovens, and oxidation furnaces. In these systems, a unit heats the gas and a blower or pressure pushes the gas to a space where it, in turn, transfers its energy to the object.

The term *radiation* describes the travel of electromagnetic energy waves through space. Radiation waves travel in vacuums as well as through gases. The sun transfers heat to the earth by radiation.

Heated Gas

Heater

Figure 8.36   Convection oven.

Heating lamps also transfer heat by radiation. Radiation is the heating method used by RTP systems. When the radiation strikes an object, the energy carried by the wave is transferred directly to the atoms of the object.

Convection ovens.   The mainstay baking oven for nonautomated fabrication lines is the convection oven (Fig. 8.36). It is a stainless steel chamber in an insulated enclosure. Either nitrogen or air is supplied by ducts surrounding the chamber and passed through a heater before being directed into the chamber. The inside of the chamber is fitted with racks for the wafer carriers. The carriers stay inside the oven for a predetermined time while the heated gas brings them up to temperature. Convection ovens used for VLSI applications have proportional band controllers and HEPA filters to maintain a clean baking environment.

There are several drawbacks to convection ovens for soft baking. One is batch-to-batch temperature variation, which arises from the amount of time the door is open for loading, the size of the load, and the variable time for all parts of the oven to reach a constant temperature. Within the oven there are different locations that heat up at various rates, depending on the gas flow. A process problem associated with convection heating is the tendency of the top layer of resist to "crust," trapping solvents in the resist (Fig. 8.37).

Hot N2 ⟶

Crust
Photoresist
Wafer

Figure 8.37   Crusting effect of ovens.

Figure 8.38 (a) Manual hot plate; (b) in-line continuous hot plate; (c) in-line single-wafer hot plate.

Manual hot plates. In manual and laboratory operations, a simple hot plate is often used for the soft bake. The wafers are placed on an aluminum holder which has a dial thermometer set in a hole drilled into it. The wafers are put on the holder, which is set on the hot plate (Fig. 8.38). The operator monitors the rise in temperature on the thermometer and removes the holder when the proper temperature is reached. With a well-controlled hot plate, an effective soft bake can be achieved. A process advantage to a hot plate is the backside heating. With the bottom of the wafer being heated by conduction, the surface on the resist layer does not crust over, a situation that can trap solvents in the resist. A hot plate process is operator-dependent and has low productivity.

In-line, single-wafer hot plates. The backside advantage of hot plate heating can be gained in track systems. A single-wafer hot plate is built into the spinner system. Wafers leaving the spinner are positioned on the hot plate and clamped to it with a vacuum. The wafer and resist are heated for a predetermined time, the vacuum released, and the wafer transferred to a carrier. These in-line systems are connected to the facility exhaust system for the removal of the solvent vapors.

Moving-belt hot plates. A constraint on the productivity of single-wafer hot plates in an integrated system is the total time of the spin step. A typical spin time is 25 to 40 s, which means the soft bake would have to be completed in that amount of time to keep the wafers

Figure 8.39 Moving-belt infrared (IR) heating.

moving in a continuous flow. For some resists and for some processes, this is too short a time. A way around the problem is a moving-belt hot plate. The wafers are placed on a heated moving steel belt, and the temperature and belt speed are set to meet the soft bake requirements and process the wafers in a continuous flow.

Moving-belt infrared ovens. The desire for fast, uniform, and noncrusting soft bake methods led to the development of infrared (IR) radiation sources (Fig. 8.39). Infrared baking is much faster than conduction baking and heats from the bottom up. Bottom up baking is the principle of conduction hot plate baking. The infrared waves pass through the resist layer without heating it, much as sunlight will pass through a window without heating it. The wafer, however, absorbs the energy, gets hot, and in turn heats the resist layer from the bottom.

Microwave baking. Microwaves as a soft bake heating source have the advantage of infrared heating but at a much faster rate due to the higher energy carried in a microwave. Soft bake temperatures can be well under 1 min. This brief time lends itself to on-chuck soft baking. Immediately after spinning, a microwave source is directed at the wafer, completing the soft bake (Fig. 8.40).

Vacuum baking. Vacuum ovens offer several advantages for a number of process steps. A vacuum oven is configured similarly to a convec-

Figure 8.40 Microwave heating.

tion oven, but is fitted to a vacuum source. Vacuum is particularly efficient for evaporation processes because the reduced pressure aids the evaporation of the solvents, reducing the reliance on the temperature. However, for soft baking the wafers must be heated to a uniform temperature. A problem arises because heating in a vacuum oven is by radiation from the heated chamber walls to the wafers. This heat transfer method is sometimes called line-of-sight, because for uniformity, each wafer must have a clear line-of-sight to the heat source. In a chest-type vacuum oven packed with carriers of vertically held wafers, this condition cannot be met. The result is poor temperature uniformity in most vacuum ovens.

The benefits of vacuum and uniform hot plate heating can be achieved with in-line, single-wafer systems. Figure 8.41 is a table summarizing the different soft bake methods.

Alignment and Exposure

Alignment and exposure (A&E) is, as the name implies, a two-purpose photomasking step. The first part of the A&E step is the positioning or alignment of the required image on the wafer surface. The second part is the encoding of the image in the photoresist layer from an exposing light or other radiation source. It was previously mentioned that the photoresist is the heart of the masking process. More correctly, the photoresist is the "material" heart of the process. A&E is the equipment heart of the process. Correct alignment of the image patterns and establishment of the precise image dimensions in the resist are absolute requirements for the functioning of the devices and circuits. Further, the wafers spend 60 percent of the process time in the lithography area.

Consider the challenge of building a typical ULSI megabit RAM circuit. The circuit can contain up to tens of millions of individual devices in an area about 0.25 in.[2] The individual parts of the devices will have feature sizes in the 1-μm range and have to "fit" or overlay within or next to each other within tolerances of one-third the nominal feature size. This tolerance has to be maintained as each of the level patterns is exposed on the wafer. Advanced processes may require up to 15 to 20 separate patterns. The allowable variation in wafer placement must be maintained over all of the pattern alignments.

Aligner system performance capabilities

An aligner system is composed of two major subsystems. One is the subsystem required to correctly position the pattern(s) on the wafer surface. The different aligner types have different alignment subsystems, which are explained in the equipment sections. The second part

Method	Bake Time Min.	Temperature Control	Productivity Type	Rate Waf/Hr.	Queuing
Hot Plate	5-15	Good	single to small batch	60	Yes
Convection Oven	30	Average - Good	Batch	400	Yes
Vacuum Oven	30	Poor - Average	Batch	200	Yes
I.R. Moving Belt	5-7	Poor - Average	Single	90	No
Conductive Moving Belt	5-7	Average	Single	90	No
Microwave	0.25	Poor Average	Single	60	No

Figure 8.41 Soft bake chart.

Alignment Selection Criteria
• Resolution Capability/Limit • Alignment Accuracy • Contamination Level • Reliability • Productivity • Overall cost of ownership (COO)

Figure 8.42 Aligner selection criteria.

is the exposure subsystem. This subsystem includes an exposure source and a mechanism for directing the radiation rays to the wafer surface.

Aligners are selected and compared by several criteria (Fig. 8.42) that relate to their ability to produce the required images in a consistent and productive manner. Perhaps the most important parameter is the *resolution capability,* or the ability of the machine to produce a particular size image. The higher the resolution capability, the better the machine. In addition to the resolution of the required image size, the aligner must be capable of placing the images in their correct position to each other. This performance parameter is called the *registration capability* of the aligner. These two factors must be performed over the entire wafer, a factor called *dimensional control.* The final performance factor is cost of ownership, which includes initial purchase cost, wafer throughput (the time required to load, align, expose, and unload the wafer), maintenance cost, and the uptime of the machine. These factors are discussed in Chap. 15 on manufacturing technology.

Alignment and exposure systems

Up to the mid-1970s, the photoresist engineer had the choice of only two A&E systems, contact and proximity aligners. Today the choice has expanded to include both optical and nonoptical aligners (Fig. 8.43). Optical aligners use an ultraviolet light source, while nonoptical systems use exposure sources from other parts of the electromagnetic spectrum. The systems in use today were developed to keep pace with the reduction of feature size, increased circuit density, and required defect reductions of the ULSI era.

Exposure sources

While aligners are very sophisticated machines, they operate on several basic optical principles. Consider a shadow image of a fork on a wall made by shining a flashlight onto a real fork held near the wall.

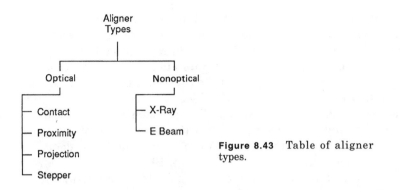

Figure 8.43   Table of aligner types.

By semiconductor standards, the image of the fork is quite imprecise. But there are ways to improve the image. One way would be to replace the flashlight with a narrower wavelength light. The white light that comes from a flashlight contains many different wavelengths (colors). At the edge of the fork a phenomenon called diffraction takes place. Diffraction is the bending of the light rays at an opaque edge (or going through a narrow slit). The amount of bending depends on the wavelength and with many wavelengths (white light) there are many rays scattering from the edge and fuzzing up the image. Using a shorter wavelength or a single wavelength source minimizes the diffraction. Another way to improve the image is to get all of the rays traveling along the same path. In normal white light, the rays coming from the light bulb are traveling in many directions and leaving the edge of the fork in many directions, again fuzzing the image. With mirrors and lenses, the light rays can be *collimated* into a band of parallel rays, improving image quality. The sharpness of the image and its dimensions are also influenced by the distance of the light behind the fork and the distance of the fork to the wall. Closing up both of these distances sharpens the image. These techniques, narrower or single wavelength exposure sources, collimated light, and strict control of the distances are all used in aligners to produce the required images.

Exposure sources are chosen to match the spectral response characteristic of the resist and the feature size of the images (see section on "Exposure speed, sensitivity, and exposure source"). Most optical aligner systems use a high-pressure mercury lamp, which produces light as a current is passed through a tube of mercury. A high-pressure atmosphere allows a higher level of stimulation of the mercury without evaporation.

The energy from the mercury comes out in bundles of waves grouped in ranges. Some resists are designed to react to the entire range of the wavelengths and some to specific wavelengths. Other resists are designed to respond to the specific high-energy peaks of a mercury lamp. The three peaks are at the 365-, 405-, and 436-nm wavelengths. They are referred to as of the I, H, and G lines, respectively. Steppers used for advanced imaging often have filters to expose the resist to the G or I lines. I-line exposure has been the source of choice in the submicron era. The I line has a wavelength of 365 nm (or 0.365μm), which is near the image size for 0.350 products. The problem is the physical difficulty of resolving images smaller than the wavelength of the light. Schemes for imaging in the sub-0.35 era include going to deep UV sources and other methods described in Chap. 10.

The advantage of confining the exposure source to a narrow wavelength is to foster complete and faster exposure of the resist. Narrowband exposure limits resolution problems associated with partial exposure of the resist at the clear/opaque edge on the mask. The exposure time is faster for a specific spectral resist since the peaks represent higher-energy portions of the spectrum.

Resolution capabilities of the aligner and the resist are primarily a function of the wavelength of the exposing light. The smaller or narrower the exposing wavelength, the higher the resolution capability. This is due to the diffraction effect (Fig. 8.44). The longer the wavelength, the more the diffraction effect, which ultimately acts as a limit on the resolution capability of the system. Shorter wavelengths of light also carry more energy, allowing shorter exposure times and in turn limiting poor resolution coming from scattering of the light in the resist and from the wafer surface (Fig. 8.45).

Figure 8.44 Diffracting reduction of image in resist.

Mask Image

Resist

Resist Image

Figure 8.45 Light scattering in resist film.

Shorter ultraviolet wavelength peaks are in a region of the mercury lamp output spectrum called the mid or deep ultraviolet (DUV). The peaks occur at 313 nm (mid UV) and at 254 nm (DUV). Deep- and mid-ultraviolet exposing sources are created by the use of filters with a standard ultraviolet source, such as mercury-zenon, zenon, or deuterium lamps. There are also DUV sources available from eximer lasers. Eximers are chemical compounds stable in an excited state and capable of laser action with high intensity output in the UV range. The following gas lasers and their associated wavelength are used: XeF (351 nm), XeCl (308 nm), RF (248 nm), AF (193 nm).[6]

There is interest in the possibility of focused ion beams for photoresist exposure. Their advantages are less scattering of the beam and lower energy scattering in the resist, which should allow smaller and sharper images. Also, the energy of ion beams is higher than that of electrons; this could lead to faster exposure times. Unfortunately, development has been hampered by a host of production problems, including vibration and the requirement for ion-optical systems for submicron imaging.[7]

Other Exposure Issues

Numerical aperture (NA) of a lens.   In the example of the shadow image of the fork, distance was introduced as a resolution factor. Distance is also a factor in semiconductor imaging. There are contact or proximity exposure systems where the wafer and mask are touching or in close proximity, respectively (see "Contact aligners" and "Proximity aligners"). But most VLSI and ULSI products are exposed with projection systems where the mask or reticle and wafer are separated. Projection optics present particular problems. The challenge is to project the image from the mask or reticle to the wafer surface with as little loss of resolution or dimensional control as possible. Small image sizes require the use of short wavelengths. The minimum image that can be resolved on the wafer is constrained by the physical attributes of the projection optical system. One attribute is the numerical aperture (NA) of the projection lens. The NA is the ability of the lens to gather light. The relationship is:

$$\sigma = \frac{k\lambda}{NA}$$

where σ = the minimum feature size
  k = a constant
  λ = the wavelength of the exposure source
  NA = the numerical aperture

The formula suggests that decreasing the wavelength and/or increasing the NA are ways to print a smaller image size. Using deep UV, x rays or e-beams are all ways to reduce the wavelength. However, there are limits to increasing the NA. That is because of a tradeoff parameter called depth of focus or depth of field. In regular photography we run into depth of field problems when the foreground is in focus and the background is out of focus or vice versa. At the wavelengths used in patterning, even the micron-sized topography on a wafer is enough to cause depth of field problems. Images both on the top surface and at the bottom of valleys must have good resolution and the correct dimensions. Another tradeoff with increasing the NA is a decrease in the field of view. This is the same phenomenon experienced going to a higher magnification with a zoom lens. At the higher powers the breadth of view is narrowed. Field of view becomes a production limit with steppers. A narrower field requires more time to complete exposure of the entire wafer.

All of these technical problems are complicated by the increasing amount of information that must be printed on advanced circuits. The technology driver product is the dynamic random access memory (DRAM). This product evolves about every three years to a new generation with four times more information. With feature size decreases, chip size increases, more stacking of components, and better designs, three times more information per square centimeter is required of the masking process.[8] This trend pressures alignment makers, especially steppers, to develop more sophisticated lens systems, exposure sources, resists, and other products such as phase shift masks (see Chap. 10).

Off-axis illumination.   Shifting the direction of the exposure beam from the perpendicular (off-axis) interrupts the interference pattern that causes standing waves in the resist.

Alignment criteria

The first mask is aligned by positioning the axis of the mask at a 90° angle to the major wafer flat on the wafer (Fig. 8.46). Subsequent masks

Figure 8.46   First mask alignment.

Figure 8.47 Types of alignment marks.

Figure 8.48 Misalignment types. (*a*) *x* direction; (*b*) rotational; (*c*) run-out.

are aligned to a previously patterned mask with the use of alignment marks (also called *targets*). These are special patterns (Fig. 8.47) located in an easily found position on the edge of each chip pattern or in the scribe line. Alignment is accomplished by the operator positioning a mark on the mask to a corresponding mark in the wafer pattern. For automatic systems (see "Steppers"), the alignment marks serve the same purpose. Alignment marks become a permanent part of the chip surface through the etching process. They are then in place for the alignment of the next layer.

Alignment errors, called *misalignment,* fall into several categories (Fig. 8.48). A common one is a simple misplacement in the *x-y* directions. Another common misalignment is rotational, where one side of the wafer is aligned but the patterns become increasingly misaligned across the wafer. A third rotational misalignment comes about when the die pattern is rotated on the mask or reticle.

Other misalignment problems associated with masks and stepper aligners are run-out and run-in. These problems arise when the chip patterns are not formed on the mask on constant centers or are placed on the chip off center. The result is that only a portion of the mask chip patterns can be properly aligned to the wafer patterns. The pattern becomes progressively misaligned across the wafer.

A rule of thumb is that circuits with micron or submicron feature sizes must meet registration tolerances of one-third the minimum feature. An *overlay budget* is calculated for the total circuit. It is the allowable accumulated alignment error for the entire mask set. For a 0.35-μm product the allowable overlay budget (Fig. 8.2) is about 0.1 μm.[9]

Aligner types

Contact aligners. Until the mid-1970s, the contact aligner was the workhorse aligner of the semiconductor industry. The alignment part of the system uses a full-wafer-size photomask positioned over a vacuum wafer chuck. The wafer is mounted on the chuck and viewed through a split-field objective microscope (Fig. 8.49). The microscope presents the operator with a simultaneous view of each side of the mask and wafer. The chuck is moved left, right, and/or rotated (x, y, and z movement) by manual controls until the wafer is aligned to the mask pattern.

Once the mask and wafer are aligned properly, a piston pushes the wafer into contact with the mask. Next, the collimated ultraviolet rays coming from a reflection and lens system pass through the mask and into the photoresist.

Contact aligners are used in production for discrete devices and circuits with SSI and MSI densities and feature sizes of approximately 5 μm and above. They also are used for flat panel displays, infrared sensors, device packages, and multichip modules (MCMs).[10] In these roles they rank as an older technology. Actually, a contact aligner is capable of submicron imaging with the proper resist and a well-tuned process. Its replacement by other systems is more related to yield losses associated from the contact of the mask and wafer.

Figure 8.49 Contact aligner system. (*a*) Alignment stage; (*b*) contact stage.

In the contact mode, particulate contamination can damage the soft resist layer, the mask, or both. Dirt adhering to the clear portions of the mask blocks light during exposure, causing an unwanted defect in negative photoresists or tearing holes in any type of resist. With contact aligners, epitaxial layer spikes on bipolar wafers can degrade the mask. Mask damage is so prevalent that masks have to be removed and discarded or cleaned every 15 to 25 exposures, a process that limits the productivity of the method. Dirt between the mask and wafer will cause resolution problems in the immediate area of the piece of dirt, due to the forced local separation of the mask and wafer. Alignment of larger-diameter wafers presents a light uniformity problem that causes image size variations and alignment problems.

The high defect levels of contact aligners, cost of masks, resolution accuracy, and alignment limitations led to the development of higher-yield systems for advanced products.

Proximity aligners. Proximity aligners were a natural evolution of contact aligners. The systems are essentially contact aligners but with mechanisms that either hold the wafer in near or soft contact with the mask. Sometimes proximity aligners are called *soft-contact machines.*

The performance of a proximity aligner is a tradeoff between resolution capability and defect density. With the wafer in soft contact with the mask, there is always some scattering of the light, which fuzzes the definition of the image in the resist. On the other hand, the soft contact also greatly reduces the number of defects associated with mask and resist damage. Even with the improved defect density, proximity aligners do not find much use in VLSI photomasking processes.

Scanning projection aligners. The end of contact aligners was in sight for years and development work was ongoing in the search for an alternative. The search centered on the concept of projecting the mask image onto the wafer surface much as a slide (the mask) is projected onto a screen (the wafer) (Fig. 8.50). While simple in concept, the technique requires an excellent optical system to expose the resist in the exact dimensions of the mask. The system became a reality with the introduction in the early 1970s of the Perkin Elmer Company scanning projection aligner.

Perkin Elmer Company avoided the problems of a full mask projection exposure in favor of a scanning technique. It used a mirror system with a slit blocking part of the light coming from the light source. The slit allows a more uniform portion of the light to shine on the mirror system, which is in turn projected onto the wafer (Fig. 8.51).

Figure 8.50 Concept of projection exposure.

Figure 8.51 Perkin Elmer light projection system. (*Courtesy of Perkin Elmer Corp.*)

Since the size of the slit is smaller than the wafer, the light beam is scanned across the wafer. With this system a new parameter, scan speed, became a parameter requiring control.

The early projection systems were sold for up to $100,000 each, at a time when a fully loaded contact aligner could be had for about $20,000. However, the increase in yields gained by the physical separation of the mask and wafer (lower damage-induced defects and reduced mask costs) made quick payback possible. The early scanning projection aligners required manual alignment of the mask and wafer. They are called 1:1 aligners, since the image dimensions on the mask are the same size as the intended image dimensions on the wafer surface.

Steppers. Scanning projection aligners were a great leap forward over contact aligners for production work, but they still had some limitations, such as alignment and overlay (registration) problems associated with full-size masks, image distortion, and mask-induced defects from dust and glass damage.

By the mid-1970s there were thoughts of stepping images (Fig. 8.52), from a reticle directly onto the wafer surface, the same technique used to make masks. A reticle, carrying the pattern of one or several chips, is aligned and exposed, then is *stepped* to the next site and the process is repeated. A reticle is of higher quality than a full-size mask, so fewer defects occur. There is better overlay and alignment because each chip is individually aligned. The procedure of stepping allows precise matching of larger-diameter wafers. Other advantages are resolution improvements because a smaller area is being exposed each time and a lessened vulnerability to dust and dirt. Some steppers are 1:1, that is, the image on the reticle has the same dimensions as those required on the wafer. Others use reticles with 5 to 10 times the final dimensions. These are called reduction steppers. Making an oversize reticle is easier, and any dirt or small glass distortions are reduced out of existence during the exposure

Figure 8.52 Step and repeat reduction system (*From Wolf and Tauber, Silicon Processing for the VLSI Era.*)

(Fig. 8.53). Generally 5× reduction is preferred. A 5× reticle is smaller and easier to make than a 10×. Also, a 5× reticle projects a larger (up to 20 mm × 20 mm) field onto the wafer, resulting in faster wafer throughput.[11]

The key to production use of steppers was the development of automatic alignment systems. There is no way that an operator could individually align several hundred die on a wafer at a productive rate. Automatic alignment is accomplished by passing low-energy laser beams through alignment marks on the reticle and reflecting them off corresponding alignment marks on the wafer surface. The signal is analyzed and information is fed to the x-y-z wafer chuck controls by a computer, which moves the wafer around until the wafer and reticle are aligned. The images are placed in the photoresist by sequentially exposing each die pattern across and down the wafer (Fig. 8.54).

An alternative to laser signal control is a vision system. These systems use a camera to capture a vision of the die and compare it to a data base. The wafer is moved until it and the mask or reticle image match the data base.

Most production steppers have UV exposure sources with G- or I-line capabilities. Steppers intended for small feature sizes are fitted with laser sources operating in the DUV range.[12] The maintenance of the correct image size during the exposure part of the process requires tight humidity and temperature control. Most steppers are

Figure 8.53 5× mask pattern transfer.

Figure 8.54 Step and repeat die alignment and exposure.

enclosed in an environmental chamber that controls these important parameters and keeps the wafers clean.

Step and scan aligners. Larger die sizes would normally drive the development of lens with larger fields of vision. Increasing the field of vision keeps the time of alignment and exposure lower. However, larger lens become expensive. An alternative is a stepper with a smaller lens and the capability to scan a smaller field over the required area (see Fig. 8.55).

X-ray aligners. The desire for higher-resolution exposure sources inevitably led to the consideration of x-ray sources (Fig. 8.56). X rays are high-energy photons with wavelengths of 4 to 50 Å.[13] This range of wavelengths is capable of very small image sizes (down to the 0.1-μm level) due to the lack of diffraction effects. X-ray aligners are projection systems using a full-size mask (1:1 mask to wafer image). They generally have higher output because of shorter exposure times. Reflection and scattering in the resist is minimized and there are few depth of focus problems. X-ray-exposed wafers show a lower level of defects from dust and organic matter on the mask because the x rays pass through the spots.

A number of difficulties have surfaced with production-level, x-ray aligners. A major problem has been with the development of x-ray blocking masks. Because x rays pass through conventional chrome and glass masks, a process that requires gold as the blocking layer and other materials that stand up to the high energy of the x rays needs to be developed (see "Maskmaking," Chap. 10).

While development work goes on in defining x-ray equipment, it also goes on in developing x-ray resists. This work is complicated because there is no standard x-ray source and the resists must show

Figure 8.55 Step and scan comparison. (*a*) step and repeat requires 42.4 mm diameter lens field for 9 cm² die; (*b*) step and scan for same die requires 23.7 mm lens field.

Figure 8.56 X-ray exposure system.

high sensitivity to x rays and also be good etch barriers. These last two factors have proven difficult to balance in resist chemistries.

X-ray sources include standard x-ray tubes or laser-driven sources as point sources or synchrotrons. Point sources are fitted to individual machines. Synchrotrons are large, expensive machines that can accelerate electrons in an orbit. The orbiting electrons give off x rays in a process called *synchrotron radiation*. While expensive, synchrotrons have the capability to supply x rays to several aligners.

Electron beam aligners. Electron beam lithography is a mature technology used in the production of high-quality masks and reticles. The system (Fig. 8.57) consists of an electron source that produces a small-diameter spot and a "blanker" beam on-off control capable of turning the beam on and off. The exposure must take place in a vacuum to prevent air molecules from interfering with the electron beam. The beam passes through magnetic and electrostatic plates capable of directing (or steering) the beam in the x-y direction on the mask, reticle, or wafer. This system is functionally similar to the beam-steering mechanisms of a television set. Precise direction of the beam requires that the beam travel in a vacuum chamber in which there is the electron beam source, support mechanisms, and the substrate being exposed.

There is no mask or reticle used to generate the pattern. The blank-

Figure 8.57 Electron beam exposure system.

ing and steering functions are controlled by a computer that has in its memory the wafer pattern directly from the CAD design stage. The beam is directed to specific positions on the surface by the deflection subsystem and the beam turned on where the resist is to be exposed. Larger substrates are mounted on an *x-y* stage and are moved under the beam to achieve full surface exposure. This alignment and exposure technique is called *direct writing*.

The pattern is exposed in the resist by either raster or vector scanning (Fig. 8.58). Raster scanning is the movement of the electron beam side to side and down the wafer. The computer directs the movement and activates the blanker in the regions where the resist is to be exposed. One drawback to raster scanning is the time required for the beam to scan, since it must travel over the entire surface. In vector scanning, the beam is moved directly to the regions that have to be exposed. At each location, small square or rectangular shaped areas are exposed, building up the desired shape of the exposed area.

Alignment and overlay parameters are very good with electron-beam systems because no distortions are introduced from masks or

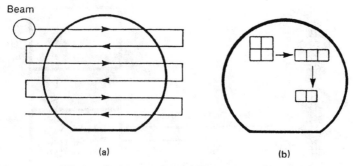

Figure 8.58 Electron beam scanning. (*a*) Raster scan; (*b*) vector scanning.

from optical effects such as diffraction. Resolution is also good, with current machines capable of 0.25-µm feature sizes.[14] Drawbacks to full use of electron-beam systems to wafer production are speed and cost. Current electron-beam aligners can expose less than ten 6-in-diameter wafers per shift and cost several million dollars per system. A factor in the slowness of the system is the time required to create the vacuum and release it in the exposing chamber.

Mix and match aligners.   Small-geometry imaging is expensive. Fortunately, only certain layers of a product mask set are critical enough to require advanced imaging techniques. For advanced circuits, fully 50 percent of the layers may be noncritical.[15] The other, less critical layers can be imaged with more established techniques using projection scanners or less expensive steppers. Mix and match aligners will probably be part of fabrication operations that use x-ray or e-beam technology.

Aligner system comparison

The table in Fig. 8.59 compares aligner system features and performance parameters.

Postexposure bake (PEB)

Standing waves are a problem that occurs with optical exposure and positive resists (see Chap 10). One technique for minimizing the effect of standing waves is to bake the wafers after exposure. The bake method could be any one of the ones described earlier. The time and temperature specifications for the PEB are a function of the baking method, exposure conditions, and resist chemistry.

Aligner System	Mask	Reticle	Exposure Sources	Resolution (microns)	Throughput 150 mm waf./hr.
Contact/Proximity	X		Hg	0.25-0.50	30-120
Scanning Projection	X	X	Hg	0.9-1.25	30-100
Step and Repeat		X	Hg/ExL/KrF/DUV	0.35-0.80	65-90
Step and Scan		X	Hg/ExL/KrF/DUV	0.25-0.40	50*
X-Ray	X	X	X-ray	0.10-	20+
E-Beam	Direct	write	Electron beam	0.25-	2-10

Figure 8.59 Aligner tool comparison table. (*Source: Solid State Technology, April 1993, p. 26.*)

The End of Optical Lithography?

Optical lithography's demise has been predicted with each new generation of feature size smaller than 1 μm. Each time, the benefits of x-ray and e-beam techniques were dusted off and a future without optical imaging confidently forecast. Unfortunately, these two techniques are expensive and have significant technical barriers to overcome before reaching a comfortable production confidence level. But the biggest deterrent to their installation in fabrication lines has been the continual "tweaking" and development of all the components of optical lithography. Resists have gotten better, exposure sources have been narrowed and controlled, system optics, and a host of process techniques (see Chap. 10) have pushed the limits to the 0.25-μm level. These evolutionary improvements may continue to make 0.18-μm products possible with optical means.

Today the industry generally concurs that beyond the 0.18-μm barrier different techniques will be needed. Stay tuned.

Key Concepts and Terms

Alignment systems	Photoresist performance factors
Alignment and exposure	Positive-resist performance
Exposure sources	Resist spin
Misalignment	Soft bake methods
Negative-resist performance	Surface preparation methods
Photoresist components	Ten-step process systems

Review Questions

1. Draw cross-sectional diagrams of the ten-step process with a positive resist and dark-field mask.

2. Make a list of the major components in photoresist and explain the role of each.

3. What type of polymers are in negative resist?

4. What type of polymers are in positive resist?

5. Describe the changes in positive and negative resist during exposure to light.

6. Name the exposure and alignment systems that require a mask or reticle.

7. Which exposure system(s) are used for ULSI processes? Explain why.

8. Name ten factors that affect the image size.

9. What is the result of too low a soft bake temperature?

10. What process steps are required to complete the first image transfer?

References

1. David J. Elliott, *Integrated Circuit Fabrication Technology*, McGraw-Hill, New York, 1976, p. 168.
2. "Photoresists for Microlithography," *Solid State Technology*, PenWell Publishing Company, June 1993, p. 42.
3. Ibid., p. 71.
4. Ibid., p. 116.
5. "Yield Engineering," LPZ Product Bulletin, 1982.
6. S. K. Ghandhi, *VLSI Fabrication Principles*, John Wiley and Sons, New York, 1994, p. 687.
7. Ibid., p. 693.
8. H. Jeong, D. Markle, G. Owen, F. Pease, A. Grenville, and R. von Bunau, "The Future of Optical Lithography," *Solid State Technology*, PenWell Publishing Company, April 1994, p. 39.
9. K. Simon, et al., "Abstract-Alignment Accuracy Improvement by Consideration of Wafer Processing Impacts," *SPIE Symposium on Microlithography*, 1994, p. 35.
10. E. Cromer, "Mask Aligners and Steppers for Precision Microlithography," *Solid State Technology*, PenWell Publishing Company, April 1993, p. 24.
11. S. Wolf and R. N. Tauber, *Silicon Processing*, vol. 1, Lattice Press, Sunset Beach, CA, 1986, p. 473.
12. E. Cromer, "Mask Aligners and Steppers for Precision Microlithography," *Solid State Technology*, PenWell Publishing Company, April 1993, p. 26.
13. David J. Elliott, *Integrated Circuit Fabrication Technology*, McGraw-Hill, New York, 1976, p. 82.
14. Ibid., p. 384.
15. J. Greeneich, "Mixing Critical and Noncritical Steppers," *Solid State Technology*, PenWell Publishing Company, October 1994, p. 79.

Photolithography—
Developing to
Final Inspection

Overview

In this chapter the methods used for photoresist developing through final inspection (steps 5 through 10 of the basic process) are explained. The end of the chapter examines the processes used for mask making and includes a discussion of alignment error budgets.

Objectives

Upon completion of this chapter, you should be able to:

1. Draw a cross section of a wafer before and after developing.
2. Make a list of the developing methods.
3. Explain the purpose and methods of hard bake.
4. List at least five reasons a wafer can be rejected at develop inspect.
5. Draw a diagram of the develop-inspect-rework loop.
6. Explain the methods and relative merits of wet and dry etch.
7. Make a list of the resist strippers used to strip photoresist from oxide and metal films.
8. Explain the purpose and methods of final inspection.

Development

After the wafer completes the alignment and exposure step, the device or circuit pattern is coded (as a latent image) in the photoresist as regions of exposed and unexposed resist (Fig. 9.1). The pattern is "developed" in the resist by the chemical dissolution of the unpolymerized resist regions. Development techniques are designed to leave in the resist layer an exact copy of the pattern that was on the mask or reticle. Problems that can result from a poor developing process (Fig. 9.1) are underdevelopment (which leaves the hole incompletely developed to the correct dimensions) and a coved sidewall. In some cases, the development will not be long enough (incomplete) and will leave a layer of resist in the hole. The third problem is overdevelopment, which removes too much resist from the image edge or top surface.

Negative and positive resists have different developing characteristics and require different chemicals and processes (Fig. 9.2).

Figure 9.1 Photoresist development. (a) Process; (b) problems.

	Positive Resist	**Negative Resist**
Developer	Sodium hydroxide (NaOH) Tetramethyl amonium hydroxide (TMAH)	Xylene Stoddard solvent
Rinse	Water (H_2O)	n-Butylacetate

Figure 9.2 Resist developer and rinse chemicals.

Negative resist development

The successful development of the image coded in the resist is dependent on the nature of the resist's exposure mechanisms. Negative resist, upon exposure to light, goes through a process of polymerization which renders the resist resistant to dissolution in the developer chemical. The dissolving rate between the two regions is high enough so that little of the resist layer is lost from the polymerized regions. The chemical preferred for most negative-resist-developing situations is xylene, which is also used as the solvent in negative resists formulas. The development step is done with a chemical developer followed by a rinse. For negative resists, the rinse chemical is usually n-butyl acetate, because it neither swells nor contracts the resist. Either of these actions can change pattern dimensions. Mixtures of alcohol and trichlorethylene (TCE) can function as a negative resist rinse, but they cause some changes in the resist film dimensions. For wafers that have been patterned with a stepper, a milder-acting Stoddart solvent may be used.

The action of the rinse is twofold. First, it rapidly dilutes the developer chemical to stop the developing action. Although the polymerized resist is developer-resistant, there is always a transition region at the exposed edge (Fig. 9.3) that contains partially polymerized molecules. If the developer is allowed to stay in full strength on the wafer, it dissolves into this region and changes the image dimensions. The second action of the rinse is to remove partially polymerized pieces of resist from the open regions in the resist film.

An all-aqueous negative resist development system has been an industry goal. Eliminating the organic solvents from the fabrication area reduces the need for the safety systems and lowers the cost of treating or removing the spent chemicals. To date, little success with an acceptable all-aqueous system has been reported.

Mask

Negative Resist
Layer

Wafer

◇ Polymerized Resist

╱ Partially Polymerized Resist

▣ Unpolymerized Resist

Figure 9.3 Transition region at resist image edge.

Positive resist development

Positive resists present a different developing condition. The two regions, polymerized and unpolymerized, have a dissolving rate difference of about 1:4. This means that during the developing step some resist is always lost from the polymerized region. Use of developers that are too aggressive or that have overly long developing times may result in an unacceptable thinning of the resist film, which in turn may cause it to lift or break down during the etch step.

Two types of chemical developers are used with positive resist, alkaline-water solutions and nonionic solutions. The alkaline-water solutions can be sodium hydroxide or potassium hydroxide. Since both of these solutions contain mobile ionic contaminants, they are not desirable in processing sensitive circuits. Most positive resist fabrication lines use a nonionic solution of tetramethylammonium hydroxide (TMAH). Sometimes a surfactant is added to break down the surface tension (that is, make the solution more wetable) to the wafer surface. The aqueous nature of positive developers makes them more environmentally attractive than the solvent developers required for negative resists.

The positive-resist developing process is more sensitive than negative processes.[1] Factors influencing the outcome are the soft bake time and temperature; degree of exposure; developer concentration; and time, temperature, and method of developing. The development process parameters are determined by matrix testing of all the variables. The effect on line width for a particular process is shown in Fig. 9.4. Tight control of the development and rinse process is critical for dimensional control when using a positive resist. The rinse chemical for positive resist developers is water. It serves the same role as

Figure 9.4 Developer temperature and exposure relationship versus line-width change.

negative resist rinsers but is cheaper, safer to use, and easier to dispose of.

Wet development methods

There are several methods used to develop resist films (Fig. 9.5). The selection of a method is dependent on the resist polarity, the feature size, defect density considerations, the thickness of the layer to be etched, and productivity.

Immersion. Immersion is the oldest development method. In its simplest form, the wafers, in a chemically resistant carrier, are immersed in a tank of the developer solution for a specific time before being transferred to a second tank of the rinse chemical (Fig. 9.6). The problems associated with such a simple wet procedure are

1. The surface tension of the liquids can prevent the chemicals from penetrating into small openings.
2. Partially dissolved pieces of resist can cling to the wafer surface.
3. The tanks can become contaminated as hundreds of wafers are processed through them.
4. The wafers can become contaminated as they are drawn through the liquid surface.
5. Developer chemicals (especially positive developers) can become diluted through use.
6. Frequent changing of solutions to eliminate problems 1, 2, and 3 raise the cost of the process.

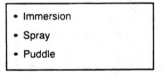

- Immersion
- Spray
- Puddle

Figure 9.5 Development methods.

Developer   Rinse   Dry

Figure 9.6 Immersion developer steps.

7. Fluctuations in the room temperature can cause changes in the developing rate of the solution.

8. The wafers have to be quickly transferred to a drying process step, which introduces a third step.

Additions are often made to the immersion tanks to improve the development process. Uniformity and penetration of small openings are aided by mechanical agitation of the bath by stirring or rocking mechanisms. A popular stirring system is a Teflon-encapsulated magnet that is coupled to a rotating magnetic field outside the tank.

Agitation is also achieved by passing ultrasonic or megasonic waves through the liquid. The waves cause a phenomenon called *cavitation*. The energy in the waves causes the liquid to separate into tiny cavities which immediately collapse. The rapid generation and collapse of the millions of microscopic cavities create a uniformity of development and help the liquid penetrate into small openings. Uniform development rates are also enhanced by the addition of heaters and temperature controllers to the bath.

Spray development. The preferred method of chemical development is by spray. In fact, spray processing is generally preferred over an immersion system for any wet process (clean, develop, etch) for a number of reasons. For instance, there is a major reduction in chemical use with spray systems. Process improvements include better image definition due to the mechanical action of the spray pressure in defining the resist edge and removing partially polymerized pieces of resist. Spray systems are always cleaner than immersion systems because each wafer is developed (or etched or cleaned) only with fresh chemicals.

Spray processing is done in either single or batch systems. In the single-wafer configuration (Fig. 9.7) the wafer is clamped on a vacuum chuck and rotated while the developer and rinse are sequentially sprayed onto the surface. Drying takes place immediately after the rinse cycle by increasing the rotational speed of the wafer chuck. In appearance and design, a spray developer system for single wafers is the same as a resist spinner but plumbed for different chemicals. Single-wafer spray developers offer the advantage of track automation by integrating the developing and hard bake processes. A major process advantage of these systems is the increased uniformity from the direct impingement of the chemical spray on the wafer.

Spray development had been a standard process for negative resist for years because negative resist is fairly insensitive to the temperature of the developer. For the temperature-sensitive positive resists,

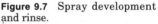

Figure 9.7   Spray development and rinse.

spray development is less effective. The problem is the phenomenon of rapid cooling of a fluid dispensed through an orifice under pressure. Called *adiabatic cooling,* it is the same phenomenon that causes a can of pressurized household cleaner to cool during dispensing. Spray developers used for positive resist often have a heated wafer chuck or a heated spray nozzle to control the develop temperature. Other problems encountered with the spray development of positive resist are machine deterioration when alkaline developers are used and foaming as the water-based developer comes out of the nozzle under pressure.

Batch developers come in two versions, single-boat and multiple-boat. These machines are the spin-rinse-dryers described in Chap. 7. As developers they require additional plumbing to accommodate the developing chemicals. Batch developing systems are in general less uniform than direct-spray, single-wafer systems because the spray does not impinge directly on each wafer surface and temperature control for positive resist processing is more complicated.

Puddle development. Spray development is very attractive for its uniformity and productivity. A process variation used to gain the advantages of spray for development of positive resist is the *puddle procedure.* This system uses a standard single-wafer spray unit. The difference between regular spray development and the puddle procedure is in the application of the developer chemical to the wafer. The process starts with the deposit of enough developer on the static wafer to cover the surface (Fig. 9.8). Surface tension holds the developer in a puddle on the wafer. The puddle sits there for some required time, usually on a chuck-heated wafer, causing the majority of the

Form Puddle   Spray during   Dry during
                     Rotation   High Speed
                                           Rotation

Figure 9.8 Puddle-spray development.

development to take place. Puddle development is, in effect, a single-wafer, front-side-only immersion process. After the required puddle time, the wafer surface is sprayed with more developer and rinsed, dried, and passed on to the next step.

Plasma descum. A particularly difficult form of incomplete development is a condition called *scumming*. The scum may be undissolved pieces of resist or dried developer[2] left on the surface. The film is very thin and is hard to detect with visual inspection. In reaction to this problem, advanced VLSI lines with micron-size openings will remove (descum) the film from the wafers in an oxygen-rich plasma chamber after a chemical develop.

Dry (or plasma) development

It is the desire of the industry to eliminate liquid processes. They are difficult to integrate into automated lines and the chemicals are a substantial expense to purchase, store, control, and remove or dispose of. The goal is to replace liquid chemical developers with a plasma etch process. *Dry plasma etch* is a well-established process for etching wafer surface layers (see "Dry Etch"). In a plasma etcher, ions energized by a plasma field chemically dissolve (etch away) exposed layer surfaces. Dry resist development requires a photoresist chemistry that leaves either the exposed or unexposed portions of the resist layer readily removable by plasma energized oxygen. In other words, only one part of the pattern is oxidized off the wafer surface. A dry development process, called DESIRE, using silyation and plasma O_2 is described in Chap. 10.

Hard Bake

Hard bake is the second heat treatment operation in the masking process. Its purpose is essentially the same as the soft bake step: the evaporation of solvents to harden the resist. For hard bake, however, the goal is exclusively to achieve good adhesion of the resist to the wafer surface. As such, this step is sometimes called preetch bake or prebake.

Hard bake methods

Hard bake is similar to soft bake in the equipment and methods used. Convection ovens, in-line and manual hot plates, infrared tunnel ovens, moving-belt conduction ovens, and vacuum ovens are all used for hard baking. The track systems are preferred for automated lines. See the "Soft bake" section in Chap. 8.

Hard bake process

The exact time and temperature of the hard bake are determined much the same as in the soft bake process. The starting point is the process recommended by the resist manufacturer. After that, the process is fine-tuned to achieve the adhesion and dimensional control required. Nominal hard bake temperatures are from 130 to 200°C for 30 min in a convection oven. Temperatures and times vary in the other methods. The minimum temperature is set to achieve good adhesion of the resist image edge to the surface. The heat-caused adhesion mechanism is dehydration and polymerization. The heat drives water out of the resist, at the same time further polymerizing it, thereby increasing its etch-resistant properties.

The upper temperature limit of the hard bake is set by the flow point of the resist. Resist is a plasticlike material that softens and flows when heated (Fig. 9.9). When the resist flows, the image dimensions are changed. Resist flow is first evident as a thickening of the resist edge when viewed with a microscope. Extreme flow exhibits itself as a series of fringe lines around the image. The fringes are an optical effect from the slope left in the resist after the flow.

Hard bake takes place either immediately after the developing step or just before the etching step, as shown in Fig. 9.10. In most production situations, the hard bake is performed in a tunnel oven that is in-line with the developer. When this procedure is used, it is important

Normal temperature   High temperature

Figure 9.9 Resist flow at high temperature.

Hard Bake Process Flows

Develop   Develop/Bake   Develop/Bake

Inspect   Inspect   Inspect

Hard Bake   Etch   Rebake

Etch   Etch

Figure 9.10 Hard bake process flow options.

that the wafers be stored in a nitrogen atmosphere and/or be processed through the develop inspection step as quickly as possible to prevent the reabsorption of water into the resist film.

A goal of process engineering is to have as many common processes as possible. For hard baking that is sometimes difficult due to the different adhesion characteristics of various wafer surfaces. The more difficult surfaces, such as aluminum- and phosphorus-doped oxides, sometimes are given a higher-temperature hard bake or a second hard bake in a convection oven just prior to being etched.

Develop Inspect

The first quality check in the photomasking process is performed after developing and baking. It is appropriately called *develop inspect* or simply DI. The purpose of the inspection is to (1) identify wafers that have a low probability of passing the final masking inspection, (2) provide process performance and process control data, and (3) identify wafers for rework.

This inspection yield, that is, the number of wafers that pass this first quality cheek, is not factored into the overall yield formula, but it is a much-watched yield for two principal reasons. The critical nature of the photomasking process to the functioning of the circuit has been emphasized throughout this text. It is at develop inspect that the process engineer has the first chance to judge the performance of the process. The second importance of the develop inspect step is related to the two types of rejects made at this inspection. First, some of the wafers will have problems from previous steps that prevent their continuation in the process. These wafers are rejected at develop inspect and discarded. Other wafers have problems associated specifically with the quality of the pattern in the resist film.

Figure 9.11   Rework process loop.

These wafers can be reworked (Fig. 9.11) by removal of the photoresist and reinsertion into the dehydration process. This is one of the few places in the entire fabrication process where a general rework of mistakes is possible.

Wafers sent back into the masking process are called *reworks* or *redos*. The process engineer has a goal of keeping the rework rate as low as possible, certainly under 10 percent and preferably less than 5 percent. One reason is that wafers that have gone through a masking rework have a lower wafer sort yield at the end of the fabrication process. Reworking causes adhesion problems and the additional handling can result in contamination and breakage. If too many wafers are in the rework loop, the overall wafer sort yield will suffer. The second reason for keeping the rework rate under control is related to the additional accounting and identification required to process the rework wafers.

The develop inspect yield and rework rate vary from mask level to mask level. In general, the first levels in the masking sequence have wider feature sizes, flatter surfaces, and lower density, all of which make for a higher yield out of the mask step. By the time the wafers are at the critical contact and metal masks, the rework rate tends to rise.

Develop inspect methods

In this section, inspection methods, process flow, and inspection parameters are explained. Descriptions of the inspection equipment and methods are in Chap. 14. In general, there are four categories of wafer-surface problems that are detected at develop inspect and final inspect:

1. Deviations in the pattern dimensions (critical dimension measurements)

2. Misaligned patterns

3. Surface problems such as contamination, holes, or scratches in the photoresist; stains; and other surface irregularities

4. Patterns that are distorted in shape

Manual inspection.   The flow diagram in Fig. 9.12 shows a typical manual develop inspection sequence. The first step is a naked-eye inspection of the wafer surface. Since no magnification is used, this inspection is sometimes called a 1× (a magnification of 1 is equal to naked-eye viewing) inspection. It may take place in normal room light, but more likely the wafer will be viewed under a collimated white light or a high-intensity ultraviolet light. The wafer is viewed at an angle in the light beam. This method is surprisingly effective in showing film thickness irregularities, gross developing problems, scratches, and contamination, especially stains.

Wafers that pass the 1× visual inspection are placed on a microscope stage for a more detailed inspection. Most industrial inspection techniques use a system called *random sampling* based on probability theory. The theory determines the size of an area that must be inspected to accurately detect the level of defects on the complete wafer. Such an inspection system is based on the operator selecting sites for inspection in a random fashion.

Random sampling is not used in manual inspection. As the defect density goes down, as it must for VLSI circuits, the area of inspection

Step	Looking For:	Methods
1.	Stains/ gross contamination	Naked eye/UV light
2.	Stains/contamination Pattern irregularities Missalignment	100-400X microscope/ SEM/ AFM/ Automatic Inspection Systems
3.	Critical dimensions	microscope/SEM/ AFM

Figure 9.12 Order of develop inspection.

must go up if the defects are to be statistically sampled. However, as the die density goes up, the individual parts also get smaller, which in turn requires a higher magnification to see them. The increasing magnification narrows the field of view, which in turn increases the time for an operator to inspect a wafer. The time required to statistically sample a large-diameter, low-defect wafer is prohibitive.

The usual procedure is to select a number of specific locations for inspection. The chosen die are assumed to geographically represent the wafer surface. This method also assumes that gross problems will show up on the selected die. This method is adequate for gross wafer-wide problems and will alert the process engineer of problems in the masking process. Often, the microscopes will have motorized and/or programmable stages that automatically go to the inspection areas on the wafer.

In addition to the issue of the die-sampling plan, there is the issue of the wafer-sampling plan. The number of wafers inspected varies with the sensitivity and difficulty of the particular mask level. The more sensitive and difficult the layer, the more wafers are likely to be inspected. The number of wafers inspected per batch varies from a few to 100 percent. In addition to process defects, dimensional measurement and control is part of the develop inspection. On each mask level there is a region or set of patterns whose dimensions are critical to the functioning of the whole circuit. A representative pattern is chosen for each level as the *critical dimension* (CD). During the develop inspection the dimensions of these patterns are measured using manual or automatic microscopic techniques.

Automatic inspection. With die getting larger, dimensions getting smaller, and processes becoming more numerous and more sophisticated, manual inspection reaches a limit of effectiveness. Automatic inspection systems designed to detect surface and pattern distortion problems are finding their way onto production lines for both off-line and on-line inspections. The systems are described in Chap. 14. Automatic systems can yield greater amounts of data, which would enable process engineers to characterize and control the processes. Automatic systems also are consistent, whereas humans vary in ability and experience fatigue when doing repetitive tasks.

Causes of develop inspection reject

There are many reasons why a wafer can be rejected or sent for rework at the develop inspection step. Generally, the only defects looked for are those added to the wafer during the current photomasking step. Defects from previous masking steps are generally

overlooked under the theory that every wafer is passed on with some
defects or problems and that the wafer arrived at the current step
with acceptable quality. If there is a serious problem with a wafer
that somehow escaped previous inspections, it is pulled from the
batch.

The inspection is generally done on a *first-fail basis*. This means
that the operator or machine inspects only until a rejectable level is
reached on the wafer. First-fail systems save inspection time but do
not give the process engineers a complete evaluation of the wafer
image quality. The information for each wafer is logged on the sheet
for data accumulation and analysis (Fig. 9.13). Automatic and semiautomatic optical inspection stations have electronic memories for
accumulating and correlating this reject data. Some advanced lines
have the inspection logs programmed into on-line PCs, which speeds
the inspection step and the data analysis.

Most of the causes for rejects have been discussed. One problem not
discussed is bridging (Fig. 9.14). It is a condition where two patterns
are connected (bridged) by a thin layer of photoresist, usually at the
metal layer. If passed on to the next steps, the bridge results in a
short between the patterns. Bridging comes from an overexposure,
poor mask definition, or a resist film that is too thick. Bridging is a
particularly vexing problem in that it occurs easily with minor
process variations.

Batch Number	Broken Wafer	Scratch	Contamination	Pinholes	MA	Bridging	Lifting Resist	Under Exposed	Incomplete Develop	No Resist	Resist Running	Wrong Mask	C.D.	Reject

Figure 9.13 Typical develop inspect log.

Figure 9.14 Bridged conduction lines.

Etch

At the completion of the develop inspect step, the mask (or reticle) pattern is defined in the photoresist layer and is ready for etch. It is in the etch step that the image is permanently transferred into the surface layer on the wafer. Etching is the process of removing the top layer(s) from the wafer surface through the openings in the resist pattern. Etching processes fall into two main categories: wet and dry. The primary goal of each is an exact transfer of the image from the mask or reticle into the wafer surface. Other etch process goals include uniformity, edge profile control, selectivity, cleanliness, and minimizing cost of ownership.

Wet Etching

The historic method of etching has been by immersion techniques using wet etchants. The procedure is similar to the preoxidation clean-rinse-dry process (Chap. 7) and immersion development. Wafers are immersed in a tank of an etchant for a specific time, transferred to a rinse station for acid removal, and transferred to a station for final rinse and a spin dry step. Wet etching is used for products with feature sizes greater than 3 μm. Below that level the control and precision needed requires dry etching techniques.

Etching uniformity and process control are enhanced by the addition of heaters and agitation devices, such as stirrers and ultrasonic or megasonic waves, to the immersion tanks.

Wet bench methods are giving way to spray systems similar to those described in Chap. 7. For etching, the control of the chemical composition and timing becomes critical.

The problem of etchant contamination of the wafers is addressed by *point-of-use filters*. These are special filters fitted to automatic chemical dispensing systems to filter-clean the chemicals just prior to filling the immersion tank. This placement catches particulate contamination from the chemicals, the pumps, and the tubing systems.

Wet etchants are selected for their ability to uniformly remove the top wafer layer without attacking the underlying material (good selectivity).

Etch time variability is a process parameter influenced by temperature variations as the boat and wafers come to temperature equilibrium in the tank and the continued etching action as the wafers are transferred to a rinse tank. Generally, the process is set at the shortest time compatible with uniform etching and high productivity. The maximum time is limited to the amount of time the resist will continue to adhere to the wafer surface.

The exactness of the image transfer is dependent on several process factors. They include incomplete etch, overetching, undercutting, selectivity, and anisotropic/isotropic etching of the sidewalls.

Incomplete etch

Incomplete etch is a situation in which a portion of the surface layer still remains in the pattern hole or on the surface (Fig. 9.15). The causes of incomplete etch are too short an etch time, the presence of a surface layer that slows the etching, or an uneven surface layer that results in incomplete etch in the more thickly coated portions of the wafer. If wet-chemical etching is used, a lowered temperature or weak etch solution will cause incomplete etch. If dry plasma etching is used, a wrong gas mixture or an improperly operated system can cause the same effect.

Overetch and undercutting

The opposite condition to incomplete etch is overetch. In any etch process there is always some degree of overetch planned into the process. This allows for thickness variations in the surface layer. Planned overetch also allows for the etch to break through any slow-etching layers on the top surface.

The ideal etch leaves vertical sidewalls in the layer (Fig. 9.16). Etch techniques that produce this ideal result are said to be *anisotropic*. However, the etchant removes material in all directions. This phenomenon is called *isotropic* etching. Etching takes place at the top of

Resist
Layer

Wafer

Figure 9.15 Incomplete etch.

Anisotropic etch   Isotropic etch

Figure 9.16 Anisotropic and isotropic etch.

Normal

Over Etch

Figure 9.17 Degrees of undercutting.

Over Etch and Resist Lifting

the layer for the entire time it takes to reach the bottom of the layer.
The result is a sloped side. This action is also called *undercutting*
(Fig. 9.17) since the surface layer is undercut below the resist edge.
An ongoing goal of the etch step is the control of undercutting to an
acceptable level. Circuit layout designers take undercutting into
account when planning the circuit. Adjacent patterns must be separated a certain distance to prevent shorting. The amount of undercutting must be calculated when the pattern is designed. The anisotropic
etching available with plasma etching is preferred for advanced circuits. Reducing undercutting allows denser circuits.

Severe undercutting (or overetch) takes place when the etch time is
excessive, the etching temperature is too high, or the etch mixture is
too strong. Undercutting is also present when the adhesion bond
between the photoresist and the wafer surface is weak. This is a constant worry, and the purpose of the dehydration, prime, soft bake, and
hard bake steps is to prevent this type of failure. Failure of the resist
bond at the edge of the etch hole can result in severe undercutting. If
the bond is very poor, the resist can lift from the wafer surface, causing catastrophic undercutting.

Selectivity

Another goal in the etch step is the preservation of the surface underlying the etched layer. If the underlying surface of the wafer is partially
etched away, the physical dimensions and electrical performance of the

devices are changed. The property of the etch process that relates to preservation of the surface is *selectivity*. It is expressed as the ratio of the etch rate of the layer material to the etch rate of the underlying surface. Silicon oxide/silicon selectivity varies from 20 to 40, depending on the etching method.[3] High selectivity implies little or no attack of the underlying surface. Good selectivity becomes a problem in etching small contacts with aspect ratios greater than 3:1.[4] Selectivity also applies to the removal of the photoresist. This is more of a consideration in dry etch processes. While the top layer(s) is being removed by the etchant, some photoresist is also being removed. The selectivity factor must be high enough to ensure that the photoresist is not removed before the etched layer.

Silicon wet etching

Silicon layers are typically etched with a solution of nitric and HF acids mixed in water. The formula becomes an important factor in control of the etch. In some ratios, the etch has an exothermic reaction with the silicon. Exothermic reactions are those that produce heat, which in turn speeds up the etch reaction, which in turn creates more heat, and so on, resulting in an uncontrollable process. Sometimes acetic acid is mixed in with the other ingredients to control the exothermic reaction.

Some devices require the etching of a trough or trench into the silicon surface. The etch formula is adjusted to make the etch rate dependent on the orientation of the wafer. <111>-oriented wafers etch at a 45° angle, while <100>-oriented wafers etch with a "flat" bottom.[5] Other orientations result in different-shaped trenches. Polysilicon films are also etched with the same basic formula.

Silicon dioxide wet etching

The most common etched layer is a thermally grown silicon dioxide. The basic etchant is hydrofluoric acid (HF). HF has the advantage of dissolving silicon dioxide without attacking silicon. However, full-strength HF has an etch rate of about 300 Å/s at room temperature.[6] This rate is too fast for a controllable process (a 3000-Å layer would etch in only 10 s).

In practice, the HF (assay of 49 percent) is mixed with water or ammonium fluoride and water. The ammonium fluoride (NH_4F) acts as a buffer to the unwanted generation of hydrogen ions which accelerate the etch rate. These solutions are known as *buffered oxide etches* or BOEs. They are mixed in different strengths to create reasonable etch

Figure 9.18 Etch rate versus temperature for BOEs.

times for the particular oxide thickness (Fig. 9.18). Some BOE formulas include a wetting agent (surfactant such as Triton X-100 or equivalent) to reduce the surface tension of the etch, allowing it to uniformly penetrate into smaller openings.

Aluminum film wet etching

Selective etching solutions for aluminum and aluminum alloy layers are based on phosphoric acid. An unfortunate by-product of the reaction of aluminum and phosphoric acid are tiny bubbles of hydrogen, as shown in the reaction in Fig. 9.19. These bubbles cling to the wafer

Bridge   Snowball

Figure 9.19 Hydrogen bubble blockage etchant.

⊗ Hydrogen Bubble

surface and block the etch action. The result is either bridges of aluminum that can cause electrical shorts between adjacent leads or spots of unwanted aluminum, called *snowballs,* left on the surface. Neutralization of this problem is accomplished by use of an aluminum etching solution that contains phosphoric acid, nitric acid, acetic acid, water, and wetting agents. A typical solution of the active ingredients (less wetting agent) is 16:1:1:2.

In addition to the special formulas, a typical aluminum etch process will include wafer agitation by stirring or moving the wafer boat up and down in the solution. Sometimes ultrasonic or megasonic waves are used to collapse and move the bubbles around.

Deposited oxide wet etching

One of the final layers on a wafer is a silicon dioxide passivation film deposited over the aluminum metallization pattern. These films are known as vapox or silox films. While the chemical composition of the films is that of silicon dioxide, the same as thermally grown silicon dioxide, they require a different etch solution. The difference is in the selectivity required of the etchant.

The usual etchant for silicon dioxide is a BOE solution. Unfortunately, the BOE attacks the underlying aluminum pads, causing bonding problems in the packaging process. This condition is called *brown,* or *stained, pads.* The preferred etchant for this layer is a solution of ammonium fluoride and acetic acid mixed in a ratio of 1:2.

Silicon nitride wet etching

Another compound favored for the passivation layer is silicon nitride. It is possible to etch this layer with wet chemical means, but it is not as easy as for the other layers. The chemical used is hot (180°C) phosphoric acid. Since the acid evaporates rapidly at this temperature, the etch must be done in a closed reflux container equipped with a cooled lid to condense the vapors. The major problem is that photoresist layers do not stand up to the etchant temperature and aggressive etch rate. Consequently, a layer of silicon dioxide or some other material is required to block the etchant. These two factors have led to the use of dry etching techniques for silicon nitride.

Wet spray etching

Wet spray etching offers several advantages over immersion etching. Primary is the added definition gained from the mechanical pressure of the spray.[7] Spray etching also minimizes contamination from the etchants. From a process control point of view, spray etch is more controllable since the etchant can be instantly removed from the surface by switching the system to a water rinse. Single-wafer spinning-chuck spray systems offer considerable process uniformity advantages.

Disadvantages to spray etching are system cost, safety considerations associated with caustic etchants in a pressurized system, and the requirement of etch-resistant materials to prevent the deterioration of the machine. On the plus side, spray systems are usually enclosed, which adds to worker safety. Figure 9.20 is a table of common semiconductor films and their common etchants.

Dry Etch

The limits of wet etching for small dimensions have been mentioned in the previous section. For review they are

1. Wet etching is limited to pattern sizes of 3 μm.

2. Wet etching is isotropic, resulting in sloped sidewalls.

3. A wet etch process requires rinse and dry steps.

4. The wet chemicals are hazardous and/or toxic.

5. Wet processes represent a contamination potential.

6. Failure of the resist-wafer bond causes undercutting.

These considerations have led to the use of dry etch processes for the definition of small feature sizes on advanced circuits. Figure 9.21 is an overview of the etching techniques used.

	COMMON ETCHANT	ETCH TEMP	RATE Å /MIN	METHOD
SiO₂	HF & NH₄F (1 : 8)	Room	700	Dip & wetting agent predip
SiO₂	HF & NH₄F (1 : 8)	Room	700	Dip & wetting agent predip
SiO₂ (Vapox)	Acetic Acid & NH₄F(2 : 1)	Room	1000	Dip
Aluminum	H₃PO₄ : 16 HNO₃ : 1 Acetic : 1 H₂O : 2 Wetting Agent	40 – 50°C	2000	a) Dip & agitation b) Spray
Si₃N₄	H₃PO₄	150 – 180°C	80	Dip
POLYSi	HNO₃ : 50 H₂O : 20 HF : 3	Room	1000	Dip

Figure 9.20 Summary of wet etching process.

Figure 9.21   Guide to etch methods.

Dry etching is a generic term that refers to the etching techniques in which gases are the primary etch medium, and the wafers are etched without wet chemicals or rinsing. The wafers enter and exit the system in a dry state. There are three dry etching techniques: plasma, ion milling, and reactive ion etch (RIE).

Plasma etching

Plasma etching, like wet etching, is a chemical process but uses gases and plasma energy to cause the chemical reaction. Comparison of silicon dioxide etching in the two systems illustrates the differences. In wet etching of silicon dioxide, the fluorine in the BOE etchant is the ingredient that dissolves the silicon dioxide, converting it to water-rinseable components. The energy required to drive the reaction comes from the internal energy in the BOE solution or from an external heater.

A plasma etcher requires the same elements: a chemical etchant and an energy source. Physically, a plasma etcher consists of a chamber, vacuum system, gas supply, and a power supply (Fig. 9.22). The wafers are loaded into the chamber and the pressure inside is

Figure 9.22   Barrel plasma etch.

reduced by the vacuum system. After the vacuum is established, the chamber is filled with the reactive gas. For the etching of silicon dioxide the gas is usually CF_4 mixed with oxygen. A power supply creates a radio frequency (RF) field through electrodes in the chamber. The field energizes the gas mixture to a plasma state. In the energized state, the fluorine attacks the silicon dioxide, converting it into volatile components that are removed from the system by the vacuum system.

Earlier plasma systems were designed with circular chambers and are called *barrel etchers*. In this design, ions are energized by the plasma in a nondirectional pattern. The etching ions attack the surface layer from all directions (isotropic etching), creating a tapered sidewall. Etch uniformity is a problem. It is hard to supply a constant amount of etchant to all the wafers in the system because of the nondirectionality of the etching ions and the close packing of the wafers.

Another consideration of barrel plasma etching is radiation damage resulting from the high-energy plasma field. The high energy causes charges to build up in the wafer surface that compromise the electrical functioning of the circuit. Protection of the wafers from the radiation is provided by perforated metal cylinders that isolate the wafers from the plasma field (Fig. 9.23).

Planar plasma etching. For more precise etching, plasma planar systems are preferred. These systems contain the basic components of the barrel system, but the wafers are placed on a grounded pallet

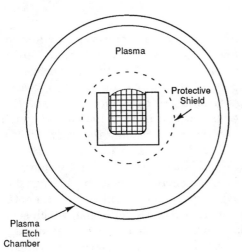

Plasma
Etch
Chamber

Figure 9.23 Barrel plasma stripper with protective shield.

Figure 9.24   Planar plasma etch.

under the RF electrode (Fig. 9.24). Etching takes place with the wafers actually in the plasma field. The etching ions are more directional than those in a barrel system, resulting in a more anisotropic etch. Almost vertical sidewalls are possible with plasma etch. Etching uniformity is increased with the rotation of the wafer pallet in the system. Planar plasma etch systems are designed in both batch and single-wafer chamber configurations. Single-wafer systems are popular because their etch parameters can be tightly controlled for uniform etching. Single-wafer systems with load-lock chambers can maintain high production rates and are compatible with in-line automation.

RF-generated parallel-plate plasma sources are giving way to new sources for 0.35-μm processing.[8] High-density, low-pressure plasma sources under development are electron cyclotron resonance (ECR), high-density reflected electron, helicon wave, inductively coupled plasma (ICP), and transformer coupled plasma (TCP).

The features of a dry etch process are: etch rate, radiation damage, selectivity, particulate generation, postetch corrosion, and lowered cost of ownership.

Etch rate.   The etch rate of a plasma system is determined by a number of factors. System design and chemistry are two of them. Others are the ion density and system pressure. Ion density (number of ions per cubic centimeter) is a function of the power supplied to the electrodes. (Power supply configurations are described in Chap. 12.) Increasing power creates more ions, which in turn increase the etch rate. Increasing ion density is similar to increasing the strength of a liquid chemical etch solution. Ion densities are in the 3×10^{10} to 3×10^{12} range.[9] System pressure influences etch rate and uniformity through a phenomenon called *mean free path*. This is the average distance a gas atom or molecule will travel before a collision with another particle. At higher pressures there are many collisions, which send the particles in many directions, which in turn causes loss of edge profile control. Low pressures are preferred, but there is a tradeoff with plasma damage, as explained in the next section. System pressures typically run in the 0.4- 50-mTorr range.[10] Etch rates vary from 600 to 2000 Å/min.[11]

Plasma/radiation damage. It would seem that the preferred system design comprises higher-density sources with low pressure. However, there is a countervailing process of radiation or plasma damage to the wafers. Within the plasma field are energetic atoms, radicals, ions, electrons, and photons.[12] These species, depending on their concentration and energy levels, cause various kinds of damage in semiconductors. The damage includes surface leakage, changes in electrical parameters, degradation of films (especially oxides), and damage to silicon, among others. There are two damage mechanisms. One is simple overexposure to the high-energy species in the plasma. The other is *dielectric wearout* from currents flowing across dielectrics during the etch cycle.[13] Higher-density sources also cause a problem for photoresist removal. The combination of the energy and low pressure tend to harden the resist to a level that is difficult to remove with conventional processes (see "Resist Stripping"). System designers are looking for plasma sources that feature high-density, low-energy ions (to reduce damage) and low-pressure operations.

Besides balancing the ion density and pressure parameters, *downstream plasma processing* is an option to reduce plasma damage. The damaging species come from the high energy applied to the gas by the plasma source. Downstream systems create the plasma field in one chamber and transport it downstream to the wafer(s). The wafers are separated from the damaging plasma. To minimize the damage, the system must allow for identification of the plasma discharge, ionic recombination, and reduction of the electron density.[14] Downstream plasma systems were developed to minimize damage during plasma resist removal. They are attracting interest for etch applications even though they add more complexity to an etch system.

Selectivity

Selectivity is a major consideration of plasma etching processes, especially when balanced against the need for overetch. Ideally, an etch time could be calculated to remove the anticipated layer thickness with just a little overetch time for safety. Unfortunately, the accumulated thickness and composition variations on multiple layer *stacks* on high-density devices present etch uniformity problems. Also, on high-density devices a phenomenon called *microloading* introduces etch rate variations. Microloading is a change in the local etch rate relative to the area of material being etched. A large area will *load* the etching process with removed material, slowing it down in that area while a smaller etch area proceeds at a faster rate. Topography issues also drive overetch considerations. A typical situation is the opening of contact holes in both thin regions and thick regions on the device or circuit (see Fig. 10.12). These factors can lead to an overetch of 50 to 80 percent[15] for metal etches and up to 200 percent for oxides and polysilicon etches.[16]

Overetch makes the issue of selectivity supercritical. There are two selectivity considerations, the photoresist and the underlying layer (usually silicon or silicon nitride). Dry etch processes have a higher resist-removal rate than wet processes. Given the thinner photoresist layers used to define small geometries and the increasing use of stacks of layers, the photoresist selectivity becomes critical. Compounding the selectivity issue are high aspect ratio patterns. Advanced devices have patterns with up to 4:1 aspect ratios. The holes are so narrow compared to their height that etching can slow up or stop near the bottom.[17]

Four methods used to control selectivity are

1. The selection of the etching gas formula

2. The etch rate

3. The dilution of the gas near the end of the process to slow down the attack of the underlying layer

4. Endpoint detectors in the system

An endpoint detector senses the presence of the etching layer material in the exhaust stream and automatically signals an end to the etch when no more material is detected.

Contamination, residues, corrosion, and cost of ownership.   Other process problems of concern, especially in the submicron range, are particulate generation, residues, post-etch corrosion, and cost of ownership factors. One approach to the reduction of particles is to use an electrostatic wafer holder in place of mechanical holders. Mechanical holders generate particles and cause wafer breakage, and the clamps cast a shadow on part of the wafer surface. The electrostatic clamp holds the wafer with a direct current (dc) potential between the wafer and the chuck.[18]

The plasma etch environment is highly reactive and many chemical reactions take place. Hydroxyl groups in the photoresist react with metal halide gases to form stable metal halide (such as AlF_3, WF_5, WF_6) and oxides such as TiO_3, TiO, and/or WO_2.[19] These residues create contamination problems and can interfere with the selective deposition of tungsten.[20]

Post-etch corrosion is caused by etch residues left on metal patterns after the etch process. The addition of copper to aluminum metal and the use of titanium or tungsten metallization increases the corrosion problem from residual chlorine after plasma etch. Minimizing this problem includes substituting fluorine-based etchants for chlorine etchants, passivating the sidewalls, and postetch processes such as removing the residual chlorine or using a native oxide to passivate the surface.[21] Other solutions include an oxygen plasma treatment, fuming nitric acid, and a wet photoresist stripper step.[22] Cost of ownership factors are detailed in Chap. 15. For etchers, a primary COO parameter is the etch rate.

Figure 9.25 lists the common gas etchants for various materials. Silicon and silicon dioxide processes favor fluorine-based etchants such as CF_4. Aluminum etching generally takes place with chlorine-based gases, such as BCl_3.

TRENDS IN PLASMA ETCH CHEMISTRY

Material Being Etched	Conventional Chemistry	New Chemistry	Benefits
PolySi	Cl_2 or BCl_3 /CCl_4 /CF_4 /$CHCl_3$ /CHF_3 } sidewall passivating gases	$SiCl_4/Cl_2$ BCl_3/Cl_2 $HBr/Cl_2/O_2$ HBr/O_2 Br_2/SF_6 SF_6 CF_4	No carbon contamination Increased selectivity to SiO_2 and resist No carbon contamination Higher etch rate
Al	Cl_2 BCl_3 } + sidewall passivating gases $SiCl_4$	$SiCl_4/Cl_2$ BCl_3/Cl_2 HBr/Cl_2	Better profile control No carbon contamination
Al-Si (1%)-Cu (0.5%)	Same as Al	BCl_3/Cl_2+N_2	N_2 accelerates Cu etch rate
Al-Cu (2%)	$BCl_3/Cl_2/CHF_3$	BCl_3/Cl_2+N_2+Al	Additional aluminum helps etch copper
W	$SF_6/Cl_2/CCl_4$	SF_6 only NF_3/Cl_2	No carbon contamination Etch stop over TiW and TiN No carbon contamination
TiW	$SF_6/Cl_2/O_2$	SF_6 only	
WSi₂, TiSi₂, CoSi₂	CCl_2F_2	CCl_2F_2/NF_3 CF_4/Cl_2	Controlled etch profile No carbon contamination
Single crystal Si	Cl_2 or BCl_3 + sidewall passivating gases	CF_3Br HBr/NF_3	Higher selectivity trench etch
SiO₂ (BPSG)	CCl_2F_2 CF_4 C_2F_6 C_3F_8	CCl_2F_2 CHF_3/CF_4 CHF_3/O_2 CH_2CHF_2	CFC alternatives
Si₃N₄	CCl_2F_2 CHF_3	CF_4/O_2 CF_4/H_2 CHF_3 CH_3CHF_2	CFC alternatives
GaAs	CCl_2F_2	$SiCl_4/SF_6$ $/NF_3$ $/CF_4$	Fluorine provides etch stop on AlGaAs
InP	None	CH_4/H_2 HI	Clean etch Higher etch rate than with CH_4/H_2

Figure 9.25   Plasma etch gas chart.

Argon Heated Cathode

Anode

Plasma
Region

Wafers

O = Argon Atoms

• = Ionized Argon Atoms

Figure 9.26 Ion beam milling.

Ion beam etching

A second type of dry etch system is the ion beam system (Fig. 9.26). Unlike the chemical plasma systems, ion beam etching is a physical process. The wafers are placed on a holder in a vacuum chamber and a stream of argon is introduced into the chamber. Upon entering the chamber, the argon is subjected to a stream of high-energy electrons from a set of cathode (−)–anode (+) electrodes. The electrons ionize the argon atoms to a high-energy state with a positive charge. The wafers are held on a negatively grounded holder which attracts the ionized argon atoms. As the argon atoms travel to the wafer holder they accelerate, picking up energy. At the wafer surface they crash into the exposed wafer layer and literally blast small amounts from the wafer surface. Scientists call this physical process *momentum transfer*. No chemical reaction takes place between the argon atoms and the wafer material. Ion beam etching is also called *sputter etching* or *ion milling*.

The material removal (etching) is highly directional (anisotropic), resulting in good definition of small openings. Being a physical process, ion milling has poor selectivity, especially with photoresist layers. Radiation damage from the ionization mechanism is also a problem.

Reactive ion etching

Reactive ion etching (RIE) systems combine plasma etching and ion beam etching principles. The systems are similar in construction to the plasma systems but have a capability of ion milling. The combination brings the benefits of chemical plasma etching along with the benefits of directional ion milling. A major advantage of RIE systems is in the etching of silicon dioxide over silicon layers. The combination etch results in a selectivity ratio of 35:1,[23] whereas ratios of only 10:1 are available with plasma-only etching. RIE systems have become the etching system of choice for most advanced product lines.

Resist effects in dry etching

For both wet and dry etching processes, a patterned photoresist layer is the preferred etch barrier. In wet etching there is almost no attack of the resist by the etchants. However, in dry etching, residual oxygen in the system attacks the resist layer. The resist must remain thick enough to stand up to the etchants without becoming so thin that pinholes are present. Some structures use deposited layers as etch barriers to avoid the problem of resist loss (see Chap. 10).

Another resist-related dry etch problem is resist baking. Within the dry etch chamber, the temperature can rise as high as 200°C, a temperature that can bake the resist to a condition that makes it hard to remove from the wafer. Another temperature-related problem is the tendency of resist patterns to flow and distort the images.

One unwanted effect of plasma etching is the deposition of *sidewall polymer* strings on the sides of etch patterns. The polymer comes from the photoresist. During a subsequent oxygen plasma resist strip step, the strings become metal oxides that are difficult to remove.[24]

Resist Stripping

After etching, the pattern is a permanent part of the top layer of the wafer. The resist layer that has acted as an etch barrier is no longer needed and is removed (or stripped) from the surface. Traditionally, the resist layer has been removed by wet chemical processing. Now, there is growing use of plasma O_2 stripping. Wet stripping follows the same flow as wet wafer cleaning or etching: strip-rinse-dry.

A number of different chemicals are used for stripping. Choices depend on the wafer surface (under the photoresist), production considerations, the polarity of the resist, and the condition of the resist (Fig. 9.27). Wafers are stripped of photoresist after a number of processes: wet etch, dry etch, and ion implantation. There are different degrees of difficulty, depending on the prior process. High-temperature hard bakes, plasma etch residues and sidewall polymers, and ion implantation crusting all present challenges for the resist removal process. Generally, the strippers are divided into the

Stripper Chemistry	Strip Temperature (Centigrade)	Surface Oxide	Metallized	Resist Polarity
Acids:				
Sulfuric acid+oxidant	125	X		+/-
Organic Acids	90-110	X	X	+/-
Chromic/Sulfuric	20	X		+/-
Solvents:				
NMP/Alkanolamine	95		X	+
DMSO/Monothanolamine	95		X	+
DMAC/Diethanolamine	100		X	+
Hydroxylamine (HDA)	65		X	+

Figure 9.27 Wet photoresist stripper chart.

categories of universal strippers and positive and negative resist-only strippers. They are also divided by type of wafer surface, metallized or nonmetallized.

Wet stripping is the most used for the following reasons:

1. It has a long process history.

2. It is cost-effective.

3. It is effective in the removal of metallic ions.

4. It is a low-temperature process and does not expose the wafers to possibly damaging radiation.

Wet chemical stripping of nonmetallized surfaces

Sulfuric acid and oxidant solutions. Solutions of sulfuric acid and an oxidant (hydrogen peroxide or ammonium persulfate[25]) are the most common wet strippers used for the removal of resist from *nonmetallic* surfaces. Nonmetallic surfaces are composed of either silicon dioxide, silicon nitride, or polysilicon. This solution strips both negative and positive resists. These are the same chemical solutions and processes used for pre-tube-cleaning wafers described in Chap. 7.

Nitric acid is sometimes used as an additive oxidant in a sulfuric acid bath. A mixing ratio of about 10:1 is typical. A drawback to nitric acid is that it turns the bath a light orange color, which can mask the buildup of carbon in the bath. Fuming nitric acid is used as a stripper in Europe and Japan. All of these solutions dissolve the resist by an oxidation mechanism.

Wet chemical stripping of metallized surfaces

Stripping from metallized surfaces is a more difficult task because the metals are subject to attack or oxidation. There are five types of wet chemicals that are used for stripping metallized surfaces. They are

1. Organic strippers
2. Chromic sulfuric acid mixtures
3. Solvent strippers
4. Solvent–amine strippers
5. Specialty strippers

Phenolic organic strippers. Organic strippers contain a combination of sulfonic acid (an organic acid) and chlorinated hydrocarbon solvents such as duodexabenzene. The formula requires phenol to create a rinseable solution. This class of stripper was introduced in the early 1960s with the invention of the stripper J-100 by Industri Chem. It was the first stripper to provide universal stripper capabilities and nonattack of metal films. Since the introduction of J-100, a number of suppliers have developed similar formulas, some designed for direct water rinsing. These were the workhorse strippers for over two decades. However, in the 1970s concern over the toxic ingredients in these formulas led to the development of sulfonic acid, nonphenolic, nonchlorinated[26] resist strippers.

Stripping the photoresist requires heating the solutions to the 90 to 120°C range. Often the process uses two or three heated strip baths. Rinsing is in two steps, first a solvent followed by a water rinse and drying step.

Chromic and sulfuric acid mixtures. These strippers are a mixture of chromium trioxide in sulfuric acid. The strippers are used at room temperature, which simplifies the process. In fact, these strippers cannot be heated or an attack of metal films will take place. The mixtures must also be kept water-free to prevent metal attack. Like the phenolic organic strippers, the chromic and sulfuric acid strippers represent an environmental problem due to the chromium trioxide, which is unacceptable in the food chain. This type of stripper is removed from the site rather than rinsed and dumped into the city drain.

Solvent positive-only strippers.   One of the advantages of positive resists is their ease of removal from the wafer surface. A positive resist layer that has not been hard-baked is easily removed from the wafer with a simple acetone soak. In fact, acetone has been the traditional positive resist stripper. Unfortunately, acetone represents a fire hazard and its use is discouraged.

Solvent–amine strippers.   Several manufacturers supply positive-only strippers based on solvent and organic amine solutions. N-methyl pyrrolidine (NMP)[27] is the solvent most used. Others are dimethylsulfoxide (DMSO), sulfolane, dimethylforamide (DMF), and dimethylacetamide (DMAC). These strippers are effective, water-rinseable, and drain-dumpable. The strippers may be heated to increase the removal rate and/or remove resist films that have been through a high temperature hard bake. Solvent and solvent–amine strippers remove the resist by a chemical dissolution mechanism.

Specialty wet strippers.   A number of wet chemical strippers have been developed to solve specific problems. One is a positive resist stripper based on hydroxylamine (HDA) chemistry.[28] The stripper removes a host of plasma etch residues and polyimide layers not removed by solvent–amine strippers. Other strippers include corrosion inhibitors.[29]

Figure 9.27 is a table of the most common wet resist strippers and their uses.

Dry stripping

Like etching, the dry plasma process can also be applied to resist stripping. The wafers are placed in a chamber and oxygen is introduced (Fig. 9.28). The plasma field energizes the oxygen to a high-energy state, which, in turn, oxidizes the resist components to gases that are removed from the chamber by the vacuum pump. In dry strippers, the plasma is generated by microwave, RF, and UV-ozone sources.[30]

The major advantage of plasma resist stripping is the elimination of wet hoods and the handling of chemicals. The principal disadvantage is its ineffectiveness in the removal of metallic ions. There is not enough energy in the plasma field to volatilize the metallic ions.

C_xH_y (resist) + O_2 (plasma energized) → CO (gas) + CO_2 (gas) + H_2O

Figure 9.28   Resist removal by plasma oxygen.

Another consideration of plasma stripping is radiation damage to the circuits from the high-energy plasma field. This problem is reduced with system designs that have the plasma chamber removed from the stripper chamber. They are called *downstream strippers* because the plasma is created downstream from the wafers. MOS wafers are more sensitive to radiation effects during stripping.

Replacement of wet stripping by dry plasma techniques has been a long-term industry desire. However, problems such as the inability of oxygen plasma to remove mobile ionic metal contamination and certain metal residues, as well as radiation damage concerns, have kept wet stripping as the mainstream photoresist removal process. Some fabrication lines use a mix-and-match approach. Plasma stripping is used to remove hardened resist layers. A follow-on wet strip step is used to remove the residuals not removed by the plasma.

Final Inspection

The final step in the basic photomasking process is a visual inspection. It is essentially the same procedure as develop inspect, with the exception that the majority of the rejects are fatal (no rework is possible). The one exception is contaminated wafers that may be recleaned and reinspected. Final inspection certifies the quality of the outgoing wafers and serves as a check on the effectiveness of the develop inspection. Wafers that should have been identified and pulled from the batch at develop inspect are called *develop inspect escapes.*

The wafers receive a first surface inspection in incident white or ultraviolet light for stains and large particulate contamination. This inspection is followed by a microscopic or automatic inspection for defects and pattern distortions. Measurement of the critical dimensions for the particular mask level is also part of the final inspection. Of primary interest is the quality of the etched pattern with underetching and undercutting being two parameters of concern. The table in Fig. 9.29 is a list of typical causes of wafer rejection found in the final inspection.

Mask Making

In Chap. 5, the steps of circuit design were detailed. In this section, the process used to construct a photomask or reticle is examined. Originally the masks were made from emulsion-coated glass plates. The emulsions were similar to those found on camera film. These masks were vulnerable to scratches, deteriorated during use, and

Possible Process Cause	Contami-nation	Misalign	Undercut	Incom-plete Etch	Wrong Mask	Pin Holes	C.D.'s	Visual Reject
Contaminated Etch	X		X	X				
Contaminated Stripper	X							
Contaminated H₂O	X							
Insufficient Rinse	X		X					
No Wet Agent				X				
Under Etch				X			X	
Over Etch			X				X	
Wrong Etch			X	X			X	
Hard Bake Too High			X	X			X	
Poor Develop				X			X	
P₂O₅ & SiO₂			X				X	
B₂O₃ & SiO₂				X			X	
Low Hard Bake			X					
Develop Inspect Escapes		X	X	X	X	X		

Figure 9.29 Final inspect rejects and process causes.

were not capable of resolving images in the sub-3-μm range. Masks for most modern work use a chrome on glass technology. This mask-making technology is almost identical to the basic wafer-patterning operation (Fig. 9.30). In fact the goal is the same, the creation of a pattern in the thin chrome layer on the glass reticle surface. The preferred materials for masks and reticles are borosilicate glass or quartz. These materials have good dimensional stability and transmission properties for the wavelengths of the exposing sources. Chrome layers are in the 1000-Å range and deposited on the glass by sputtering (see Chap. 12). Advanced masks and reticles use layers of chromium, chromium oxide, and chromium nitride.[31]

Mask- or reticle-making follows a number of different paths. There are a number of different process paths, depending on the starting exposure method (pattern generation, laser, e-beam) and the end result (reticle or mask) (Fig. 9.31). Flow A shows the process for making a reticle using a pattern generator, an older technology. A pattern generator consists of a light source and a series of motor-driven shutters. The chrome-covered mask or reticle with a layer of photoresist is moved under the light source as the shutters are moved and opened to allow precisely shaped patterns of light to shine onto the resist creating the desired pattern. The reticle pattern is transferred to the

Figure 9.30 Major steps in mask/reticle plate processing.

resist-covered mask blank by a step-and-repeat process to create a master plate. The master plate is used to create multiple working mask plates in a contact printer. This tool brings the master into contact with a resist-covered mask blank and has a UV light source for transferring the image. After each of the exposure steps (pattern generation, laser, e-beam, master plate expose, and contact print) the reticle or mask is processed through the development, develop inspect, etch, strip, and final inspection steps that transfer the pattern permanently into the chrome layer. Inspections are very critical since any undetected mistake or defect has the potential of creating thousands of scrap wafers. Reticles for this use are generally 5 to 20 times the final image size on the mask.[32]

Advanced products with very small geometries and tight alignment budgets require high-quality reticles and/or masks. The reticles and masks for these processes are made with lasers or e-beam direct write exposure (Flows A and B). Laser exposure uses a wavelength of 364 nm, making it an I-line system. It allows using standard optical resists and is faster than e-beam. Direct write laser sources are turned on and off with an acousto-optical modulator (AOM).[33] In all

Figure 9.31 Mask/reticle-making processing flows.

cases, the reticle or mask is processed to etch the pattern into the chrome.

Other mask and reticle process flows may be employed. The reticle in Flow A or the master plate may be laser or e-beam generated.

VLSI- and ULSI-level circuits require virtually defect-free and dimensionally perfect masks and reticles. Critical dimension (CD) budgets from all sources are 10 percent or better leaving the reticles with a 4 percent error margin.[34] There are procedures to eliminate unwanted chrome spots and pattern protrusions with laser "zapping" techniques. There is also a technology to fill in missing pattern parts and fill pinholes by using focused ion beams (FIB).

Summary

For VLSI and ULSI work, the resolution and registration requirements are very stringent. In 1977, the minimum feature size was

3 μm. By the mid-1980s, it had passed the 1-μm barrier. By the 1990s, 0.5-μm sizes were common with 0.35-μm technology planned for production circuits. Circuit design projections call for minimum feature sizes less than 0.1 μm.[35]

Chip manufacturers calculate several budgets for each circuit product. A *critical dimension* (CD) *budget* calculates the allowable variation in the image dimensions on the wafer surface. For products with submicron minimum feature sizes, the CD tolerances are 10 to 15 percent.[36] Also of concern is the critical defect size relative to the minimum feature size. These two parameters are brought together in an *error budget* calculated for the product. An *overlay budget* is the allowable accumulated alignment error for the entire mask set. A rule of thumb is that circuits with micron or submicron feature sizes must meet registration tolerances of one-third the minimum feature. For a 0.35-μm product, the allowable overlay budget is about 0.1 μm.[37]

Key Concepts and Terms

Develop inspect and rework	Negative resist developers
Dry etch methods	Overlay budget
Dry stripping	Plasma descum
Error budget	Positive resist developers
Etch process	Puddle develop
Final inspect	Resist development
Hard bake methods	Resist stripping
Hard bake process	Spray develop
Immersion develop	Wet etch methods
Mask making	Wet strip chemicals

Review Questions

1. Name the major methods of resist development.

2. What are the chemicals used to develop negative and positive resists?

3. What is the purpose of the hard bake step?

4. Name three methods used for hard bake.

5. What problems arise if the hard bake temperature is too low? Too high?

6. Name the preferred wet etchants for etching silicon dioxide layers, silicon nitride layers, and aluminum layers.

7. Name two advantages of dry etching over wet etching.

8. List the three principal dry etching techniques.

9. Write a wet stripping flow diagram.

10. Name the purpose and methods of final inspect.

11. List at least five problems and solutions for resolving submicron patterns.

References

1. D. Elliott, *Integrated Circuit Fabrication Technology,* McGraw-Hill, New York, 1976, p. 216.
2. S. Wolf and R. Tauber, *Silicon Processing for the VLSI Era,* Lattice Press, Newport Beach, CA, 1986, p. 530.
3. P. Singer, "Meeting Oxide, Poly and Metal Etch Requirements," *Semiconductor International,* Cahners Publishing, April 1993, p. 51.
4. Ibid, p. 51.
5. S. Wolf and R. Tauber, *Silicon Processing for the VLSI Era,* Lattice Press, Newport Beach, CA, 1986, p. 532.
6. Ibid.
7. C. Murray, "Wet etching update," *Semiconductor International,* May 1986, p. 82.
8. P. Burggraaf, "Advanced Plasma Sources: What's Working?," *Semiconductor International,* Cahners Publishing, May 1994, p. 57.
9. P. Singer, "Meeting Oxide, Poly and Metal Etch Requirements," *Semiconductor International,* Cahners Publishing, April 1993, p. 53.
10. Ibid.
11. D. Elliott, *Integrated Circuit Fabrication Technology,* McGraw-Hill, New York, 1976, p. 275.
12. S. Fonsh, C, Viswanathan, and Y. Chan, "A Survey of Damage Effects in Plasma Etching," *Solid State Technology,* PenWell Publishing Company, July 1994, p. 99.
13. Ibid.
14. C. Boitnott, "Downstream Plasma Processing: Considerations for Selective Etch and Other Processes," *Solid State Technology,* PenWell Publishing Company, October 1994, p. 51.
15. P. Riley, S. Pengm, and L. Fang, "Plasma Etching of Aluminum for ULFI Circuits," *Solid State Technology,* PenWell Publishing Company, February 1993, p. 4.
16. M. Engelhardt, "Advanced Polysilicon Etching in a Magnetically Confined Reactor," *Solid State Technology,* PenWell Publishing Company, June 1993, p. 57.
17. P. Singer, "Meeting Oxide, Poly and Metal Etch Requirements," *Semiconductor International,* Cahners Publishing, April 1993, p. 51.
18. B. Newboe, "Wafer Chucks Now Have an Electrostatic Hold," *Semiconductor International,* Cahners Publishing, February 1993, p. 30.
19. C. Cardinaud, M. Peignon, and G, Turban, "Surface Modification of Positive Photoresist Mask during Reactive Ion Etching of Si and W in SF_6 Plasm," *J. Electrochemical Society,* vol. 198, 1991, p. 284.
20. W. M. Lee, *A Proven Sub-Micron Photoresist Stripper Solution for Post Metal and Via Hole Processes,* EKC Technology, Inc, Hayward, CA, 1993.
21. F. Clayton and S. Beeson, "High-Rate Anisotropic Etching of Aluminum on a Single-Wafer Reactive Ion Etcher," *Solid State Technology,* PenWell Publishing Company, July 1993, p. 93.
22. W. M. Lee, *A Proven Sub-Micron Photoresist Stripper Solution for Post Metal and Via Hole Processes,* EKC Technology, Inc, Hayward, CA, 1993.
23. D. Elliott, *Integrated Circuit Fabrication Technology,* McGraw-Hill, New York, 1976, p. 282.

24. W. M. Lee, A Proven Sub-Micron Photoresist Stripper Solution for Post Metal and Via Hole Processes, EKC Technology, Inc, Hayward, CA, 1993.
25. EKC Technology Inc., Technical Bulletin SA-80.
26. Ibid—Nophenol 922.
27. Ibid—Posistrip Series.
28. W. M. Lee, *A Proven Sub-Micron Photoresist Stripper Solution for Post Metal and Via Hole Processes,* EKC Technology, Inc, Hayward, CA, 1993.
29. M. D. Levenson, "Wet Stripper Companies Clean Up," *Solid State Technology,* PenWell Publishing Company, April 1994, p. 31.
30. P. Burggraaf, "What's Driving Resist Dry Stripping?," *Solid State Technology,* PenWell Publishing Company, November 1994, p. 61.
31. B. Grenon, et al., "A Comparison of Commercially Available Chromium-Coated Quartz Mask Substrates," *OCG Microlithography Seminar,* INTERFACE 94, p. 37.
32. S. Wolf and R. N. Tauber, *Silicon Processing,* vol. 1, Lattice Press, Sunset Beach, CA, p. 477.
33. J. Reynolds, "Mask Making Tour Video Course," *Semiconductor Services,* Redwood City, August 1991, Segment 5.
34. J. Reynolds, "Elusive Mask Defects: Random Reticle CD Variation," *Solid State Technology,* PenWell Publishing Company, September 1994, p. 99.
35. C. Hu, "MOSFET Scaling in the Next Decade and Beyond," *Semiconductor International,* Cahners Publishing, June 1994, p. 105.
36. J. Wiley and J. Reynolds, "Device Yield and Reliability by Specification of Mask Defects," *Solid State Technology,* PenWell Publishing Company, July 1993, p. 65.
37. K. Simon, et al., Abstract—"Alignment Accuracy Improvement by Consideration of Wafer Processing Impacts," *SPIE Symposium on Microlithography,* 1994, p. 35.

10

Advanced Photolithographic Processes

Overview

The demands of decreasing feature sizes, especially into the submicron range, and the increasing need for low-defect densities have outstripped the capabilities of the basic photoresist processes. New processes and improvements to existing processes have allowed the continuation of product advances and high-density chips. These processes and the problems they solve are described in this chapter.

Objectives

Upon completion of this chapter, you should be able to:

1. Describe four exposure-related effects that cause image distortion.
2. Draw a cross-sectional flow diagram of a dual-layer resist process.
3. Describe the uses and process of a lift-off procedure.
4. List two planarization techniques.
5. State an advantage of an image-reversal process.
6. Describe how antireflective layers, contrast enhancement layers, and resist dye additives improve resolution.
7. Identify the parts of a pellicle and the advantages it offers to a resist process.

Introduction

The ten-step patterning process detailed in Chaps. 8 and 9 is a basic process based on one photoresist layer. It would be sufficient for the production of MSI circuits and some simple LSI and VLSI circuits. However, the VLSI and ULSI requirements for decreased feature sizes and defect densities strained the limits of the basic process. The limits showed up in the sub-3 μm range and became critical in the submicron era. Problems include the physical limitations associated with optical exposure equipment, the resolution limits of photoresists, and a host of surface problems, including reflective surfaces and multilevel topographies. In the mid-1970s, it was widely accepted that optical photoresist processes had a lower resolution limit of about 1.5 μm. This projection gave rise to the interest in x-ray and e-beam exposure systems. However, manipulation and improvements on the basic processes have successively lowered the usable range of optical lithography to the 0.2-μm range.[1]

In this chapter some of the limits of optical imaging and advanced process solutions are presented. The general resist resolution problem comes from the wavefront that arrives at the resist layer. Simple cross-sectional drawings show the arriving rays as uniform arrows. The actual radiation in the wavefront has a mixture of directions and energies that vary across surface and in the vertical direction. This is called the *aerial image*.[2] Resolving 0.5- and 0.33-μm images with optical lithography requires management and manipulation of the aerial image. Control methods fall into four general areas: optical resolution, resist resolution limits, surface problems, and etch definition.

Optical resolution and exposure system advances

In Chap. 8, the issue of image resolution and exposure wavelength was explored. In general, the way to expose smaller images is to use a smaller-wavelength exposure source. However, this leads to a smaller depth of field (DOF). In the sub-0.5-μm range exposure processes extend from I line to deep UV (Fig. 10.1). The DOF problem requires other refinements including variable numerical aperture lens, annular-ring illumination, off-axis illumination, and phase shift masks. In addition, there are other optical effects that come into play as the image size gets smaller and the pattern density gets larger. These issues and solutions are examined.

Variable numerical aperture (NA) lenses

The NA factor is an overall measurement of a lens's ability to gather light. The tradeoff is depth of field (DOF). A lens with a better ability

Figure 10.1 Lithography resolution process roadmap. (*Source: Semiconductor International, February 1993*).

to gather light (higher NA) suffers from a decreased DOF (see Chap. 8). Unfortunately, advanced circuits have many layers with high and low points (*surface topography*) that require a large DOF. A stepper with one lens is limited to exposing wafers within its associated DOF which may or may not include the levels on the wafer surface. Newer steppers come with a variable NA lens, allowing its use over a wider range of DOF requirements.

Contrast effects

Good resolution becomes difficult in regions of the mask where an opaque line is surrounded by a large clear area. The large amount of radiation coming around the opaque line tends to shrink the dimension of the line in the resist layer (Fig. 10.2) as the exposing rays diffract around the edge of the pattern. This problem is called a *proximity effect*.

Another contrast effect is called *subject contrast* (Fig. 10.3). This situation comes about when some exposure radiation penetrates the opaque region of a mask or reflects off the wafer surface into the resist. The result is a partially exposed region that leaves a distorted image after the development step. This is more of a problem with negative resist than with positive resist.

Masks

Phase shift masks (PSMs)

Another diffraction problem occurs as two mask patterns get closer together. At some point, normal diffraction causes the exposure rays

Figure 10.2 Proximity effects.

Figure 10.3 Subject contrast.

to start touching, leaving the patterns unresolved in the resist. The blending of the two diffraction patterns into one occurs because all the rays are in the same phase. *Phase* is a wave term that relates to the relative positions of the peaks and valleys of a wave (Fig. 10.4). The waves in (*a*) are in phase and those in (*b*) out of phase. One way to prevent the diffraction patterns from wiping out two adjacent mask patterns is to cover one of the openings (Fig. 10.5) with a transparent layer that shifts one of the sets of exposing rays *out of phase,* which will null the blending. This approach is called *alternating phase shift mask* or *alternating aperture phase shift mask* (AAPSM). It requires the deposition of a layer of silicon dioxide on the mask or reticle and a photomasking process to remove the oxide layer from alternate patterns. Covering every other clear opening works well for repeated array patterns such as those found in memory products.

Another solution is the addition of phase-shifting layers to the edges of the mask or reticle patterns. This process also requires an

In phase
(a)

Out of phase
(b)

Figure 10.4 Wave phases.

Figure 10.5 Light intensity patterns. (*a*) without phase shifting, (*b*) with phase shifting. (*Source: VLSI Fabrication Principles by Ghandhi*)

oxide deposition and full masking process. There are several variations on this approach, for example, subresolution (or outrigger) and rim phase shift masks.[3]

Optical proximity corrected (OPC) masks

It is known that in the sub-0.5-μm range a perfect image on the mask or reticle can be distorted by proximity effects, resulting in a defective pattern in the resist. OPC masks attempt to reverse this situation by putting a distorted image on the mask or reticle that is designed to produce a perfect image in the resist. A computer is used to analyze exposure process conditions (contrast effects) and design the mask's pattern shape.[4]

Illumination

Annular ring illumination

Annular ring illumination is a technique first introduced in Perkin Elmer scanning projection aligners. One of the guns in the resolution arsenal is a more uniform exposing light. Unfortunately, conventional optical exposure sources produce a light spot that is too nonuniform for small image exposure. However, within the spot there are areas (rings) of more uniform energy. An annular ring illuminator blocks off all but a *ring* portion of the spot, directing a more uniform wave of exposing radiation to the wafer.

Off-axis illumination (OAI)

Off-axis illumination is an approach to greater resolution similar to AAPSM. In the mask case, on-board phase shifters are used to put the rays that arrive at the projection lens of a stepper out of phase. OAI moves the incident ray angle off the usual 90°. The angle is chosen such that rays arrive out of phase, thus canceling diffraction effects.

Pellicles

The development of projection exposure systems (projection aligners and steppers) brought with them an increased mask and reticle lifetime. With the increased lifetime came the incentive to make higher-quality masks and reticles. In a production line where masks are used for a long time, wafer-sort yield loss comes from dirt and scratches picked up during handling and use. One source of damage comes from mask and reticle cleaning steps. This situation is a "Catch-22." The masks and reticles become dirty during the process and require cleaning. The cleaning procedure then itself becomes a source of contamination, scratches, and breakage.

A solution to these problems is a pellicle (Fig. 10.6). A *pellicle* is a thin layer of an optically neutral polymer stretched onto a frame. The frame is designed to fit onto the mask or reticle. The pellicle is fitted to the mask or reticle after the mask is made and cleaned. Once in place, the pellicle membrane is the surface collecting any dirt or dust in the environment. The height of the membrane above the mask surface holds the dirt particles out of the focal plane of the mask. In effect, the particles are transparent to the exposing rays.

Another benefit of a pellicle is the elimination of scratches from the mask surface since the surface is covered. A third benefit is that once pelliclized, a mask or reticle does not need in-process cleaning. For some applications, the pellicle membrane is given an antireflective coating which assists in the imaging of small geometries, especially on reflective wafer surfaces. These benefits translate into wafer-sort yield increases of 5 to 30 percent.[5]

Pellicle membranes are made from either nitrocellulose (NC) or cellulose acetate (AC). NC films are used in exposure systems with

Figure 10.6 Pellicle.

broadband exposure sources (340 to 460 nm)[6] while AC films are used in mid-ultraviolet applications. The membranes are thin (0.80 to 2.5 μm) and must show a high transmission rate for the rate exposure wavelengths. A typical pellicle will exhibit over a 99 percent transmission rate for the peaks of the exposing wavelengths. Pellicle effectiveness requires stringent thickness control, on the order of ± 800 Å, and control of particles to less than 25 μm in diameter.

A pellicle membrane is made by a spin casting technique. The pellicle material is dissolved in a solvent and spun onto a rigid substrate, such as a glass plate. This is the same technique used to spin photoresist onto wafers. Thickness of the membrane is controlled by the viscosity of the solution and the spin speed of the spin coater. The membrane is removed from the substrate and fixed onto the frame. Frame shapes are determined by the size and shape of the mask or reticle. Cleanliness control requires class 10 or better clean rooms and antistatic packaging.

Surface Problems

The resolution of small images is affected by several conditions on the wafer surface. Reflections off the surface layers, increasing variation of the topography, and the etching of multilayer stacks all require special process steps or "tweaking" of the process.

Resist light scattering

In addition to light radiation reflecting off the wafer surface, the radiation tends to diffuse into the resist, causing poor image definition.

The amount of diffusion is in proportion to the resist thickness. Some additives put in the photoresist to increase radiation absorption also increase the amount of radiation diffusion, thus reducing image resolution.

Subsurface reflectivity

The high-intensity exposing radiation ideally is directed at a 90° angle to the wafer surface. When this ideal situation exists, exposing waves reflect directly up and down in the resist, leaving a well-defined exposed image (Fig. 10.7). In reality, some of the exposing waves are traveling at angles other than 90° and expose unwanted portions of the resist.

This subsurface reflectivity varies with the surface layer material and the surface smoothness. Metal layers, especially aluminum and aluminum alloys, have higher reflectivity properties. A goal of the deposition processes is a consistent and smooth surface to control this form of reflection.

Reflection problems are intensified on wafers with many steps, also called a varied topography. The sidewalls of the steps reflect radiation at angles into the resist, causing poor image resolution. A particular problem is light interference at the step that causes a "notching" of the pattern as it crosses the step (Fig. 10.8).

Antireflective Coatings

Antireflective coatings (ARCs) spun onto the wafer surface before the resist (Fig. 10.9) can aid the patterning of small images. The ARC layer brings several advantages to the masking process. First is a planarizing of the surface, which makes for a more planarized resist layer. Second, an ARC cuts down on light scattering from the surface into the resist, which helps in the definition of small images. An ARC can also minimize standing wave effects and improve the image contrast. The latter benefit comes from increased exposure latitude with a proper ARC.

Figure 10.7   Subsurface reflectivity.

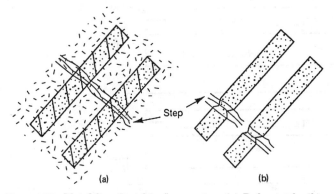

Figure 10.8 Metal line "notching" over step. (*a*) Before etch; (*b*) after etch.

An ARC is spun onto the wafer and baked. After the resist is spun on top of the ARC, the wafer is aligned and exposed. The pattern is developed in both the resist and the ARC. During the etch, the ARC acts as an etch barrier. To be effective, an ARC material must transmit light in the same range as the resist. It must also have good adhesion properties with the wafer surface and the resist. Two other requirements are that the ARC must have a refractive index that matches the resist, and that the ARC must develop and be stripped with the same chemicals as the resist.

There are several penalties associated with the use of an ARC. One is an additional layer requiring a separate spin and bake. The resolution gains offered by an ARC can be offset with poor thickness control and/or with an ill-controlled developing step. The time of exposure can increase 30 to 50 percent, increasing the wafer throughput time. ARC layers may also be used as the intermediate layer in a trilayer resist process or used on the top of the photoresist, called a top antireflective coating or TAR.

Standing waves

In "Subsurface reflectivity," it was mentioned that the ideal exposure situation is when the radiation waves are directed to the wafer surface at 90°. This is true when only reflection problems are under consideration. However, 90° reflection causes another problem in positive photoresists, the creation of standing waves. As the radiation wave reflects off the surface and travels back up through the resist, it interferes constructively and destructively with the incoming wave, creating regions of varying energy (Fig. 10.10). The result after development is a rippled sidewall and a loss of resolution. A number of solutions are used to moderate standing waves, including dyes in the resist and separate

Figure 10.9 Antireflective process sequence.

Figure 10.10   Standing-wave effect. (a) During exposure; (b) after develop.

antireflective coatings directly on the wafer surface. Most positive resist processes include a postexposure bake (PEB) step before development of the resist layer used to minimize standing waves.

Planarization

Increased surface density and more surface layers made possible the advancement of circuits to VLSI levels. As the various layers are etched into patterns, the surface becomes one of high and low plateaus. The plateaus are delineated by steps, each of which have different reflection qualities. The resolution of submicron (especially 0.5 μm and smaller) images on this type of surface is not possible in a simple one-resist layer process. One problem is depth of field. A limited DOF lens cannot resolve images on both the high and low plateaus. Another is light reflection at the steps, causing the notching of metal lines (see Fig. 10.8). A nonlithography problem is the thinning of deposited metal layers at the steps. A number of techniques are used to offset the effects of a varied wafer topography. The techniques of multilayer resist processing, planarization layers, reflow, and chemical-mechanical polishing are collectively known as planarization techniques.

Multilayer resists and surface imaging

There are a number of multilayer resist processes. The choice of a process depends on the size of the resist opening and the severity of the surface topography. While multilayer resist processes add the penalty of more process steps in some situations, they are the only reliable means of creating the desired images. A multilayer resist process features a thicker bottom layer that serves to fill in the val-

Incident Radiation

Mask

Resist

Metal

Substrate

Figure 10.11 Light reflection at steps.

leys and planarize the surface. The image is first formed in a layer of photoresist on the top of the planarizing layer(s). This *surface imaging* allows small-dimension imaging to avoid DOF problems because the surface is flat, i.e., away from reflecting steps and in the same plane.

A dual multilayer resist process uses two layers of photoresist, each with a different polarity. The process is suited to resolve small geometries on wafers with a varied topography. First a relatively thick layer of resist is applied and baked to the thermal flow point (Fig. 10.12). A typical thickness is three to four times the highest step height on the wafer. The goal is a planar top resist surface. A typical multilayer process will use a positive-acting polymethylmeth acrylate (PMMA) resist sensitive to deep ultraviolet radiation.

Next, a thin layer of positive resist sensitive to just ultraviolet radiation is spun on top of the first layer and processed through the development step. The thin top layer allows the resolution of the pattern without the adverse effects encountered with thick resist layers or reflections from steps in the surface. Since the top layer conforms to the shape of the bottom layer, it is referred to as a *conformal layer* or *portable conformal layer.* This top layer of resist acts as a radiation block, leaving the bottom layer unpatterned. Next, the wafer is given a blanket or flood (no mask) deep ultraviolet exposure, which exposes the underlying positive resist through the holes in the top layer, thus extending the pattern down to the wafer surface. A development step completes the hole resolution and the wafer is ready for etch.

Considerations in the choice of photoresists are compatibility of the two resists through the process, reflection problems from the subsurface, standing waves, and sensitivity problems with PMMA resists.[7] In addition, the two resists must have compatible bake processes and independent developing chemistries.

Variations on the basic dual-level resist process include dyes in the PMMA and the use of antireflection layers under the first resist layer.

Start

1. Apply 1st layer.

2. Bake 1st layer to cause slight flow.

3. Apply 2nd layer and process to develop.

4. Flood expose 1st layer.

5. Develop 1st layer.

6. Etch and strip.

Figure 10.12 Dual-layer photoresist processing.

Many variations on the basic dual-level process are practiced. One use of a dual-layer resist process is as a lift-off technique. By adjusting the development of the bottom layer, an overhang can be created that assists in the clean definition of a deposited metal line on the surface (Fig. 10.13).

A trilevel resist process (Fig. 10.14) incorporates a "hard" layer between the two resist layers. The hard layer may be a deposited layer of silicon dioxide or other developer-resistant material. As in the dual-layer process, the image is formed in the top photoresist layer. Then the image is transferred into the hard layer by a conventional

Figure 10.13 Dual-layer resist lift-off process.

etching step. The finishing step is the formation of the pattern in the bottom layer, using the hard layer as an etch mask. The use of a hard intermediate layer makes possible the use of nonphotoresist bottom layers, such as a polyimide layer.

DESIRE process

A novel approach to surface imaging is the DESIRE process (Fig. 10.15).[8] Like other multilayer resist processes, the concept is to planarize the wafer and form the image in a surface layer. The DESIRE (Diffusion Enhanced SIlylating REsist) process uses a layer exposed by standard UV exposure. In this process, the exposure is confined to the top layer of the planarizing layer. Next the wafer is placed in a chamber (see vapor prime baking)[9] for exposure to HMDS, called a *silyation process.* During this step silicon becomes incorporated into the exposed areas. The silicon-rich areas become a hard mask that allows the dry development and removal of the underlying material with an anisotropic RIE etch (see Chap. 9). During the etch step, the silyated areas are converted to silicon dioxide (SiO_2), forming a more resistant etch mask.

Polyimide planarization layers

Polyimides have been used for years in printed circuit board manufacturing. For semiconductor use, the polyimides offer the dielectric strength of deposited silicon dioxide films and the process advantage of application to the wafer with the same spinning equipment used for photoresist.[10]

Start

1. Apply 1st layer

2. Bake 1st layer to
   cause slight flow

3. Deposit "hard" layer

4. Spin, expose and develop
   thin top layer of resist

5. Etch hard layer

6. Flood expose, develop
   planarizing layer

7. Etch surface layer

Figure 10.14   Trilevel resist process.

Figure 10.15 DESIRE process. (*Source: Solid State Technology, June 1987*).

Once applied to the wafer, the polyimides flow over the surface, making it more planar. After application and flow, the polyimide can be covered with a hard layer and patterned with chemicals much like a photoresist. A popular use of polyimide layers is as an interdielectric layer between two layers of conducting metal. The planarizing effect of the polyimide makes the definition of the second metal layer easier.

Chemical mechanical polishing (CMP)

All of the planarization methods described fall short of guaranteeing an absolutely flat wafer surface. The only process that provides a *global planarization* of an entire wafer is chemical mechanical polishing. CMP is the same basic process used to polish wafers after crystal growth and slicing (see Chap. 3). The wafers are mounted upside down on a holder and rotated above a pad sitting on a pallet, rotating in the other direction (Fig. 10.16). A slurry containing a chemical that dissolves the wafer layer and an abrasive that physically removes it is flowed between the wafer and the platen.

CMP is most often used to flatten the profiles that build up in multimetal interconnection schemes (Chap. 13). In this use, the process takes on the added challenges of removing oxides and metals. Furthermore, a typical flatness specification across the wafer must be in the 1500-Å range.[11] Achieving this flatness must take place without contaminating or scratching the wafer. Also, the process must have rigorous endpoint control to avoid polishing away portions of the functioning circuit parts.

A particular challenge is developing a slurry that is noncontaminating, balances chemical removal with abrasive aggressiveness, polishes at an acceptable production rate, and still produces a planar sur-

Figure 10.16 CMP system. (*Source: SpeedFam CMP-V system, Semiconductor International, May 1993*).

face. This challenge becomes more difficult when the surface has both oxide and metal layers.

In addition to planarization for lithography improvements CMP is used for removing top layer metals after a plug deposition[12] (see Chap. 13).

Reflow

Some device schemes use a hard planarizing layer or layers. A popular layer is a deposited silicon dioxide doped with about 4 to 5 percent boron, called boron silicate glass (BSG). The presence of the boron causes the glass to flow at a relatively low temperature (less than 500°C), creating a planarized surface.

Another hard planarizing layer used is a spin-on-glass (SOG) layer. The glass is a mixture of silicon dioxide in a solvent that evaporates quickly. After spin application the glass film is baked, leaving a planarized silicon dioxide film. The glass as spun is brittle[13] and some formulas contain between 1 and 10 percent carbon to increase its resistance to cracking.

Photoresist Process Advances

The ten-step lithography process detailed in Chaps. 8 and 9 is based on a single photoresist layer and assumes that the layer can resolve the required images without pinholes or failure during the etch process. The submicron era has forced the development of advances on the single resist layer process.

Double masking

For the patterning of smaller geometries the resist engineer prefers thinner resist layers and shorter exposure times. Unfortunately, these solutions give rise to higher densities of pinholes in the resist. A process variation that resolves this problem is double masking.

Double masking starts with a thin layer of resist that is processed through the development step (Fig. 10.17). Within the thin resist, small image sizes can be resolved to specification. After the first resist layer, a second layer of resist is spun onto the wafer. The wafer goes through another alignment and exposure, but with an oversize mask. The oversize mask makes for an easy alignment and protects the previously patterned image from being decreased by the light. The second resist layer fills in any pinholes in the first layer, providing protection during the etch step. The drawback to this process is the requirement of two masks and the longer time to process the wafer.

Process Step	Cross Section

1. 1st layer processed to develop

2. 2nd photoresist layer applied and exposed with oversize mask

3. 2nd layer developed

4. After etch and strip

Figure 10.17 Double masking.

Image reversal

The preference for positive resists over negative resists for small-geometry patterning has been discussed. One of the advantages with positive resists is the use of dark-field masks for the imaging of holes. Dark-field masks offer a lower-defect process because the majority of the surface is covered by hard chrome which does not damage like glass. However, some mask levels require the printing of islands rather than holes. Metal mask levels are island patterns. Unfortunately, the printing of an island with a positive resist requires the use of a clear field mask with its glass damage potential.

A process that allows the printing of islands with positive resists and dark-field masks is image reversal, which involves the formation of the image in the resist with a dark-field mask by conventional masking steps (Fig. 10.18). At the conclusion of the exposure step there is an image in the resist that is reversed from the desired image. That is, if the resist was developed, a hole rather than an island would be formed.

The image reversal process involves exposing the resist-covered wafer to amine vapors in a vacuum oven. The vapors penetrate the

Figure 10.18 Image reversal.

resist, reversing its polarity. Upon removal of the wafers from the oven, they are given a flood exposure which completes the reversal process. The effect of the amine bake and flood exposure is to change the relative dissolution rates of the exposed and unexposed regions, thus reversing the original image when the resist is finally developed. This process is capable of the same resolution capabilities as a nonreversed positive process.[14]

Contrast enhancement layers

Optical projection system resolution is approaching the limits imposed by the constraints of the lens and the wavelength of the exposing radiation. The two set up a condition where the resist *contrast threshold* becomes the limiting factor. This is because at short exposure times and ultraviolet and deep ultraviolet energies, the energy of the exposing wave varies in intensity. Thus the image formed in the resist is fuzzy.

A method used to decrease this threshold is a contrast enhancement layer (CEL), which is a layer spun on top of the resist that is initially opaque to the exposing radiation (Fig. 10.19). During the exposure cycle, the CEL becomes bleached (transparent) and allows the radiation to pass into the underlying photoresist. The CEL responds first to the higher intensities before turning transparent, in effect storing the lower intensities before turning transparent. The result is that the resist receives a uniform exposure of high-intensity radiation which improves its resolution capability. Another way to imagine the role of the CEL is as the top layer of a dual-layer resist system with the image being formed in the thin top layer.

1. Spin Coat Photoresist — 1.6 μm AZ 1400 Series Resist

2. Spin Coat CEL — 0.25 μm Typical Thickness

3. Expose Pattern

4. Remove CEL

5. Develop Resist

Figure 10.19 Contrast enhancement layer process flow.

Before the resist is developed, the CEL is removed by a chemical spray and development of the resist proceeds by normal processing. Positive resist processes normally capable of 1.0-μm resolution can achieve a 0.5-μm image with a CEL.

Dyed resists

Various dyes may be added into the resist during manufacture. A dye may have one or several effects during the exposure step. One possible effect is the absorption of radiation, thereby attenuating the reflected radiation and minimizing standing wave effects. Another is a change of the dissolution rate of the resist polymer during development. This effect creates a cleaner developed line (increased contrast).[15] An important use of dyes is to eliminate the notching that occurs in thin lines of deposited material crossing over surface steps. Addition of a dye to a resist can cause an increase in exposure time of 5 to 50 percent.[16]

Improving Etch Definition

Forming the correct image in the photoresist is a critical step, but not the only one that defines the image in the wafer surface. Etching must also be controlled and precise.

Lift-off process

The final dimensions of the images in the surface layer are the result of variations in both the exposure step and the etch step. In processes where etch undercutting (resist adhesion) is a problem, such as aluminum etching, the etch component of the dimensional variation can be the dominant one.

A patterning variation that eliminates the etch variation component is lift-off (Fig. 10.20). In this variation the wafer is processed through the development step, leaving a hole in the resist layer where a deposited layer is to be located. The exposure and development steps are adjusted to create a negative slope in the sidewall of the hole.

Next, the wafer receives the deposited layer, which covers the entire wafer and fills in the hole. Definition of the pattern comes when the wafer is processed through a photoresist removal step that lifts off the resist and unwanted metal layer. Usually the removal step is assisted by ultrasonic agitation. This helps form a clean break of the deposited film at the resist edge. After resist and film removal, the desired pattern is left on the wafer surface.

Self-aligned structures

Overetching has the effect of placing two structures closer together than intended. There is always some amount of overetching, but the alignment of some structures is absolutely critical. One solution is the *self-aligned structure,* such as a MOS gate (Chap. 16). The definition of the gate structure also defines the neighboring source or drain regions

Process Step	Cross Section
1. Photoresist process up to developing	
2. Deposit metal	
3. Photoresist removed	

Figure 10.20 Lift-off.

Gate

Source and Drain Self Aligned to Gate

Figure 10.21   Self-aligned silicon gate structure (SAG).

(Fig 10.21). Opening the source or drain regions is a simple procedure of dip etching the oxide off the source or drain regions. The thinner oxide on the source/drain region allows a short etch time that does not allow enough time to etch the sides of the gate structure. The subsequent source or drain doping places the dopants next to the gate. This technique uses a differential oxide thickness and dip etch to define or etch other structures. The design uses the gate structure as a doping block.

Etch profile control

In Chap. 9 the concept of anisotropic etching was explained as a way of producing vertical (or nearly vertical) etched sidewalls. The problem becomes more complicated when the etched layer is actually a stack of different materials (Fig. 10.22). Use of an etchant that has poor selectivity can produce a stack with varying widths (Fig. 10.22*b*). Profile control for multilayer etches involves a tradeoff among the etchant chemistry, power levels, and the system pressure and design.

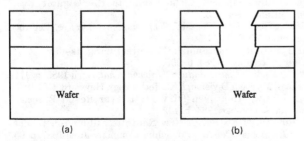

Figure 10.22   (*a*) Anisotropic and (*b*) isotropic etch of surface layer "stack."

Key Concepts and Terms

Antireflective layers	Multilayer resist processes
CMP	Pellicle
Contrast enhancement	Planarization
Image contrast	Self-aligned structures
Image reversal	Standing wave effect
Lift-off	Standing waves
Light scattering	Subsurface reflectivity

Review Questions

1. What is a contrast effect and how does it affect an image?

2. Name two problems resulting from subsurface reflection.

3. What is the purpose of the top layer of a multilayer resist process?

4. What is the purpose of the bottom layer of a multilayer resist process?

5. What is flood exposure and at what step is it used in a multiresist process?

6. What basic resist process step is eliminated in a lift-off process?

7. What advantages come from a planarization surface?

8. How does a contrast enhancement layer improve resolution?

9. How does a pellicle improve the wafer-sort yield?

10. Explain the advantage of a self-aligned structure and give an example.

References

1. C. Mack, Lithography, "Forecast 1993, Fitting the Pieces Together," *Semiconductor International,* Cahners Publishing, January 1993, p. 31.
2. M. D. Levenson, "Extending Optical Lithography to the Gigabit Era," *Microlithography World,* Autumn 1994, p. 5.
3. J. Reynolds, "Mask Making Tour Video Course," *Semiconductor Services,* Redwood City, August 1991, Segment 10.
4. C. Spence, "Optical Proximity Photomask Manufacturing Issues," *OCG Microlithography Seminar Proceedings,* 1984, p. 255.
5. R. Ixcoff, "Pellicles 1985: an Update," *Semiconductor International,* April 1985, p. 111.
6. Micropel product data sheet, Micropel Division, EKC Technology, Hayward, CA, 1988.
7. C. H. Ling and K. L. Liauw, "Improved DUV Multilayer Resist Process," *Semiconductor International,* November 1984, p. 102.
8. F. Coopmans and B. Roland, "DESIRE: A New Route to Submicron Optical Lithography," *Solid State Technology,* PenWell Publishing Company, June 1987, p. 93.
9. B. Moffatt, Private conversation, Yield Engineering Systems, San Jose, CA, 1985.

10. K. Skidmore, "Techniques for planarizing device topography," *Semiconductor International,* April 1988, p. 116.
11. P. Singer, "Chemical-mechanical Polishing: A New Focus on Consumables," *Semiconductor International,* Cahners Publishing, February 1994, p. 49.
12. R. Iscoff, "CMP Takes a Global View," *Semiconductor International,* Cahners Publishing, May 1994, p. 74.
13. K. Skidmore, "Techniques for planarizing device topography," *Semiconductor International,* April 1988, p. 117.
14. E. Aling and G. Stauffer, "Image Reversal of Positive Photoresist," *SPIE Journal,* vol. 539, March 11–12, 1985, p. 194.
15. J. F. Boland, H. F. Stanford, and S. A. Fine, "Dye Effects on Exposure and Development of Positive Resists," *Micro Manufacturing and Test,* August 1985, p. 18.
16. J. Housley, R. Williams, and I. Horiuchi, "Dyes in Photoresists: Today's View," *Semiconductor International,* April 1988, p. 142.

11

Doping

Overview

Conductive regions and N-P junctions are essential parts of semiconductor devices. They are formed in and on the wafer by diffusion or ion implantation techniques. In this chapter, the definition of a junction and the principles and processes of diffusion and ion implantation are explained in detail.

Objectives

Upon the completion of this chapter, you should be able to:

1. Define an N-P junction.
2. Draw a flow diagram of a complete diffusion process.
3. Describe the differences between the deposition and drive-in steps.
4. List the three types of deposition sources.
5. Draw a typical concentration vs distance curve for a deposition and drive-in.
6. List the major parts of an ion implanter.
7. Describe the principle of an ion implanter.
8. Compare the advantages and disadvantages of diffusion and ion implant processes.

Definition of a Junction

The structure that makes transistors and diodes work is an N-P or P-N junction. A *junction* is the separation between a region rich in negative electrons (N-type region) and a region rich in holes (P-type region). The exact location of a junction is where the concentration of electrons equals the concentration of holes. This concept is explained more fully in the section called "Formation of a Doped Region and Junction by Diffusion."

The usual way to form junctions in the surface of a semiconductor wafer is by thermal diffusion or by ion implantation.

Concept of Diffusion

In the early days of semiconductor processing, junctions were formed in the wafer surfaces by alloying dots (Fig. 11.1) of metal into the wafer surface. To form a junction, the dot material had to be of an opposite conductivity type from the wafer conductivity type. This nonplanar method was unreliable because it produced junctions that varied in concentration, depth, and area.

A major advance in junction formation was the development of diffusion doping techniques. *Diffusion,* the movement of one material through another, is a natural chemical process with many examples in everyday life. Two conditions are necessary for a diffusion to take place. First, one of the materials must be at a higher concentration than the other. Second, there must be sufficient energy in the system for the higher-concentration material to move into or through the other.

An example of a gas-state diffusion is the action of a common pressurized spray can (Fig. 11.2). Inside the can is some material, a room

Figure 11.1 Alloy junction transistor.

Figure 11.2 Examples of diffusion.

deodorant or such, under pressure. The pressure allows the storage of the material at a higher concentration than it exists in the air outside the can. When the nozzle is depressed, the material leaves the can and moves into the surrounding air. Initially the movement out of the can is driven by the pressure. Thereafter, movement of the gas into the room proceeds by the process of diffusion. The movement takes place while the nozzle is depressed and continues after it is closed. The diffusion will continue as long as the advancing spray is at a concentration higher than that in the air. As the material moves away from the can the concentration of the material is progressively less. This is a characteristic of a diffusion process. Diffusion will continue until the concentration is even throughout the room.

An example of a liquid-state diffusion is represented when a drop of ink is dropped into a glass of water. The ink is more concentrated than the surrounding water and immediately starts diffusing into the glass of water. The diffusion will continue until the whole glass of water is the same color. This example can be used to illustrate the influence of energy on the diffusion process. If the water in the glass is heated (giving the water more energy) the ink will spread into the water faster.

The same diffusion phenomenon takes place when a doped wafer is exposed to a concentration of atoms of an opposite concentration type.

Formation of a Doped Region and Junction by Diffusion

The formation of a doped region and junction is illustrated by examining the doping of a wafer in a diffusion process. The starting condition is depicted in Fig. 11.3. The wafer illustrated is from a P-type crystal. The +'s in the diagram represent the P-type dopants that were incorporated into the crystal during the crystal-growing process. They are uniformly distributed throughout the wafer.

In many texts and articles dopant atoms are referred to as *impurities*. This term comes from the field of chemistry which contributed the mathematics of diffusion. Unfortunately, the term impurity is

Figure 11.3   P-type wafer ready for diffusion.

confusing since the industry spends so much time and energy eliminating contaminating impurities from the process. In this text, the terms *dopant atoms* or *dopants* are used in reference to the creation of the needed N- and P-type regions and *impurities* is reserved for unwanted dopants and contaminants.

The wafer receives a thermal oxidation and a patterning process that leaves a hole in the oxide layer. In a diffusion tube, the wafer is exposed to a concentration of N-type dopants at a high temperature (the −'s in Fig. 11.4). The N-type dopants diffuse through the hole in the oxide layer.

The effect in the wafer is illustrated by examining what happens at different levels in the wafer. The conditions in the diffusion tube are set such that the number of N-type dopant atoms that diffuse into the wafer surface are greater than the number of P-type atoms in layer 1. In the illustration there are seven more N-type atoms than P-type atoms and that level is electrically an N-type layer. This process has converted the top layer from P type to N type.

The diffusion process proceeds with N-type atoms diffusing from the first level down to the second level (Fig. 11.5). At the second level there are again more N-type dopants than P-type dopants, converting

+ = P Type Dopant Atoms   − = N Type Dopant Atoms

Figure 11.4   Start of diffusion process.

Figure 11.5   Cross section of wafer at conclusion of diffusion.

Layer	# N's (−)	# P's (+)	Net (N − P)	Layer
1	12	5	7	N
2	10	5	5	N
3	8	5	3	N
4	5	5	0	Jct
5	3	5	−1	P
6	0	5	−5	P

Figure 11.6 Dopant amounts and level conductivity type.

level 2 to N type. In the table (Fig. 11.6) is an accounting of the number of N-type and P-type atoms at each level. This process goes deeper into the wafer.

The N-P junction

At level 4 there are exactly the same number of N-type and P-type atoms. This level is the location of an N-P junction. The definition of an N-P junction is the location in the wafer where the number of N-type and P-type dopant atoms are equal. Note that below the junction at level 5 there are three N-type atoms, which are not enough to convert that layer to N type.

The term *N-P junction* indicates that there is a higher concentration of N-type dopants in the doped region. A P-N junction would indicate a higher P-type concentration in the doped region. The behavior of electrical currents across semiconductor junctions gives rise to the particular performance of individual semiconductor devices and is the subject of Chap. 14. In this chapter the emphasis is on the formation and character of doped regions in a wafer.

Goals of solid-state diffusion

The goals of a diffusion process are threefold:

1. The creation of a specific number (concentration) of dopant atoms in the wafer surface

2. To create an N-P or P-N junction at a specific distance below the wafer surface

3. To create a specific distribution (concentration) of dopant atoms in the wafer surface

Lateral diffusion

The diffusion doping process depicted in Fig. 11.5 shows the incoming dopant atoms traveling straight down into the wafer. In reality, the dopant atoms move in all directions. An accurate cross section

Figure 11.7 Side-diffused N-type dopants.

(Fig. 11.7) would show that some of the atoms have moved in a lateral direction, forming a junction under the oxide barrier. This movement is also called *lateral* or *side diffusion*. The amount of the lateral or side diffusion is approximately 85 percent of the vertical junction depth. The effects of side diffusion on circuit density are discussed in the introduction to ion implantation.

Same-type doping

Some devices call for a doping with a dopant type the same as the host region. In other words, an N-type dopant will be put into an N-type wafer or a P-type dopant will be put into a P-type wafer (Fig. 11.8). When this situation happens, the added dopant atoms simply increase the concentration of the dopant atoms in the localized region. No junction is formed.

Graphical representation of junctions

In a cross section (Fig. 11.9) of a semiconductor device, the N-P junctions are indicated simply as regions in the device. There is no graphical convention to represent N- or P-type areas. The drawings just

Figure 11.8 P-type doing of P-type wafer.

Ideal

Lateral Diffusion

Figure 11.9 Cross-sectional representation of diffuse junctions.

show the relative location of the doped region and the junction. This type of graphical representation gives little information about the concentration of the dopant atoms and only approximates the actual dimensions of the regions. A drawing of a 2-μm-deep junction in a 20-mm-thick wafer, scaled to an 8-ft wafer, would have the junction depth only be 0.4-in thick!

Concentration versus depth graphs

Another two-dimensional graphical representation of a doped region is the concentration versus depth graph. This type of graph has the concentration on the vertical axis and the depth into the wafer on the horizontal axis. An example of such a graph is illustrated in Fig. 11.10. The illustration uses the data from the doping example shown in Fig. 11.6. First the P-type dopant concentration is plotted. In the example there are exactly five P-type dopant atoms at every level, resulting in a straight horizontal line on the graph (Fig. 11.10b). Next the number of N-type dopant atoms is plotted. Since the number of atoms decreases deeper into the wafer, the plotted line slopes down and to the right. At level 4, the number of N- and P-type dopants is equal and the lines cross. This is a graphical representation of the location of the junction.

The graph of an incoming dopant concentration versus depth profile for an actual process is not a straight line. They are curved lines. The shape of the curve is determined by the physics of the dopant technique. The actual shapes are discussed in the deposition and drive-in sections.

In some devices, a third doping will take place into the same region as the first one shown in Fig. 11.11. A cross-sectional and concentration versus depth representation is shown. When the dopant types are alternated, two different junctions are created.

Diffusion Process Steps

The use of *solid-state diffusion* to create junctions in semiconductor wafers requires two steps. Step one is called *deposition* and step number two is called *drive-in–oxidation*. Both steps take place in a horizontal or vertical tube furnace. The equipment is the same as that described in Chap. 7 for oxidation.

Deposition

The first step of a diffusion process is called *deposition;* it is also called *predeposition, dep, or predep*. It takes place in a tube furnace,

(a)

(b)

(c)

Figure 11.10 Construction of concentration versus depth curve. (a) Axes; (b) P-type dopant; (c) N-type and P-type dopant.

Figure 11.11 Doping and layering. N-type doping and reoxidation of source drain.

Figure 11.12 Deposition.

with the wafers placed in the flat zone of the tube. A source of dopant atoms is located in the source cabinet and their vapors are transferred into the tube at a required concentration (Fig. 11.12).

In the tube, the dopant atoms proceed to diffuse into the exposed wafer surface as explained in "Formation of a Doped Region and Junction by Diffusion." Within the wafer, the dopant atoms move by two different mechanisms. In the vacancy model (Fig. 11.13a) the

Figure 11.13 Diffusion models. (a) Vacancy model; (b) interstitial model.

dopant atoms move by filling empty crystal positions, the vacancies. The second (Fig. 11. 13*b*) model relies on interstitial movement of the dopant.[1] In this model the dopant atom moves through the spaces between the crystal sites, that is, intersite.

A deposition process is controlled or limited by several factors. One is the *diffusivity* of the particular dopant. Diffusivity is a measure of the rate of movement of the dopant through the particular wafer material. The higher the diffusivity, the faster the dopant moves through the wafer. Diffusivity increases with temperature.

Another factor is the *maximum solid solubility* of the dopant in the wafer material. Maximum solid solubility is the maximum concentration of a particular dopant that can be put into the wafer. A familiar analogy is the maximum liquid solubility of sugar in coffee. The coffee can only dissolve a certain amount of sugar before collecting in the bottom of the cup as a solid. The maximum solid solubility limit increases with increasing temperature.

A semiconductor deposition is performed with a concentration of dopant in the tube that exceeds the maximum solid solubility of the dopant in the wafer material. This situation adds to the control of the deposition process by eliminating the dopant concentration as a variable. The amount of dopant entering the wafer surface is a function of the temperature only, and the deposition is said to take place at solid solubility. Maximum solubility levels for various dopants in silicon are shown in Fig. 11.14.

The concentration of dopant atoms at each level in the wafer is an important factor in the performance of junction diodes and transistors. A dopant concentration versus depth curve for a deposition is shown in Fig. 11.15. The shape of the curve is specific; it is known in mathematics as an *error function*. An important factor in device performance is the concentration of the dopant at the wafer surface. This is called the *surface concentration* and is the quantity indicated where the error function curve intersects the vertical axis. Another deposition parameter is the total quantity of atoms diffused into the wafer. This amount increases with the time of the deposition. Mathematically, the quantity of atoms (Q) is represented by the area under the error function curve.

Deposition steps

A deposition process requires four steps. They are

1. Preclean and etch
2. Tube deposition

Figure 11.14 The solid solubility of impurities in silicon.

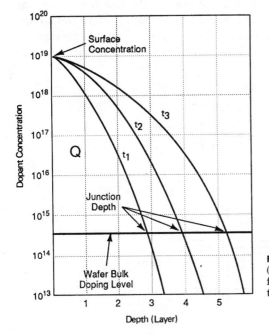

Figure 11.15 Typical deposition (error function) dopant profile for three different deposition times.

3. Deglaze

4. Evaluation

Preclean and etch. Wafers coming to the deposition station first receive a preclean to remove particulates and stains. The chemicals and processes are the same as in the preoxidation process cleaning. After the preclean, the wafers will be chemically etched in an HF or HF and water solution to remove any oxide that may have grown on the exposed silicon surface. An oxide can form on the silicon from exposure to the air and from the chemical preclean process. Removal of the oxide is essential to allow the dopants to diffuse unimpeded into the wafer surface. The etch time and concentration must be balanced to prevent the surface-blocking oxide from being removed or thinned too much.

Deposition. The deposition process, like oxidation, requires a minimum of three cycles. The first cycle is the loading cycle, which takes place in a nitrogen atmosphere. The second cycle is the actual doping cycle. The third cycle is the exit cycle, which also takes place in a nitrogen atmosphere.

Wafers are positioned on the boats at right angles (Fig. 11.16) or parallel to the tube axis. Right-angle placement is the highest packing density, but it can cause uniformity problems since the wafers act as baffles to the gas flow. For uniform doping the gas must mix uniformly between all of the wafers. Parallel placement offers the advantage of more uniform doping since the doping gas proceeds unimpeded through the wafer boat. The disadvantage is a lower loading density.

Figure 11.16 Boat loading patterns. (a) Parallel; (b) perpendicular.

In both placement systems dummy wafers are placed on the outsides and/or the front and back positions of the boat to create uniform doping of the interior device wafers.

Deglaze.   During the deposition cycle, a thin oxide can form on the exposed wafer surface. This oxide is doped and can act as an unwanted source of dopant during the drive-in-oxidation step. Also, a deposition-created oxide can be unetchable, causing incomplete etch in a subsequent masking process. The oxide is removed from the surface by immersion in a diluted HF solution, followed by a water rinse and a drying step.

Evaluation.   Test wafers are added to the boat of device wafers that go into the deposition tube. A deposition test wafer has no patterns and has a conductivity type opposite that of the dopant. They are placed on the deposition boat in locations to sample the deposition distribution throughout the wafer batch. After deglazing the test wafers are evaluated. The primary in-line test is of the sheet resistance with a four-point probe or contactless apparatus. The concept of sheet resistance and the measurement techniques are explained in Chap. 14. The junction depth after deposition is very thin and is generally not measured at this point.

Certification of process and tube cleanliness is determined by oxidizing test wafers and performing a capacitance-voltage measurement for mobile ionic contaminants (Chap. 14). Device wafers are 100 percent inspected or sample-inspected for surface contamination and/or stains with high-intensity ultraviolet lamps or microscopes.

Dopant sources

A deposition depends on the presence of a concentration of dopant atom vapors in the tube. The vapors are created from a dopant source located in the source cabinet of the tube furnace and passed into the tube with a carrier gas. Dopant sources are in either liquid, gaseous, or solid states. Several dopant elements are available in more than one state (Fig. 11.17).[2]

Liquid sources.   Liquid sources of dopants are chlorinated or brominated compounds of the desired element. Thus, a boron liquid source is boron tribromide (BBr_3) and a phosphorus liquid source is phosphorus oxychloride ($POCl_3$). The liquid sources are held in temperature-controlled quartz flasks (Fig. 11.18). An inert gas, such as nitrogen, is bubbled through the heated liquid, and the gas becomes saturated with the dopant vapors. The flask is connected to a gas manifold

Type	Element	Compound Name	Formula	State	Diffusion Reactions*
N	Antimony	Antimony Trioxide	Sb_2O_3	Solid	
	Arsenic	Arsenic Trioxide	As_2O_3	Solid	$2AsH_3 + 3O_2 \longrightarrow As_2O_3 + 3H_2O$
		Arsine	AsH_3	Gas	
	Phosphorus	Phosphorus Oxychloride	$POCl_3$	Liquid	$4POCl_3 + 3O_2 \longrightarrow 2P_2O_5 + 6Cl_2$
		Phosphorus Pentoxide	P_2O_5	Solid	$2PH_3 + 4O_2 \longrightarrow P_2O_5 + 3H_2O$
		Phosphine	PH_3	Gas	
P	Boron	Boron Tribromide	BBr_3	Liquid	$4BBr_3 + 3O_2 \longrightarrow 2B_2O_3 + 6Br_2$
		Boron Trioxide	B_2O_3	Solid	$B_2H_6 + 3O_2 \longrightarrow B_2O_3 + 3H_2O$
		Diborane	B_2H_6	Gas	
		Boron Trichloride	BCl_3	Gas	$BCl_3 + 3H_2 \longrightarrow 2B + 6HCl$
		Boron Nitride	BN	Solid	
	Gold	Gold	Au	Solid (Evap.)	
	Iron		Fe		
	Copper		Cu		
	Lithium		Li		Undesirable impurities
	Zinc		Zn		from contamination
	Manganese		Mn		
	Nickel		Ni		
	Sodium		Na		

*Note: Only selected diffusion reactions are listed

Figure 11.17 Deposition source table.

Inert Gas   Carrier Gas   Dopant

Figure 11.18 Liquid dopant source.

whose valves are controlled by a microprocessor. The nitrogen carries the dopant vapors into the tube and maintains a gas volume sufficient to create a laminar flow condition in the tube. Laminar flow is required to prevent the gas from spiraling in the tube, creating nonuniform doping. Another device used to create laminar flow is a baffle (Fig. 11.19) at the tube inlet. Baffles break up the incoming gas stream into a laminar form.

Figure 11.19 Use of baffle to create laminar flow.

Also connected to the manifold is a reaction gas that is required to create the elemental dopant form in the tube. For BBr_3, the reaction gas is oxygen, which creates the boron oxide (B_2O_3), as shown in Fig. 11.20. At the wafer surface, a boron trioxide layer deposits on the silicon, and the boron diffuses from the oxide into the surface. Liquid sources offer the advantages of low to moderate cost and consistent doping. The disadvantages are uniformity problems (especially for larger-diameter wafers), safety considerations, and the potential of contamination associated with the opening of the flasks for recharging. Several vendors supply liquid dopants in pluggable sealed ampules that minimize contamination and safety problems.

Gas sources. Many wafer manufacturers prefer gas dopant sources. These are hydrated forms of the dopant atom such as arsine (AsH_3) and diborane (B_2H_6). The gases are mixed in different dilutions in pressurized containers and connected directly to the gas manifold (Fig. 11.21). Gas sources offer the advantage of precise control through pressure regulators and are favored for deposition on larger-diameter wafers. The processes are in general cleaner, since the pressurized sources last longer than liquid sources. On the downside, unwanted chemical reactions in the manifold can create silica dust that can contaminate the tube and wafers.

Solid sources. The original deposition sources were solid. An oxide powder of the desired dopant was placed on a quartz holder, called a *spoon,* and placed in a source tube furnace attached to the main deposition tube (Fig. 11.22a). In the source furnace, the oxide gives off source vapors that are carried into the deposition tube where the dif-

$$4BBr_3 + 3O_2 \rightarrow 2B_2O_3 + 6Br_3$$

Figure 11.20 Reaction of liquid source BBR_3 and oxygen in a deposition tube.

Figure 11.21   Gas source manifold.

fusion takes place. This setup is called a *remote solid source*. Remote solid sources are economical but nonuniform. They are used primarily in the doping of discrete devices where high precision is not required.

A more popular solid source is the planar source wafer (Fig. 11.22*b*). These are wafer-size "slugs" that contain the desired dopant. Boron slugs are a compound of boron and a nitride (BN). Slugs are also available for arsenic and phosphorus diffusions.

Remote.

(a)

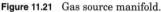

Neighbor

(b)

Figure 11.22   Solid source. (a) Remote furnace; (b) neighbor source.

The slugs are stacked on the deposition boat, with one slug between every two device wafers. This arrangement is called a *solid neighbor source*. In the tube the dopant diffuses out of the slug, crosses the short distance to the wafer, and diffuses into the surface. This system provides good uniformity for larger-diameter wafers since the slug is the same size as the wafer. Slugs are safer to use, having no toxic vapors at room temperature. Drawbacks to the slug process include breakage of the slugs, lower productivity (the slugs take up space in the tube), and the necessity of cleaning the slugs. Some slug compounds require bake steps to maintain the dopant activity.

The third solid dopant source is a conformal layer spun directly on the wafer surface. The sources are powdered oxides (same as remote sources) mixed in solvents. They are spun onto the wafer surface using photoresist-type spinners and baked to evaporate the solvents. Left on the surface is a layer of doped oxide which conforms to the wafer surface. The wafers are placed on a boat and positioned in the deposition tube where the heat drives the dopant out of the oxide and into the wafer.

Spin-on dopants have the potential of high uniformity and offer high productivity, and like the slugs are safe to handle. Problems with the system are the distribution of the dopant in the oxide layer; thickness variations, especially over surface steps; and the costs of the additional spin and bake steps.

Closed-tube deposition

Some manufacturers use the closed-tube system, which was developed by IBM (Fig. 11.23). The wafers and a powdered dopant source are placed in a quartz capsule (or ampule) that is evacuated of air to a very low pressure and sealed. The capsule is placed in a tube furnace where the heat causes the source powder to give off dopant vapors that diffuse into the wafers. The diffusion is uniform due to the absence of an atmosphere in the capsule. This system has not been widely accepted due to the high cost of the capsule (which is broken open and discarded after each use), and the expensive vacuum pumps required.

Figure 11.23 Capsule diffusion.

Drive-in–Oxidation

The second major part of the diffusion process is the drive-in–oxidation step. It is also variously known as *drive-in, diffusion, reoxidation,* or *reox.* The purpose of this step is twofold:

1. The redistribution of the dopant deeper into the wafer. During the deposition a high-concentration and shallow layer of dopants are diffused into the surface. In the drive-in there is no dopant source. The heat alone drives the dopant atoms deeper and wider into the wafer just as material from a spray can will continue to spread into the room after the nozzle is released. During this step the total amount of atoms (Q) from the deposition step remains constant. The surface concentration is reduced and the distribution of atoms takes a new shape. The distribution after the drive-in is described by mathematicians as a gaussian distribution (Fig. 11.24). The junction depth increases. Generally the drive-in–oxidation step takes place at a higher temperature than the deposition step.

(a)

(b)

Figure 11.24 Drive in-oxidation. (a) Cross section of wafer; (b) dopant concentration in wafer.

2. The second purpose of drive-in–oxidation is the oxidation of the exposed silicon surfaces. The atmosphere in the tube is oxygen or water vapor, which performs the oxidation simultaneously as the drive-in of the dopants is taking place.

The setup, process steps, and equipment for the drive-in–oxidation step are the same as an oxidation process. After the completion of the drive-in, the wafers are again evaluated. The test wafers (from the deposition step) are again measured for surface concentration with a four-point probe and the device wafers checked for cleanliness. The test wafers are measured for junction depth and perhaps for mobile ionic contamination. After some diffusion steps, a test structure in the engineering die may be electrically probed for junction parameters.

Oxidation effects

The oxidation of the silicon surface affects the final distribution of the dopants.[3] The effects are related to the relocation of the top-level dopants after the oxidation. Recall that the silicon in the silicon dioxide film is consumed from the wafer surface. The question to ask is, What happened to the dopants that were in the top level? The answer to that question depends on the conductivity type of the dopant.

If the dopant is an N-type, an effect called *pile-up* (Fig. 11.25a)

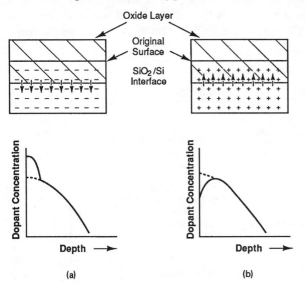

Figure 11.25 Pile-up and depletion of dopants during oxidation. (a) Pile-up of N-type dopants; (b) depletion of P-type dopants.

occurs. As the oxide-silicon interface advances into the surface, the N-type dopant atoms segregate into the silicon rather than the oxide. The effect is to increase the number of these dopants in the new top layer of the silicon. In other words, the N-type dopants pile up in the wafer surface and the surface concentration of the dopant is increased. Pile-up changes the electrical performance of the devices.

If the dopant is the P-type boron, an opposite effect occurs. The boron atoms are more soluble in the oxide and are drawn up into it (Fig. 11.25b). The effect on the wafer surface is a lowering of the concentration of boron atoms, which lowers the surface concentration and also affects the electrical performance of the devices. A summary of the deposition and drive-in–oxidation steps are listed in Fig. 11.26. The effect of the deposition and drive-in steps on the diffusion electrical parameters is shown in Fig. 11.27.

Introduction to Ion Implantation

The advent of high-density circuits requires smaller feature sizes and closer spacing of the circuit components. Thermal diffusion places a limit on the production of advanced circuits. Five problems are lateral diffusion, ultrathin junctions, poor doping control, surface contamination interference, and dislocation generation. Lateral diffusion occurs during deposition and drive-in and also continues every time the wafer is heated up to a range in which diffusion movement can take place (Fig. 11.28). The circuit designer must leave enough room

	Deposition	Drive-In
Goals	Introduction of Dopant	1. Redistribution of Dopant 2. Reoxidation
Variables		1. Surface Composition 2. Junction Depth 3. Time 4. Diffusivity 5. Temperature 6. Quantity of Atoms
Source Conditions	Continuous Source	No Source
Temperature Range	900 – 1100°C	1050 – 1200°C
Oxidation	No	Yes

Figure 11.26 Summary of deposition and drive-in steps.

Change	Parameter	Deposition	Drive-in/ Oxidation
Increase Temperature (same time)	Dopant concentration (Dc)	Increases	Same
	Surface concentration (Co)	Increases	N type-Lower P type- Higher
	Junction depth (Xj)	Deeper	Deeper
	Sheet Resistance	(Rs1) Lower	(Rs2) N type-Lower P type- Higher
Increase Time (same temp.)	Dopant concentration (Dc)	Increases	Same
	Surface concentration (Co)	Same	N type-Lower P type- Higher
	Junction depth (Xj)	Deeper	Deeper
	Sheet Resistance	(Rs1) Lower	(Rs2) N type-Lower P type- Higher

Figure 11.27 Doping processes parameter changes.

Required spacing

Figure 11.28 Side diffusion.

between adjacent regions to prevent the laterally diffused regions from touching and shorting. The cumulative effect for a dense circuit can be a greatly increased die area. Another problem with high temperatures is crystal damage. Every time a wafer is heated and cooled, crystal damage from dislocations occurs. A high concentration of these dislocations can cause device failure from leakage currents. One goal of an advanced process sequence is a decreased thermal budget to reduce these two problems.

The advent of MOS transistors presented two different problems, low dopant concentration control and ultrathin junctions. Gate regions with dopant concentrations below 10^{15} atoms/cm^2 are required for efficient MOS transistors. However, this level is difficult to achieve consistently with a diffusion process. Scaling down MOS transistors smaller to achieve higher packing densities also requires thin junction depths in the source and drain areas.[4] At the 0.18-μm design rule level junctions will be in the 40 nm range.[5] The third problem is imposed by the physics of a diffused region. As illustrated in Figs. 11.15 and 11.24, the majority of the dopant atoms are located near the wafer surface. This puts the majority of the current traveling near the surface where the dopant atoms are located. Unfortunately, this is the same location where contaminants are found (in and on the wafer surface) that interfere or degrade the current flow.

A fourth problem comes from the high temperatures required of the diffusion process. Every time a wafer is heated and cooled, crystal damage from dislocations occurs. A high concentration of these dislocations can cause device failure from leakage currents.

Ion implantation overcomes these limits of diffusion and also adds additional benefits. During the ion implant process there is no side diffusion, the process takes place at close to room temperature, the dopant atoms are placed below the wafer surface, and a wide range of doping concentrations is possible. With ion implantation there is greater control of the location and number of dopants put in the wafer. Also, photoresist and thin metal layers can be used as doping barriers along with the usual silicon dioxide layers. Given the benefits, it is not surprising that the majority of doping steps for advanced circuits are done by ion implantation. Diffusion still finds use in the doping of less critical layers and will continue to be used for lower-integration-level circuits.

Concept of Ion Implantation

Diffusion is a chemical process. Ion implantation is a physical process, that is, the act of implanting does not rely on chemical interaction between the dopant and the wafer material. An analogy that demonstrates the concept of ion implantation is a cannon firing balls into a wall (Fig. 11.29). Given enough momentum from the powder in the cannon, the balls will penetrate the wall and come to rest below the surface of the wall. The same events take place in an ion implantation machine. Dopant atoms are ionized and isolated, accelerated (gain momentum), formed into a beam, and swept across the wafer. The dopant atoms physically bombard the wafer, enter the surface, and come to rest below the surface (Fig. 11.30b).

Ion Implantation System

An ion implanter is a collection of very sophisticated subsystems (Fig. 11.30a), each performing a specific action on the ions. Ion implanters

Figure 11.29 Ion implantation analogy.

Figure 11.30 Ion implantation. (a) Block diagram of ion implanter; (b) distribution of implanted atoms in wafers.

come in a variety of designs used for advanced research and/or high-volume production. All of the machines have the same major subsystems as described below.

Implant sources

The same dopant elements used in diffusion processes are used in ion implant processes. In diffusion processes the dopants originate in either liquid, gas, or solid sources. For ion implantation, only gas and solid sources are used.

Gases are favored for ion implantation because of their ease of use and higher control. The gases are mostly fluorine based, such as PF_5, AsF_5, $B^{11}F_3$, SbF_3, and PF_3. The gas cylinders are connected to the ion source subsystem through mass flow meters, which offer more control of the gas flow than normal flow meters.

Certain solid sources are used for special applications. One is phosphorus pentoxide (P_2O_5). In this system the powder is heated to create vapors that are carried into the ion source chamber.

Ionization chamber

The name "ion implant" implies that ions are a part of the process. Recall that ions are atoms or molecules with a negative or positive charge. The ions implanted are ionized atoms of the dopants. The ionization occurs in a chamber that is fed by the source vapors. The chamber is maintained at a low pressure (vacuum) of about 10^{-3} torr. Inside the chamber is a filament that is heated to the point where electrons are created from the filament surface. The negatively charged electrons are attracted to an oppositely charged anode in the chamber. During the travel from the filament to the anode, the electrons collide with the dopant source molecules and create a host of positively charged ions from the elements in the molecule. The results of the ionization of the source BF_3 are shown in Fig. 11.31.

Another ionization method uses a cold-cathode technique to generate the electrons. A high-voltage electric field is created between a cathode and anode, which creates the electrons in a self-sustaining process.

Ion selection

At the top of the list in Fig. 11.31 is a lone boron ion. This is the atom desired in the wafer surface. The other species resulting from the ionization of the boron trifluoride are not wanted in the wafer. The boron ion must be *selected* from the group of positive ions. This process is called *analyzing, mass analyzing, selection,* or *ion separation.*

Selection is accomplished in a mass analyzer. This subsystem was first developed during the Manhattan Project for the atomic bomb. The analyzer (Fig. 11.32) creates a magnetic field. The species leave the ionization subsystem with voltages of 15 to 40 keV (thousand electron-volts). In other words, they are traveling at a relatively high

$e^- \rightarrow$

B^+

BF^+

$BF_2{}^+$

BF_3

F^+

$F_2{}^+$

$B^{11}{}^+$

$B^{11}F^+$

$B^{11}F_2{}^+$

$BF_3{}^+$

Figure 11.31 Ionization species of BF_3.

Figure 11.32   Analyzing magnet.

speed. In the field, each of the positively charged species is bent into an arc with a specific radius. The radius of the arc is dictated by the mass of the individual species, its speed, and the strength of the magnetic field. At the exit end of the analyzer is a slit that will allow only one species to exit. The magnetic field is adjusted to match the path of the boron ion to the exit slit position. Thus only the boron ion leaves the analyzing subsystem.

In some systems, the analyzing also takes place after acceleration (Fig. 11.30). After acceleration, analysis will be necessary if the implant species molecule is the type that might separate in the acceleration process and/or an uncontaminated beam is necessary.

Acceleration tube

Upon leaving the analyzing section, the boron ion moves into an acceleration tube. The purpose is to accelerate the ion to a high enough velocity such that it will have sufficient momentum to penetrate the wafer surface. *Momentum* is defined as the product of the mass of the atom multiplied by its velocity.

The required velocity is achieved by taking advantage of the fact that negative and positive charges attract each other. The tube is based on a linear design with annular anodes along its axis. Each of the anodes has a negative charge. The charge amount increases down the tube. As the positively charged ion enters the tube, it immediately starts to accelerate down the tube. The voltage value is selected based on the mass of the ion and the momentum required at the wafer end of the implanter. The higher the voltage, the higher the momentum, the faster and deeper the dopant ion can be implanted. Voltages range from 5 to 10 keV for low-energy implanters to 0.2 to 2.5 MeV (million electron-volts)[6] for high-energy implanters.

Ion implanters are classified into the following categories: medium- and high-current machines, high-energy implanters, and oxygen ion implanters. The stream of positive ions exiting the tube is actually an electric current. The beam current level translates into the number of ions implanted per minute. The higher the current, the more atoms implanted. The amount of atoms implanted is called the *dose*. Medium-current machines produce currents in the 0.5 to 1.7 mA (milliampere) range at energies from 30 to 200 keV. High-current machines generate beam currents of about 10 mA at energies up to 200 keV.[7] High-energy implanters are used in several CMOS doping applications, including wells, channel stops, and deep buried layers (see Chap. 16).

Wafer charging

One problem with high-current machines is an unacceptable degree of electrical charging (*wafer charging*) on the wafer surface. The high-current beam carries excess positive charges that charge the wafer surface. The positive charge draws neutralizing electrons from the surface, the bulk, and the beam. High charge levels can degrade or destroy surface dielectric layers. Wafer charging is a particular problem on thinner MOS gate dielectrics (less than 10 nm).[8] Methods used to neutralize or reduce the charge are *flood guns* specifically designed to provide electrons, the plasma bridge method of providing low-energy electrons,[9] and control of the electron path with magnetic fields.[10]

The relationships of current and energy for production-level implanters is shown in Fig. 11.33. High energy machines accelerate

Figure 11.33 Ion implanters. The conventional descriptions of ion implanters were based on applications. However, some of today's more advanced implanters cannot be classified easily; several systems are capable of a broader process window than the conventional descriptions indicate. (*From Ref. 4*)

ions in the 10 keV to 3.0 MeV range, with beam currents up to 1.0 mA. Oxygen implant machines are high-current implanters used to implant oxygen in SOI applications (see Chap. 16).

A successful ion implant relies on the implantation of only the desired dopant atom. Single-dopant implantation requires that the system be maintained at a low pressure, greater than 10^{-6} torr. The danger is that any residual molecules in the system (such as air) can become accelerated and end up in the wafer surface. Either oil-diffusion or cryogenic high-vacuum pumps are employed to reduce the pressure. The operation of these systems is described in Chap. 12.

Beam focus

Upon exiting the acceleration tube, the beam separates due to repulsion of like charges. The separation (or defocus) causes uneven ion density and nonuniform layers in the wafer. For successful implantation the beam must be focused. Electrostatic or magnetic lenses are used to focus the ions into a small diameter beam or a band of parallel beams.[11]

Neutral beam trap

Despite the vacuum removal of the majority of the air in the system, there are still some residual gas molecules in the vicinity of the ion beam. Collisions between the ions and the residual gas atoms result in a neutralization of the dopant ion.

$$P^+ + N_2 \rightarrow P^0 \text{ (neutral)} + N_2^+$$

In the wafer, these "neutrals" cause nonuniform doping and, since they cannot be "counted" by evaluation equipment, they result in incorrect counting of the amount of dopants in the wafer. Suppression of the neutral beam is accomplished by bending the ion beam (Fig. 11.34) with electrostatic plates, leaving the neutral beam to travel straight ahead away from the wafers.

Figure 11.34 Deflection of ion beam from neutral beam.

Beam scanning

Given a 1-cm-diameter beam or a narrow band, a procedure must be provided to uniformly dope an entire wafer. Three methods are used: beam scanning, mechanical scanning, and shuttering, alone or in combination.

A beam-scanning system (Fig. 11.35) has the beam pass between a number of electrostatic plates. Negative and positive charges can be controllably changed on the plates to attract and repel the ionized beam. By manipulating the charges in two dimensions, the beam can be swept across the entire wafer surface in a raster scan pattern.

Beam sweeping is used primarily in medium-current machines for single-wafer implanting. The procedure is fast and uniform. A drawback is the requirement that the beam be moved completely off the wafer to make the turns. For a large-diameter wafer, this procedure can lengthen the implant time by 30 percent or more. Another problem occurs with high-current machines where the high density of ions causes discharges (called *space charge forces*) that destroy the electrostatic plates. Wide beams are swept across the wafer. In some systems the wafer is rotated 90° after each sweep of the beams to ensure uniformity.[12]

Mechanical scanning approaches the scanning problem by holding the beam in one position and moving the wafer(s) in front of it. Mechanical scanning is used primarily on high-current machines. One advantage is that there is no wasted time to turn the beam and the beam speed is constant. A number of different designs are used (Fig. 11.36). The procedure is essentially a batch process, which is

Figure 11.35 Electrostatic beam scanning.

Figure 11.36   Mechanical scanning.

efficient, but there are higher maintenance and alignment chores. If the wafer is at an angle to the beam, nonuniform implant depths can result. But in some cases the wafer will be oriented at an angle to the beam. Beam shuttering employs either an electronic field or a mechanical shutter to turn the beam on when it is on the wafer and off when it is not. Most systems use a combination of beam sweeping and mechanical movement.

Target chamber

The actual implantation takes place in the target chamber, sometimes called an end chamber. It includes the scanning system and the entrance and exit mechanisms. There are several tough requirements for end stations. Wafers must be loaded into the chamber; a vacuum created; wafers individually placed on the holder; the implant completed; and the wafers dismounted, loaded into a cassette, and removed from the chamber. The mechanical motions can take longer than the implant itself. Improvements include load locks to allow loading without breaking the chamber vacuum. A big challenge is to keep particulate generation low during all of the mechanical movements in the chamber.[13] Installation of antistatic devices in the chamber is critical. Electrostatic handlers (no mechanical clamps) are an option.[14]

Wafer breakage can cause contamination from wafer chips and dust, which in turn requires time-consuming cleaning. Contamination on the wafers causes shadowing that blocks the ion beam. Production speed must be maintained with a system that quickly evacuates the chamber for implanting and returns it to room pressure for exit and reloading. The target chamber may house a detector (called a *Faraday cup*) to "count" the number of ions impacting the surface.

These detectors can automate the process by allowing beam contact with the wafer until the correct dose is achieved.

High-current implantation can cause the wafer to heat up, and these machines often have cooling mechanisms on the wafer holders. These machines may also have an *electron flood gun* (Fig. 11.37) designed to minimize a buildup of a charge on the wafer surface that can electrostatically attract contamination.

Ion implant masks

A major advantage of ion implantation is the variety of masks that are effective blocks to the ion beam. In diffusion processing the only effective mask is silicon dioxide. Most films employed in the semiconductor process can be used to block the beam, including photoresist, silicon dioxide, silicon nitride, aluminum, and other thin metal films. Figure 11.38 compares the thicknesses required to block a 200-keV implant for various dopants.

The use of resist films as a beam block rather than an etched opening in an oxide layer offers the same dimension control advantage as the lift-off process; the etch step and its variability are eliminated. Use of resist layers also is more productive. Options to the use of silicon dioxide increase overall yield by minimizing the number of heating steps the wafers undergo.

Dopant Concentration in Implanted Regions

The distribution of the ions in the wafer surface is different than the distribution after a diffusion process. The number and location of the

Figure 11.37 Electron flood gun.

Figure 11.38 Barrier thickness required to block ion beam.

dopant atoms in a diffusion process is determined by the diffusion laws and time and temperature. In an ion implant process, the number of atoms (dose) implanted is determined by the beam current density (ions per square centimeter) and the implant time.

The location of the ions in the wafer is a function of the incoming energy of the ions, the orientation of the wafer, and the stopping mechanism of the ion. The first two factors are physical ones. The heavier the incoming ion and/or the higher its energy, the deeper into the wafer it will move. The wafer orientation influences the stopping position because different crystal planes have different atomic densities and the incoming ions are stopped by the wafer atoms.

Within the wafer, the ions are slowed and stopped by two mechanisms. The positive ions are slowed by electronic interactions with the negatively charged electrons in the crystal. The other interaction is the physical collision with the nucleus of the wafer atoms. All of the stopping factors are variable; the ions have a distribution of energies, the crystal is not perfect, and the electronic interactions and collisions vary. The net result is that the ions come to rest at various depths in the wafer (Fig. 11.39). They are centered about a depth called the *projected range* and fall off in density on each side of it. Additional implants create similar distribution patterns. Projected ranges for different dopants are shown in Fig. 11.40. The mathematical shape of the ion distribution is a gaussian curve. A junction between the implanted ions and the bulk doping in the wafer takes place where the ion concentration equals the bulk doping concentration.

Crystal damage

During the process of implantation the wafer crystal structure is damaged by the colliding ions. There are three types of damage: lattice damage, damage cluster, and vacancy-interstitial.[15]

Lattice damage occurs when the ions collide with host atoms and displace them from their lattice site. A damage cluster occurs when

Figure 11.39 Dopant concentration profile after ion implant.

Figure 11.40 Projected range of various dopants in silicon. (*After Blanchard, Trapp, and Shepard.*)

displaced atoms in turn displace other substrate atoms, creating a cluster of displaced atoms. The most common implant-produced defect is a vacancy-interstitial. This defect comes about when an incoming ion knocks a substrate atom from a lattice site and the displaced atom comes to rest in a nonlattice position (Fig. 11.41).

Light atoms, such as boron, produce a small number of displaced atoms. The heavier atoms, phosphorus and arsenic, generate a large

Figure 11.41   Vacancy-interstitial damage mechanism.

number of displaced atoms. With prolonged bombardment, the regions of dense disorder may change to an amorphous (noncrystal) structure. In addition to the structural damage to the wafer from ion implantation, there is an electrical effect. The damaged regions do not have the required electrical characteristics because the implanted atoms do not occupy lattice sites.

Annealing

Restoration of the crystal damage and electrical activation of the dopants can be achieved by a thermal heating step. The temperature of the anneal is below the diffusion temperature of the dopant to prevent lateral diffusion. A typical anneal in a tube furnace will take place between 600 and 1000°C in a hydrogen atmosphere.

RTP techniques are also used for post-implant annealing. RTP offers fast surface heating that restores the damage without the substrate temperature rising to the diffusion level. Additionally, the anneal can take place in seconds while a tube process takes 15 to 30 minutes. If the wafer has a heavy subsurface amorphous layer, the damage may be restored with a second implant of a light atom, such as oxygen or neon.

Channeling

The crystalline structure of the wafer presents a problem during the ion implantation process. The problem comes about when the major axis of the crystal wafer is presented to the ion beam. Ions can travel down the channels, reaching a depth as much as 10 times the calculated depth. An ion concentration profile of a channeled cross section (Fig. 11.42) shows a significant amount of additional dopants. Channeling is minimized by several techniques: a blocking amorphous surface layer, misorientation of the wafer, and creating a damage layer in the wafer surface.

The usual blocking amorphous layer is simply a thin layer of grown silicon dioxide (Fig. 11.43). The layer randomizes the direction of the ion beam so that the ions enter the wafer at different angles and not

Figure 11.42 Effect of channeled ions on total dose.

Figure 11.43 Implant through an amorphous oxide layer.

Figure 11.44 Misorient the beam direction to all crystal axes.

directly down the crystal channels. Misorientation of the wafer 3 to 7° off the major plane also has the effect of preventing the ions from entering the channels (Fig. 11.44). Predamaging the wafer surface with a heavy silicon or germanium implant creates a randomizing layer in the wafer surface (Fig. 11.45). The method increases the use of the expensive ion-implant machine. Channeling is more of a problem with low-energy implants and heavy ions.[16]

Evaluation of Implanted Layers

The evaluation of implanted wafers is essentially the same as for diffused layers. Four-point probes are used to determine the sheet resis-

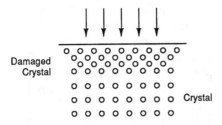

Damaged
Crystal

Crystal

Figure 11.45 Predamage on the crystal surface.

tance of the layer. Spreading resistance techniques and capacitance-voltage techniques determine the concentration profile, dose, and junction depth. Junction depths can also be determined by bevel and decoration methods. These procedures are explained in Chap. 14.

For implanted layers, a special structure, called the *Van Der Pauw structure,* is sometimes used in place of a four-point probe (Fig. 11.46). The structure allows a determination of sheet resistance without the contact resistance problems of a four-point probe. Variation in the implanted wafers can come from many sources: the beam uniformity, variations in voltage, scanning variations, and problems in the mechanical systems. These potential problems have the possibility of causing a wider sheet resistance surface variation than a diffusion process. To detect and control the sheet resistance across the wafer surface, mapping techniques are popular and, for critical implants, required. Wafer surface maps (Fig. 11.47) are drawn from four-point probe measurements that are computer-corrected for proximity and edge effects.

Figure 11.46 Van Der Pauw test pattern.

Run 2   Run 4

Figure 11.47 Four-point probe surface measurement pattern. (*Courtesy of Prometrics.*)

A measurement technique unique to ion implantation is optical dosimetry. The technique requires spinning a glass disk with a layer of photoresist. Before being placed in the ion implanter, the resist film is scanned with a dosimeter that measures the absorption of the film. The information is stored in a computer. The wafer and film receive the same implant as the device wafers. The resist absorbs the ion dose and darkens. After the implant, the film is again scanned. The computer subtracts the before-implant value for each location and prints out a contour map of the surface. The spacing of the contour lines is an indication of the uniformity of the dopants in the surface (Fig. 11.48).

Uses of Ion Implantation

An ion implantation can be substituted for any deposition process. The greater control and lack of side diffusion make it the preferred doping technique for dense and small-feature-size circuits. A predeposition application in CMOS devices is the creation of the deep P-type wells (see Chap. 16).

One of the most important uses of ion implantation is for MOS gate threshold adjustment (Fig. 11.49). A MOS transistor consists of three

Figure 11.48 Contour map of ion dose.

Gate Region

Figure 11.49 Ion doping of MOS gate region.

parts: a source, a drain, and a gate. During operation, a voltage is applied between the source and drain regions. However, no current can flow between the regions until the gate becomes conductive. The gate becomes conductive when a voltage applied to it causes a conductive channel to form in the surface and connects the source and drain. The amount of voltage required to first form the connecting channel is called the *threshold voltage* of the device. The threshold voltage is very sensitive to the dopant concentration in the wafer surface under the gate. Ion implantation is used to create the required dopant concentration in the gate region. Also in MOS technology, ion implantation is used to alter the field dopant concentration. However, in this use the purpose is to set a concentration level that prevents current flow between adjacent devices. In this application the implanted layer is part of a device *isolation* scheme.

In bipolar technology, ion implantation is used to create all of the various transistor parts. The "custom" dopant profiles available from implantation can improve device performance. One particular application is arsenic buried layers. When the buried layer is diffused, the high concentration of arsenic atoms affects the quality of a subsequent epitaxial layer deposited on the surface. By using an arsenic ion implant, a high concentration of arsenic is possible and an anneal process restores the damage, allowing a higher-quality epitaxial layer to be deposited.

Resistors for both MOS and bipolar circuits are good candidates for ion implantation. Diffused resistors vary in uniformity from 5 to 10 percent, whereas ion-implanted resistors have variations of less than 1 percent or better.

The Future of Doping

Since the mid-1970s the uses and applications of ion implantation have grown steadily. For advanced circuit fabrication, it has become the preferred method of doping for the following reasons:

- Precise dose control from 10^{10} to 10^{16} atoms/cm^2
- Uniform topping of large areas
- Dopant profile through energy selection
- Relative ease of implanting all dopant elements
- Minimal side diffusion
- Implanting of nondopant atoms
- Can dope through surface layers
- Choice of dopant barriers for selective doping

Ion implantation also has its drawbacks. The equipment is expensive and complex. Training and maintenance take longer than for comparable diffusion processes. The machines present new dangers in the form of high voltages and more toxic gas use. From the process perspective, the biggest worry is over the ability of the annealing processes to completely eliminate the implant-induced damage. However, despite the drawbacks, ion implantation is the preferred doping process for advanced circuits. In addition, many new structures are only possible with the unique advantages offered by ion implantation.

Ion implantation may run out of steam beyond the shallow junction range of 40 nm.[17] A new technique under investigation is *plasma doping,* also called PLAD. With this technique, the wafer is placed in a plasma field (similar to ion milling or sputtering) in the presence of dopant atoms. With proper charges on the dopant atoms and the wafer, the dopant atoms accelerate to the wafer surface and penetrate much like an ion implantation. The difference is the lower energy of the plasma field yielding less wafer charging and giving more control of the shallow junctions.[18]

Key Concepts and Terms

Channeling	Drive-in–oxidation
Closed-tube deposition	Implant analogy
Crystal damage	Implant mask types
Deposition	Implant sources
Diffusion	Implanted dopant profiles
Diffusion dopant profiles	Implanted layer uses
Diffusion process steps	Implanter subsystems
Dopant source states	Ion implantation
Dopant source types	Open-tube deposition

Review Questions

1. Explain the difference between a deposition and a drive-in–oxidation.

2. Name the three types of sources used in a diffusion process.

3. List two evaluation measurements made after the drive-in–oxidation.

4. Is the junction depth deeper after deposition or the drive-in–oxidation?

5. What type of source is BCl_3?

6. Name the subsystems in an ion implantation.

7. In what subsystem is the dopant ion selected?

8. What is the difference between a diffusion and an ion-implanted dopant profile? Sketch an example of each.

9. Name two materials used as an ion implant doping barrier.

10. Why is an anneal required after ion implantation?

References

1. P. B. Griffin and J. D. Plummer, "Advanced Diffusion Models for VLSI," *Solid State Technology,* May 1988, p. 171.
2. K. T. Robinson, "A Guide to Impurity Doping," *Micromanufacturing and Test,* April 1986, p. 52.
3. P. Guise and R. Blanchard, *Modern Semiconductor Fabrication Technology,* Reston Books, Reston, VA, 1986, p. 46.
4. S. Felch et al., "A Comparison of Three Techniques for Profiling Ultrashallow p⁺-n Junctions," *Solid State Technology,* PenWell Publishing Company, January 1993, p. 45.
5. N. Cheung, "Ion implantation," *Semiconductor International,* Cahners Publishing, January 1994, p. 44.
6. P. Burggraaf, "Ion Implanters: Major Trends," *Semiconductor International,* April 1986, p. 78.
7. R. Iscoff, "Are Ion Implanters the Newest Clean Machines?" *Semiconductor International,* Cahners Publishing, October 1994, p. 65.
8. N. Cheung, "Ion implantation," *Semiconductor International,* Cahners Publishing, January 1993, p. 35.
9. J. England et al., "Charge Neutralization During High-Current Ion Implantation," *Solid State Technology,* Penwell Publishing, July 1994, p. 115.
10. Japan Report, *Semiconductor International,* Cahners Publishing, November 1994, p. 32.
11. Eaton Corp., Product Video, NV8200P, 1993.
12. Ibid.
13. R. Iscoff, "Are Ion Implanters the Newest Clean Machines?" *Semiconductor International,* Cahners Publishing, October 1994, p. 65.
14. "Wafer Handler for Ion Implanters, Varian Semiconductor Equipment," *Solid State Technology,* Penwell Publishing, July 1994, p. 131.
15. J. Hayes and P. Van Zant, *Doping Today Seminar Manual,* Semiconductor Services, San Jose, CA, 1985.
16. D. Zrudsky, "Channeling control in ion implantation," *Solid State Technology,* July 1988, p. 73.
17. N. Cheung, "Ion Implantation," *Semiconductor International,* Cahners Publishing, January 1993, p. 35.
18. P. Singer, "Plasma Doping: An Implant Alternative?" *Semiconductor International,* Cahners Publishing, May 1994, p. 34.

12

Deposition

Overview

In addition to the doped regions, it takes various other layers of semiconductors, dielectrics, and conductors to complete the device or circuit. Many are put on the wafer by chemical vapor deposition (CVD) techniques (CVD). This chapter describes the most commonly used CVD techniques and the types of materials deposited with each.

Objectives

Upon completion of this chapter, you should be able to:

1. Name the basic parts of a CVD reactor.
2. Describe the principle of chemical vapor deposition.
3. List the conductor, semiconductor, and insulator materials deposited by CVD techniques.
4. Know the difference between atmospheric CVD, LPCVD, hot-wall, and cold-wall systems.
5. Explain the difference between epitaxial and polysilicon layers.

Introduction

Advances in photomasking technology have allowed the fabrication of VLSI and ULSI circuits. But as the circuits have shrunk they also have grown in the vertical direction, through increased numbers of deposited layers. In the 1960s bipolar devices had two layers deposited by chemical vapor deposition (CVD), an epitaxial layer, and a top side passivation layer of silicon dioxide (Fig. 12.1), while early MOS devices

Figure 12.1   Cross section of bipolar circuit showing epitaxial layer and isolation.

had just a passivation layer (Fig. 12.2). By the 1990s advanced devices featured four or more levels of metal interconnects requiring numerous deposited layers. The added layers function in the devices as either semiconductors, dielectrics (insulators), or conductors. Figure 12.3 lists common OVD techniques. There are two primary techniques for layer deposition: chemical vapor deposition (CVD) and physical vapor deposition (PVD). The PVD techniques of evaporation and sputtering are explained in Chap. 13. The uses of the particular films, while indicated in this chapter, are detailed in Chaps. 16 and 17. Chemical vapor deposition, the subject of this chapter, is practiced in a number of atmospheric and low pressure techniques.

Film standards

Device layers must meet general and specific parameters. The specific parameters are noted in the sections on individual layer materials. General criteria that all films must meet for semiconductor use include:

- Uniform thickness
- Smooth and flat surface
- Uniform composition and grain size
- Stress-free films
- Purity
- Integrity

Uniform thickness is required of films to meet both electrical and mechanical specifications. Deposited films must be continuous and

"Old" MOS

"Modern" MOS

Figure 12.2 Evolution of MOS layers.

Atmospheric Pressure (AP)	Low Pressure (LP) and Ultra High Vacuum (UHV)
Cold wall	Hot wall
• Horizontal	Plasma enhanced
• Vertical	Vertical isothermal
• Barrel	Molecular beam epitaxy (MBE)
• Vapor phase epitaxy	
Metalorganic CVD	

Figure 12.3 Overview of deposition systems.

free of pinholes to prevent the passage of contamination and to prevent shorting of sandwiched layers. This is of great importance for thin films. Epitaxial films have shrunk from 5-μm levels to submicron thicknesses. Recall that the thickness of a layer is one of the factors contributing to its resistance. Also, thinner layers tend to have more pinholes and less mechanical strength. Of particular concern is the maintenance of thickness over steps (Fig. 12.4). Excessive thinning at a step can cause electrical shorts and/or unwanted induced charges in the device.

Surface flatness is as important as the thickness. In Chap. 10 the effect of steps and surface roughness on image formation was detailed. Deposited films must be flat and as smooth as the material and deposition method will allow to minimize steps, cracking, and subsurface reflections.

Deposited films must be of the desired uniform composition. Some of the reactions are complex and it is possible that films will be deposited with different than the intended composition. *Stoichiometry* is the methodology by which the quantities of reactants and products in chemical reactions are determined. In addition to chemical composition, grain size is important. During deposition the film materials tend to collect or grow into grains. Varying grain size within films of the same composition and thickness will yield different electrical and mechanical properties. This is because electrical current flow is affected as it passes through the grain interfaces. Mechanical properties also change with the size of grain interface area.

Stress-free films are another requirement. A film deposited with excess stress will relieve itself by forming cracks. Cracked films cause surface roughness and can allow contamination to pass through to the wafer. In the extreme they cause electrical shorts.

Purity, that is, no unwanted chemical elements or molecules in the film, is required for the film to carry out its intended function. For example, oxygen contamination of an epitaxial film will change its electrical properties. Purity also includes the exclusion of mobile ionic contaminants and particulates.

(a)   (b)

Figure 12.4 Thinning of deposited layer (*b*) at step.

Chemical Vapor Deposition Basics

Not surprisingly, the growth in the number and kinds of deposited films has resulted in a number of deposition techniques. Whereas the process engineer of the 1960s had no choice but to use atmospheric chemical vapor deposition (CVD), today's engineer has many more options (Fig. 12.3). These techniques are described in the following sections.

Thus far the terms *deposition* and *CVD* have been used without explanation. In semiconductor processing, deposition refers to any process in which a material is physically deposited on the wafer surface. Grown films are those, such as silicon dioxide, that form from the material in the wafer surface. The majority of films are deposited by a CVD technique. In concept the process is simple (Fig. 12.5). Chemicals containing the atoms or molecules required in the final film are mixed and reacted in a deposition chamber to form a vapor. The atoms or molecules deposit on the wafer surface and build up to form a film. Figure 12.5 illustrates the reaction of silicon tetrachloride ($SiCl_4$) with hydrogen to form a deposited layer of silicon on the wafer. Generally, CVD reactions require the addition of energy to the system, such as heating the chamber or the wafer.

The chemical reactions that take place fall into the four categories of pyrolysis, reduction, oxidation, and nitridation (Fig. 12.6). *Pyrolysis* is the process of chemical reaction driven by heat alone. *Reduction* causes a chemical reaction by reacting a molecule with hydrogen.

$$SiCl_4 + 2H_2 \longrightarrow Si + 4HCl \uparrow$$

wafer

Figure 12.5 Chemical vapor deposition of silicon from silicon tetrachloride.

Pyrolysis	$SiH_4 \rightarrow Si + 2H_2$
Reduction	$SiCl_4 + 2H_2 \rightarrow SI + 4HCL$
Oxidation	$SiH_4 + O_2 \rightarrow SiO_2 + 2H_2$
Nitridation	$3SiH_2Cl_2 + 4NH_3 \rightarrow Si_3N_4 + pH + 6H_2$

Figure 12.6 Examples of CVD reactions.

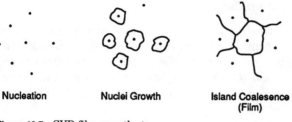

Nucleation   Nuclei Growth   Island Coalesence
                                                  (Film)

Figure 12.7 CVD film growth steps.

Oxidation is the chemical reaction of an atom or molecule with oxygen. *Nitridation* is the chemical process of forming silicon nitride.

Deposited film growth proceeds in several distinct stages (Fig. 12.7) The first stage, *nucleation,* is very important and critically dependent on substrate quality. Nucleation occurs as the first few atoms or molecules deposit on the surface. These first atoms or molecules form islands that grow into larger islands. In the third stage the islands spread, finally coalescing into a continuous film. This is the transition stage of the film growth with a typical thickness of several hundred angstroms. The transition region film has chemical and physical properties much different than the final, thicker "bulk" film.[1]

After the transition film is formed, the growth of the bulk begins. Processes are designed to produce three bulk film structures: amorphous, polycrystalline, and single crystal (Fig. 12.8). These terms have been defined before. A poorly defined or controlled process can result in a film with the wrong structure. For example, attempting to grow a single-crystal epitaxial film on a wafer with islands of unremoved oxide will result in regions of polysilicon in the bulk film.

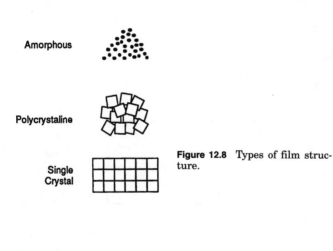

Amorphous

Polycrystaline

Single
Crystal

Figure 12.8 Types of film structure.

Basic CVD system design

CVD systems come in a wide variety of designs and options. Understanding the many variations is helped by an examination of the basic subsystems common to most CVD systems (Fig. 12.9). In most respects, a CVD system has the same basic parts as a tube furnace (described in Chap. 7): source cabinet, reaction chamber, energy source, wafer holder (boat), and loading and unloading mechanisms. In some cases the CVD system *is* a tube furnace identical to those used for oxidation and diffusion. The source chemicals are housed in a source section. Vapors are generated from pressurized gas cylinders or liquid source bubblers. Gas flow control is maintained by pressure regulators, mass flow meters, and timers.[2]

The actual deposition takes place on the wafers in a reaction chamber. Most reactions require an energy source that heats the chamber or the wafers directly. Energy sources are either heat (conduction-convection), induction RF, radiant, plasma, or ultraviolet. Energy sources are explained in the sections on particular systems. Temperatures range from room temperature to 1250°C depending on the reaction, film thickness required, and the growth parameters.

The fourth basic part of the system is the wafer holder. Different chamber configurations and heat sources dictate the style and material of the holders. Most production-level systems for VLSI work are automated from wafer load to unload. A full production system will have an associated cleaning section or station and a loading area.

Figure 12.9 Basic CVD subsystems.

CVD Process Steps

A CVD process follows the same steps as an oxidation or diffusion process. For review, the steps are preclean (and etch, if required), deposition, and evaluation. Cleaning processes are those already described to remove particulates and mobile ionic contaminants. Chemical vapor deposition, like an oxidation, take place in cycles. First the wafers are loaded into a chamber, usually containing an inert atmosphere. Next the wafers are brought to temperature. Chemical vapors are introduced for as long as required to deposit the film. Finally the chemical source vapors are flushed out and the wafers removed. Evaluation of the films are for thickness, step coverage, purity, cleanliness, and composition. Evaluation techniques are explained in Chap. 14.

CVD System Types

CVD systems (Fig. 12.3) are divided into two primary types: atmospheric pressure (AP) and low pressure (LP). There are a number of atmospheric pressure CVD systems (APCVD). Most advanced device films are deposited in systems where the pressure has been lowered. These are called low pressure CVD or LPCVD.

Another differentiation is cold wall or hot wall. Cold-wall systems directly heat the wafer holder or wafers, with induction or radiant heating. The walls of the chamber remain cold (or cooler). Hot-wall systems heat the wafers, the wafer holder, and the chamber walls. The advantage of cold-wall CVD is that the reaction occurs only at the heated wafer holder. In a hot-wall system the reaction occurs throughout the chamber, leaving reaction products on the inside chamber walls. The reaction products build up, necessitating rigorous and frequent cleaning to prevent contaminating the wafers.

CVD systems are operated by two principal energy sources, thermal and plasma. Thermal sources are tube furnaces, hot plates, and RF induction. Plasma-enhanced chemical vapor deposition (PECVD) in combination with lower pressure offers the unique advantage of lowered temperatures and good film composition and coverage.

A specialty CVD used to deposit compound films such as GaAs is vapor phase epitaxy (VPE). A newer technique used to deposit metals is a metalorganic CVD (MOCVD) source in a VPE system. The last deposition method that will be described is the non-CVD molecular beam epitaxy (MBE) used for low-temperature deposition of thin films in a very controlled process.

Atmospheric Pressure CVD Systems

As the name implies, atmospheric CVD system reactions and deposition take place at atmospheric pressure. There are a number of system designs that fall under this category (Fig. 12.10).

Horizontal tube–induction-heated APCVD

The first widespread use of CVD was for the deposition of silicon epitaxial films for bipolar devices. The basic system design is still used (Fig. 12.11). It is essentially a horizontal tube furnace, but with some significant differences. First the tube has a square cross section. The major difference, however, is in the heating method and wafer holder.

Wafers are arranged on a flat graphite slab and positioned in the tube. Surrounding the tube are copper coils that are connected to an RF generator. The RF waves traveling in the coils pass through the quartz tube and the flowing gas in the tube without heating them. This is the cold-wall aspect of the system. When the radiant waves

Figure 12.10 CVD reactor designs (*from Wolf and Tauber, Silicon Processing for the VLSI*)

Figure 12.11 Cold-wall induction APCVD with horizontal susceptor.

reach the graphite wafer holder, they "couple" with the molecules of the holder causing the graphite to heat up. This heating method is called *induction*.

The heat of the holder is passed to the wafers by conduction. The film deposition takes place at the wafer surface (and at the holder surface). One problem with this type of system is downstream depletion of the reactants in the laminar gas flow. Laminar flow is required to minimize turbulence. But if the wafers are laid flat in the chamber the layer of gas closest to the wafers becomes depleted. This results in successively thinner films along the wafer holder. A wafer holder tilted in the tube corrects the problem (Fig. 12.12).

Barrel radiant–induction-heated APCVD

Larger-diameter wafers laid horizontally on a holder in a horizontal system have a low packing density. And larger wafer holders strain the uniformity capabilities of the system.

The development of the barrel radiant-heated system (Fig. 12.13) solved these problems. The reaction chamber of the system is a cylindrical stainless steel barrel with high-intensity quartz heaters placed about the inside surface. The wafers are placed on a graphite holder that rotates in the center of the barrel. The rotation of the wafers produces a more uniform film thickness compared to horizontal systems.

Radiant heat from the lamps heats the wafer surface, where the deposition takes place. While some heating of the chamber walls occurs, the system is close to a cold-wall deposition. Direct radiant heating produces a very controlled and even film growth. In an induction-heated system, the wafers are heated from the bottom, and as the film grows there is some small but measurable drop in temperature at the film surface. In the barrel system the wafer surface is always facing the lamps and receives a more uniform temperature and film growth rate.

In 1987, Applied Material introduced a jumbo barrel system for large-diameter wafers featuring an induction heating system.[3] A principal advantage of the barrel reactor is an increased productivity based

Figure 12.12   Cold-wall induction APCVD with tilted susceptor.

Gas Flow

Graphite Wafer Holder

Radiant Heaters

Figure 12.13 Cylindrical or barrel system.

on the increased number of wafers per cycle. This system configuration is popular for deposition of epitaxial silicon in the 900 to 1250°C range.

Pancake induction-heated APCVD

The pancake or vertical-flow APCVD system has been a favorite for small fabrication lines and R&D labs (Fig. 12.14). The wafers are held on a rotating holder of graphite and heated by induction-conduction from an RF coil below the holder. The reaction gases are fed through a tube exiting above the wafers. Vertical gas flow offers the advantage of a continuous supply of fresh reactants to the wafers, thus minimizing downstream depletion. The combination of the rotation and vertical flow of the gases produces good film uniformity. Productivity in

Figure 12.14 Rotating pancake APCVD.

Gas

smaller systems is restricted, as in the horizontal tube system, by the number of wafers that the pancake can accommodate.

A production-level variation of the pancake design is produced by Gemini Research. The reactor features radiant-resistance heating and a large-capacity holder with robot autoloading.[4]

Continuous-conduction-heated APCVD

Two horizontal conduction-heated APCVD systems feature mixing the gases outside the chamber and showering them onto the wafers. In one design a heated hot plate wafer holder moves back and forth (Fig. 12.15) under a series of nozzles that dispense a vapor of the desired material. Another version of the system (Fig. 12.16) has wafers moving on a belt under a plenum that dispenses the reacted gases.

Horizontal-conduction-heated APCVD

One of the original CVD designs is the horizontal conduction-heated APCVD system (Fig. 12.17) used to deposit the silicon dioxide passivation layers. Wafers are mounted on a removable hot plate and placed in a stainless steel chamber. The hot plate heats the wafers and chamber walls (hot-wall system). Reaction gases are fed into the chamber.

Low-Pressure CVD (LPCVD) Systems

Uniformity and process control within atmospheric pressure CVD systems rely on temperature control and the flow dynamics in the

Figure 12.15 Moving hot-plate APCVD with shuttle.

Figure 12.16 Continuous hot-plate APCVD.

Horizontal   Stainless Steel
                  Reaction tube

Gas Flow.

Resistance
Heated Wafer and Holder

Figure 12.17   Hot-plate APCVD.

system. A factor influencing film uniformity and step coverage is the mean free path of the molecules in the reaction chamber. The mean free path is the average distance a molecule will travel before colliding with an object in the chamber be it another molecule, the wafers, or wafer holder. Collisions change the direction of the particles. The longer the mean free path, the higher the uniformity of the film deposition. A major determiner of the length of the mean free path is the pressure in the system. Lowering the pressure in the chamber increases the mean free path and the film uniformity. Decreasing the pressure also allows a lowering of the deposition temperature.

These benefits became available to the industry in 1974 when Unicorp, under license to Motorola Inc., introduced an LPCVD system that operated at a few hundred millitorr.[5] The complete list of LPCVD system advantages includes:

- Lower chemical reaction temperature
- Good step coverage and uniformity
- Vertical loading of wafers for increased productivity and lower exposure to particles
- Less dependence on gas flow dynamics
- Less time for gas phase reaction particles to form
- Can be performed in standard tube furnaces

A vacuum pump must be added to the system to reduce the pressure in the chamber. A discussion of the types of pumps used with LPCVD systems is in Chap. 13.

Horizontal conduction–convection-heated LPCVD

A standard production-level LPCVD system is an adaption of a horizontal tube furnace (Fig. 12.18), with three major exceptions. The tube is connected to a vacuum pump that pulls the system down to a

Pressure Sensor

Water Cooled
End Cap
Assembly

Quartz Reactor Tube

3 Zoned Resistance Heater

Vacuum Pump

Figure 12.18 Horizontal hot-wall LPCVD system.

Gas Inlets

pressure range of 0.25 to 2.0 Torr.[6] A second change is a ramping of the temperature in the center zone to offset reaction depletion down the tube. The third change may be special injectors at the gas inlet end to improve gas mixing and deposition uniformity. In some systems, the injectors are positioned directly over the wafers. Disadvantages of this system design are particles formed on the inside wall surface (hot-wall reactions), uniformity along the tube axis, the use of cages around the wafers to minimize particle contamination, and the higher downtime required for frequent cleaning.

These systems are most often used for polysilicon, silicon dioxide, and silicon nitride films with typical thickness uniformities of ± 5 percent. The primary deposition variables are temperature, pressure, gas flow, gas partial pressure, and wafer spacing. These variables are carefully balanced for each deposition process. The deposition rates are somewhat lower (100 to 500 Å/min) than AP systems, but productivity is enhanced by the vertical wafer-loading densities that can approach 200 wafers per deposition.

Vertical isothermal LPCVD

A novel LPCVD system was introduced by Anicon in the mid-1980s. Called a *vertical isothermal reactor* (Fig. 12.19), the reaction takes place in a reduced-pressure bell-jar-type reaction chamber. The wafers, which are held vertically in a quartz boat, are heated from below, creating an isothermal (uniform) temperature at each wafer. The system features vertical gas flow.

Gases are metered separately into a tube that allows the mixing just above the wafers. The deposition takes place in the vertical plane. Film uniformity is improved since there is no downstream depletion of the gas and there is a continuous supply of equal amounts of "fresh" reactants. Reported uniformities are typically ± 2 percent. Wafer cleanliness is improved since the gas supply is fresh

Figure 12.19 Vertical flow isotherm LPCVD. (*Courtesy of Anicon/SVG.*)

and the cage protects the wafers from reaction particles. Cleanliness is also improved because the wafers are loaded from the bottom and not through the deposition zone.

Ultrahigh-vacuum CVD (UHVCVD)

Low-temperature deposition is desirable to minimize crystal damage and lower the thermal budget, which in turn minimizes the lateral diffusion of doped regions. One approach is the CVD of silicon and silicon germanium (SiGe) in ultralow-vacuum conditions. Lowering the pressure allows keeping the deposition temperature low. UHVCVD takes place in a tube furnace where the pressure is initially reduced to 1 to 5×10^{-9} millibar (mbar). Deposition pressure is in the 10^{-3} mbar range.[7]

Plasma-Enhanced CVD (PECVD) Systems

Replacement of silicon dioxide passivation layers with silicon nitride led to the development of PECVD techniques. A thermal silicon dioxide deposition temperature of approximately 660°C causes unacceptable alloying of aluminum interconnects into the silicon surface (see Chap. 13). A solution to this problem was a plasma enhancement of the deposition energy. The increased energy allows a temperature under the 450°C maximum level for deposition over aluminum layers. Plasma-enhanced systems are physically similar to plasma etch systems. They feature a parallel plate chamber operated at a low pressure. A radio-frequency-induced glow discharge, or other plasma source (see Chap. 9) is used to induce a plasma field in the deposition gas. The combination of low pressure and lower temperatures provides good film uniformity and throughput.

PECVD reactors have the capability of also using the plasma for etching and cleaning the wafer prior to the deposition step. This step is the same as the dry etch processes described in Chap. 9. This in situ cleaning prepares the deposition surface, eliminating the problem of added contamination picked up during the loading step.

Horizontal-vertical flow PECVD.   This system follows the design of a bottom-heated pancake vertical-flow CVD (Fig. 12.20). The plasma region is created in the top of the chamber by a radio frequency (RF) feed to an electrode or other plasma source. Wafer heating comes from radiant heaters mounted below the wafer holder, creating a cold-wall deposition system. With PECVD systems, there are several additional critical parameters to control in addition to those in a standard LPCVD reactor. They are the RF power density, the RF frequency, and the duty cycle. Film deposition speed is generally increased, but it must be controlled to prevent film stress and/or cracking.

A single-wafer chamber PECVD system (Fig. 12.21) where the chamber is small and each successive wafer is exposed to identical conditions addresses most control needs. Single-wafer systems are generally slower than batch processes. The productivity tradeoff with the larger-chamber batch machines lies in fast mechanisms to feed wafers in and out of the chamber and how quickly the vacuum can be established and released. Load-lock systems enhance the productivity

Figure 12.20 Vertical-flow pancake PECVD.

Figure 12.21 Single-chamber planar PECVD.

by moving the wafers into an antechamber, pumping it down to the required pressure, and moving the wafers into the deposition chamber as it becomes available.

Barrel radiant-heated PECVD. This system is a standard barrel radiant-heated system with low pressure and plasma capabilities. It is favored for the deposition of tungsten silicide.

Horizontal-tube PECVD. This system is based on a standard tube furnace design. The plasma region is created in the specially designed wafer boat. The boat consists of slabs of graphite, alternately connected to an RF power supply (Fig. 12.22). Each pair of slabs is in effect a mini-parallel-plate plasma generator. Wafers are placed between the plates and the deposition gases passed over and through the holder where the localized plasma enhances the deposition.

Primary advantages of this approach to PECVD are the economic savings from adapting a tube furnace and high productivity. Dis-

Figure 12.22 Plasma generating graphite wafer boat.

advantages are in the more difficult control of the film stoichiometry and the large, heavy wafer carrier.

Photochemical (Ultraviolet) LPCVD Systems

Photochemical systems utilize an ultraviolet light source to dissociate the reactant gases into the desired film material. The major advantage of the photochemical LPCVD approach is a lower reaction temperature. These systems can deposit both silicon dioxide and silicon nitride at temperatures between 50 to 200°C. The lower temperature can result in a deposited film with fewer stress cracks and very good step coverage and with less thermal damage to the wafer surface. Heating of the wafers is provided by a hot plate and operates the system at the lower end of the LPCVD pressure range, 0.3 to 1.0 torr. While offering advantages to the CVD arsenal, these systems are still under development and are not in use as a mainstream LPCVD technique. Figure 12.23 is a summary table of CVD systems by temperature range.

Vapor Phase Epitaxy (VPE)

VPE differs from the CVD systems described in its ability to deposit compound materials, such as gallium arsenide. A VPE system[8] is a combination of a standard liquid source tube furnace and a two-zone diffusion furnace (Fig. 12.24). An example is the particular arrangement in Fig. 12.24 used to deposit epitaxial gallium arsenide. The creation of the GaAs layer on the wafer in the main chamber proceeds in two stages. $AsCl_3$ is bubbled into the first section of the tube where it reacts with a solid source of gallium that is sitting in a boat. The arsenic trichloride reacts with the hydrogen in the first section to form arsenic by the reaction

Level	Temperature Range	Methods
High Temp.:	600–1250°C	R.F. Induction (Cold Wall) Radiant Heat (Cold Wall) Resistance Coils (Hot Wall)
Mid Temp.:	200–600°C	Hot Plates Plasma Enhanced LPCVD
Low Range:	22–200°C	Hot Plates P.E. CVD Photochemical

Figure 12.23 Summary table of CVD methods.

Figure 12.24 Diagram of gallium arsenide VPE deposition system.

$$4AsCl_3 + 6H_2 \rightarrow 12HCl + As_4$$

The arsenic deposits on the gallium, forming a crust. The hydrogen passing over the crust reacts in the first section to form three gases that pass into the wafer section.

$$\frac{GaAs}{(solid)} + \frac{HCl}{(gas)} \leftrightarrow \frac{GaCl}{(gas)} + \frac{\frac{1}{2}H_2}{(gas)} + \frac{\frac{1}{4}As_4}{(gas)}$$

This section is at a somewhat lower temperature and the reaction proceeds in reverse, depositing GaAs on the wafers. The technique offers the advantages of clean films, since the gallium and arsenic trichloride are available in very pure forms and have higher production rates than the MBE technique. On the downside, the film structures produced are not the quality of MBE films.

Molecular Beam Epitaxy (MBE)

Deposition rate control, low deposition temperature, and controlled film stoichiometry are always goals in film-deposition systems. Molecular beam epitaxy (MBE) has emerged out of the laboratory to claim production status as these issues have become more important. MBE is an evaporation rather than a CVD process. The system consists of a deposition chamber (Fig. 12.25) that is maintained at a low pressure to 10^{-10} torr. Within the chamber is one or more cells (called *effusion cells*) that contain a very pure sample of the target material desired on the wafer. Shutters on the cells allow exposure of the wafer to the source material(s). An electron beam[9] is directed into the center of the target material, which it heats to the liquid state. In this state, atoms evaporate out of the material, exit the cell through an opening, and deposit on the wafers. If the material source is a gas,

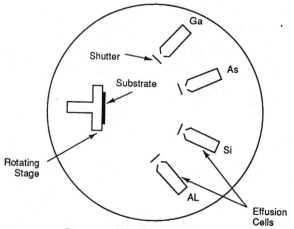

Figure 12.25   Diagram of MBE deposition system.

the technique is called *gas source MBE* or *GSMBE*. For most applications, the wafer in the chamber is heated to give additional energy to the arriving atoms. The additional energy fosters epitaxial growth and good film quality.

If the wafer surface is exposed, the depositing atoms will assume the orientation of the wafer and grow an epitaxial layer. MBE offers the intriguing option of in situ doping by the inclusion of dopant sources in the chamber. The usual silicon dopant sources are not usable in MBE systems. Solid gallium is used for P-type doping and antimony for N-type doping.[10] Phosphorus deposition is virtually impossible with MBE.[11]

The primary advantage of MBE for silicon technology is the low temperature (400 to 800°C), which minimizes autodoping and out-diffusion. Perhaps the biggest advantage of MBE is the ability to form multiple layers on the wafer surface during one process step (one pump down). This option requires the mounting of several effusion cells in the chamber and shutter arrangements to direct the evaporant beams to the wafer in the right order and for the correct time.

An advantage and disadvantage of MBE is the low film growth rate of 60 to 600 Å/min.[12] On the plus side, the films produced are very controllable. Films can be grown (and mixed) in monolayer increments. However, most semiconductor layers do not need this level of control and quality, making the low productivity and expense of the system an expensive luxury.

A bonus possibility with MBE is the incorporation of film growth and quality-analyzing instruments in the chamber. With these instruments the process can become very controlled and produce uniform films from wafer to wafer. MBE has found production use in the fabrication of special microwave devices and for compound semiconductors such as gallium arsenide.[12]

Metalorganic CVD (MOCVD)

MOCVD is one of the latest options for CVD of compound materials. Where VPE refers to a compound material deposition system, MOCVD refers to the sources used in VPE systems (Fig 12.26). There are two chemistries used, halides and metalorganic. The reactions described for the VPE deposition of gallium arsenide above is a halide process. A group III halide (gallium) is formed in the hot zone and the III–IV compound is deposited in the cold zone. In the metalorganic process[13] for gallium arsenide, trimethylgallium is metered into the reaction chamber along with arsine to form gallium arsenide by the reaction

$$(CH_3)_3Ga + AsH_3 \rightarrow GaAs + 3CH_4$$

Whereas MBE processes are slow, MOCVD processes can meet volume production requirements and accommodate larger substrates.[14] MOCVD is also capable of producing multiple layers with very abrupt changes in composition. This characteristic is critical to compound semiconductor structures. Also, MOCVD, unlike MBE, can deposit phosphorus, necessary, for example, in InGaAsP devices. Common devices made by MOCVD processes are photocathodes, high poser LEDs, long wavelength lasers, visible lasers, and orange LEDs (see Chap. 16).

MOCVD is a broad term referring to the metalorganic chemical vapor deposit of semiconductor films. When metalorganic sources are

Figure 12.26 MOCVD system. (*Source: VLSI Fabrication Principles by Ghandi*)

used in a vapor phase epitaxial system to grow epitaxial layers the method is called MOVPE.[15]

Deposited Films

The types of films deposited by CVD techniques are divided into their electrical classifications of semiconductors, dielectrics, and conductors. In the following sections the deposition of the various films is examined. For each film, its principal use(s) in semiconductor devices is presented along with the particular film properties for the described use. The uses are treated in general. (For a more detailed explanation of film roles in particular devices, see Chap. 16. Deposition methods for conductive metal films are discussed in Chap. 13.)

Deposited Semiconductors

So far in this text we have discussed the formation of wafers as the base of semiconductor devices and circuits. However, there are several drawbacks to using bulk wafers for high-quality devices and circuits. Crystal quality, doping ranges, and doping control all limit bulk wafer use. Early on, these factors placed a limit on the fabrication of high-performance bipolar transistors. A solution was found in the development of a deposited silicon layer, called an *epitaxial layer*. It is one of the major advances in the industry. Epitaxial layers were a part of the technology as early as 1950.[16] Since then deposited silicon layers have found additional uses in advanced bipolar device design (a high-quality base for CMOS circuits) and silicon epitaxial layers deposited on sapphire and other substrates (Chap. 14). Gallium arsenide and other III–1V and II–VI films are also deposited in epitaxial films. Epitaxial films that are the same material as the substrate (e.g., silicon on silicon) are *homoepitaxial*. When the deposited material is different than the substrate (e.g., GaAs on silicon) the film is called *hetroepitaxial*.

Epitaxial Silicon

The term *epitaxial* comes from the Greek word meaning "arranged upon." In semiconductor technology it refers to the single crystalline structure of the film. The structure comes about when silicon atoms are deposited on a bare silicon wafer in a CVD reactor (Fig. 12.27). When the chemical reactants are controlled and the system parameters set correctly, the depositing atoms arrive at the wafer surface with sufficient energy to move around on the surface and orient them-

Figure 12.27 Epitaxial and poly-silicon film growth.

selves to the crystal arrangement of the wafer atoms. Thus an epitaxial film deposited on a <111>-oriented wafer will take on a <111> orientation.

If, on the other hand, the wafer surface has a thin layer of silicon dioxide, an amorphous surface layer, or contamination, the depositing atoms have no structure to align to. The resulting film structure is polysilicon. This condition is useful for some applications, such as MOS gates, but is unwanted if the goal is to grow a single-crystal film structure.

Silicon tetrachloride source chemistry. A number of different sources are used for the deposition of epitaxial silicon (Fig. 12.28). Deposition temperature, film quality, growth rate, and compatibility with a particular system are factors in choosing a silicon source. An important process parameter is the deposition temperature. The higher the temperature, the faster the growth rate. Faster growth rates create more crystal defects and film cracking and stress. Higher temperatures also cause higher levels of autodoping and out-diffusion. (These effects are described in the following text.)

Silicon tetrachloride ($SiCl_4$) is the favored source of silicon for deposition. It allows a high formation temperature (growth rate) and has a

Silicon tetrachloride	$SiCl_4 + 2H_2 \leftrightarrow Si + 4HCl$
Silane	$SiH_4 + heat \rightarrow Si + 2H_2$
Dichlorosilane	$SiH_2Cl_2 \leftrightarrow Si + 2HCl$

Figure 12.28 Epitaxial silicon chemical sources.

reversible chemical reaction. In Fig. 12.28, there is a double-headed arrow, which indicates that the reaction creates silicon atoms in one direction and removes (etches) silicon in the other direction. Within the reactor these two reactions compete with each other.

Initially, the silicon surface is etched, preparing it for the deposition reaction. In the second stage, the deposition of silicon is faster than the etch, with the net result of a deposited film.

The graph in Fig. 12.29 shows the effect of the two reactions. With an increasing percentage of $SiCl_4$ molecules in the gas stream, the deposition rate first increases. At the 0.1 ratio, the etching reaction starts to dominate and slows down the growth rate. This latter reaction is actually one of the first events in the reactor. Hydrogen chloride (HCl) gas is metered into the chamber where it etches away a thin layer of the silicon surface, preparing it for the silicon deposition.

Silane source chemistry. The second-most-used silicon source chemistry is silane (SiH_4). Silane offers the advantage of not requiring a second reaction gas. It forms silicon atoms by decomposing when heated. The reaction takes place several hundred degrees lower than a silicon tetrachloride deposition, which is attractive from an autodoping and wafer warping perspective. Also, silane does not produce pattern shift (see "Epitaxial film quality"). Unfortunately, the reaction occurs at all locations in the system, creating a powdery film inside the reactor, which in turn contaminates the wafers. Silane finds more use as a source for polysilicon and silicon dioxide depositions.

Dichlorosilane source chemistry. Dichlorosilane (SiH_2Cl_2) is also a lower-temperature silicon source that is used for thin epitaxial films. The lower temperature reduces autodoping and solid-state diffusion from previously diffused buried layers and provides a more uniform crystal structure.

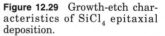

Figure 12.29 Growth-etch characteristics of $SiCl_4$ epitaxial deposition.

Epitaxial film doping.   One of the advantages of an epitaxial film is the precise doping and doping range made available by the process. Silicon wafers are manufactured in a concentration range of approximately 10^{13} to 10^{19} atoms/cm^3. Epitaxial films can be grown from 10^{12} to 10^{20} atoms/cm^3. The upper limit is close to the solid solubility of phosphorus in silicon.

Doping in the film is achieved by the addition of a dopant gas stream to the deposition reactants. The sources of the dopant gases are exactly the same chemistries and delivery systems used in deposition doping furnaces. In effect, the CVD deposition chamber is turned into a doping system. In the chamber the dopants become incorporated into the growing film where they establish the required resistivity. Both N- and P-type films can be grown on either N- or P-type wafers. The classic epitaxial film in bipolar technology is an N-type epitaxial film on a P-type wafer.

Epitaxial film quality.   Epitaxial film quality is a prime concern of the process. In addition to the usual considerations about contamination, there are a number of faults specifically associated with epitaxial growth. Contaminated systems can cause a problem called *haze*.[17] Haze is a surface problem that varies from a microscopic disruption to severe cases that are observable as a dull matte finish. Haze comes about from residual oxygen in the reactant gases or from leaks in the system.

Contaminants on the surface at the start of the deposition result in an accelerated growth known as *spikes* (Fig. 12.30). The spikes can be as high as the film thickness. They cause holes and disruption in photoresist layers and other deposited films.

During the growth, a number of crystal problems can occur. One is *stacking faults*. A stacking fault is due to the inclusion of an extra atomic plane with a corresponding "dislocation" of the atoms around the plane. A stacking fault begins at the wafer surface and "grows" to the film surface. The shape of the stacking fault depends on the orientation of the film and wafer. Faults in <111>-oriented films have a

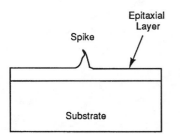

Figure 12.30   Epitaxial growth spike.

Figure 12.31   Stacking fault on <111> Si.

pyramidal shape (Fig. 12.31), while <100>-oriented wafers form rectangular-shaped stacking faults. The faults are detected by either x-ray or etching techniques.

A growth problem associated with <111> wafers is *pattern shift.* This problem occurs when the deposition rate is too high and the film planes grow at an angle to the surface. Pattern shift is a problem when alignment to a subsurface pattern relies on locating it from a film surface step (Fig. 12.32). Another major growth problem is *slip.* This condition comes about from poor control of the deposition parameters and results in a "slippage" of the crystal along plane interfaces (Fig. 12.33).

There are two issues associated with the temperature of the deposition: autodoping and out-diffusion. *Autodoping* of the growing film

Figure 12.32   Epitaxial pattern shift.

Figure 12.33   Crystal slip.

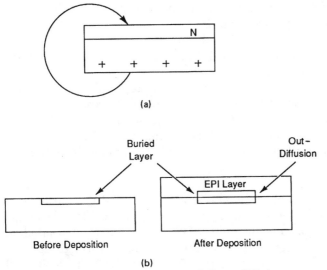

Figure 12.34 (*a*) Epitaxial autodoping and (*b*) out-diffusion.

occurs when dopant atoms from the back of the wafer diffuse out from the wafer (Fig. 12.34), mix in the gas stream, and become incorporated into the growing film. In the film they change the resistivity and the conductivity level. Autodoping in a P-type film, grown over an N-type wafer, will be less P-type than intended and have a lower P-type concentration as the autodoped atoms neutralize a number of the P-type atoms in the film.

Out-diffusion causes the same effect but at the epitaxial layer–wafer interface. The source of the out-diffused atoms is doped regions diffused into the wafer before the epitaxial deposition. In bipolar devices the regions are called *buried layers* or *subcollectors*. In the usual format the buried layer is an N-type region in a P-type wafer, over which is grown an N-type epitaxial layer. During the deposition the N-type atoms diffuse out and become incorporated into the bottom of the epitaxial film, changing the concentration. In the extreme, the buried layer can out-diffuse up into the bipolar device structure, causing electrical malfunctions.

CMOS epitaxy. Until the late 1970s, the dominant use of epitaxial films was as the collector region of bipolar transistors. The technique provided a quality substrate for device operation and a clever means of isolating adjacent devices (see Chap. 16). A newer and perhaps more dominant use of silicon epitaxial films is for CMOS circuit wafers. The need for an epitaxial layer was driven by a CMOS circuit problem called *latch-up* (see Chap. 16).

Epitaxial process.   A typical epitaxial process starts with a complete and rigorous cleaning of the wafer surface prior to loading the reactor. Within the deposition chamber a number of steps take place to correctly deposit the film. A typical $SiCl_4$ epitaxial process is shown in Fig. 12.35. The first several steps are a gas-phase cleaning of the wafer surface. Deposition follows the cleaning with a cool-down cycle at the end. During all the steps, control of the temperature and gas flows is critical.

Selective epitaxial silicon.   Advancement in epitaxial deposition systems has introduced the selective growth of epitaxial films. Whereas the epitaxial films for bipolar and CMOS substrates are deposited on the entire wafer, in selective growth they are grown through holes in either silicon dioxide or silicon nitride films. The wafer is positioned in the reactor chamber and the epitaxial film grows directly on the silicon exposed at the bottom of the hole (Fig. 12.36). As the film grows, it takes on the crystal orientation of the underlying wafer. An advantage of such a structure is that devices formed in the surfaces of

Cycle	Temperature	Gas	Purpose
1	Room	N_2	Purge air from system
2	Room	H_2	Reduce any organic contaminants of wafers in system
3	(Heating)	N_2	Bring system to deposition temperature
4	Deposition Temperature	HCl	Etch wafer to prepare surface for epi deposition
5	Deposition Temperature	Source + Dopant + Carrier	Grow epitaxial film
6	(Heat Off)	N_2	Purge system of reactant gases

Figure 12.35 Typical $SiCl_4$ epitaxial deposition process.

Figure 12.36 Steps in selective epitaxial growth.

the epitaxial regions are isolated from each other by the oxide or nitride regions.

If the deposition is allowed to continue onto the isolating surface, the structure of the film switches to a polysilicon structure. Another outcome of extended deposition is that the overlaying deposited layer becomes entirely epitaxial in nature. All of these outcomes add attractive structure options for advanced device design.

Polysilicon and Amorphous Silicon Deposition

Until the advent of silicon-gate MOS devices (Fig. 12.37) in the mid-1970s, polysilicon layers had little or no use in device structures. Silicon-gate device technology drove the need for reliable processes to deposit thin layers of polysilicon. By the mid-1980s, polysilicon seemed to be the workhorse material of advanced devices. In addition to MOS gates, uses for polysilicon include load resistors in SRAM devices, trench fills, multilayer poly in Electrically Erasable Programmable Read Only Memories (EEPROMs), contact barrier layers, emitters in bipolar devices, and part of silicide metallization schemes (see Chaps. 13 and 16).

Early processes involved simply placing the oxide-covered wafers in a horizontal APCVD system and letting the polysilicon deposit on the oxide. The major difference of early polysilicon depositions from epitaxial depositions was the use of silane sources. While silane is not favored for epitaxial film deposition, it is more than adequate for polysilicon depositions.

Typical polysilicon deposition processes take place in the 600 to 650°C range. The deposition may be from either 100 percent silane or gas streams containing N_2 or H_2. The structure of polysilicon was previously described as a total nonarrangement of the silicon atoms. In the case of deposited polysilicon, the structure is somewhat different. During the early stages of deposition, at temperatures below 575°C,

Silicon Gate

N   N

P

Figure 12.37   Cross section of silicon gate MOS transistor.

the structure is amorphous (no structure). The polysilicon structure formed by deposition techniques consists of small pockets (crystallites or grains) of single-crystalline silicon separated by grain boundaries. This structure is called *columnar poly*.

The importance of grain size and grain boundary consistency shows up in the electrical current flow characteristics of the films. Current resistance comes as the current crosses the grain boundaries. The larger the grain boundaries, the higher the resistance. The achievement of consistent current flow from device to device and within a device is dependent on a well-controlled polysilicon structure. One of the advantages claimed for the use of H_2 in the gas stream is the reduction of surface impurities and moisture, which in turn results in a reduced grain size. Moisture or oxygen impurities in the system cause the growth of silicon dioxide within the structure. The oxide increases the resistance of the film and its etchability in subsequent masking steps.

All of the system's usual operating parameters (temperature, silane concentration, pump speed, nitrogen flow, and other gas flows[18]) affect the deposition rate and the grain size. Often the wafers will receive a postdeposition anneal in the 600°C range to further crystallize the film. The process of recrystallization goes on whenever the wafers go through a high-temperature process. The grain size and electrical parameters of the polysilicon film on the finished device or circuit are never the same as the deposited film.

Also influencing the grain size is the presence of dopants in the gas stream. In many devices or circuits a strip of polysilicon functions as a conductor which requires doping to decrease its resistivity. Doping can be done by diffusion before or implantation after the deposition. In situ doping takes place by adding gas dopant sources in the source cabinet and metering them into the chamber. When diborane (boron source) is added, there is a large increase in the deposition rate. An opposite effect takes place when phosphine (phosphorus source) or arsine (arsenic source) is the dopant gas. Undesirable effects of in situ doping are a loss of film uniformity, doping uniformity, and control of the deposition rate.

Doped polysilicon film resistivities are less than those of equally doped epitaxial or bulk silicon. The lower resistivities are due to dopants being trapped in the grain boundaries.

Most polysilicon layers are deposited with LPCVD systems that provide good productivity and lower deposition temperatures. LPCVD provides good step coverage, a requirement since polysilicon layers are usually deposited later in the process and the surface has become varied in its topography. Single-chamber polysilicon LPCVD systems offer the advantage of higher deposition rates without increased temperatures.[19]

SOS and SOI

These two acronyms stand for *silicon on sapphire* and *silicon on insulator*. Both refer to the deposition of silicon on a nonsemiconductor surface. The need for such structures came about from the limits placed on some MOS devices by the presence of a semiconducting substrate under the active device. These problems are resolved by forming a silicon layer on an insulating substrate. The first substrate used for this purpose was sapphire (SOS). As different substrates were investigated, the term was expanded to the more general silicon on insulator (SOI).

One technique is a direct deposition on the substrate followed by a recrystallization process (laser heating, strip heaters, oxygen implantation) to create a usable film.[20] Another approach is a selective deposition through holes in a surface oxide with an overgrowth to form the continuous film.

Another SOI method is called SIMOX. In this process the top layer of a wafer is converted to oxide with a heavy oxygen implant. An epitaxial layer is grown on top of the oxide. There has been some exploration of *bonded wafers* as well. In this approach, two wafers are bonded together followed by the thinning (grinding and polishing) of one wafer to a thin device layer thickness.[21]

Insulators and Dielectrics

CVD is the favored method of depositing films that will function in the device or circuit as insulators or dielectrics. The two films in widespread use are silicon dioxide and silicon nitride. In general, the two films find a multiplicity of uses in device and circuit designs. While they have processing and quality differences, they must meet the same general requirements as other deposited films.

Silicon dioxide. Deposited silicon dioxide films are best known from their long-term use as a final passivation layer covering the completed wafer. In this role, they provide physical and chemical protection to the underlying circuit devices and components. Deposited silicon dioxide films used as a protective top layer are known by the proprietary terms *Vapox, Pyrox,* or *Silox.* Vapox (vapor-deposited oxide) is a term coined by Fairchild engineers. Pyrox stands for pyrolitic oxide. Silox is a registered trademark of Applied Materials, Inc. Sometimes the layer is simply called a *glass*. This protective role has expanded, and deposited silicon oxide layers are used as interdielectric layers in multimetallization schemes, as insulation between polysilicon and metallization layers, as doping barriers, as diffusion sources, and as isolation regions.

Silicon dioxide has become a major part of silicon gate structures. These structures include gate stacks, which consist of thermal oxide or silicon dioxide, or oxinitride, or silicon dioxide (TEOS deposited), and various silicon dioxide fillers for plugs in multimetal designs.[22]

CVD-deposited silicon dioxide films vary in structure and stoichiometry from thermally grown oxides. Depending on the deposition temperature, deposited oxides will have a lower density and different mechanical properties, such as index of refraction, resistance to cracking, dielectric strength, and etch rate. These factors are highly affected by the addition of dopants to the film. In many processes the deposited film will receive a high-temperature anneal, a process called *densification*. After the densification, the deposited silicon dioxide film is close to the structure and properties of a thermal oxide.

The need for a low-temperature-deposited SiO_2 was dictated by the unacceptable alloying of aluminum and silicon at temperatures above 450°C. The early deposition process used was a horizontal conduction-heated APCVD system from silane and oxygen by the reaction

$$SiH_4 + O_2 \rightarrow SiO_2 + 2H_2$$

This process produced films that were of unacceptable quality for use in advanced device designs and on larger wafers due to the poorer film quality produced by the 450°C deposition temperature.

The development of LPCVD systems made possible higher-quality films, especially for the factors of step coverage and lower stress. LPCVD processes are the preferred deposition techniques from both quality and productivity considerations. High-temperature (900°C) LPCVD of silicon dioxide is performed with a dichlorosilane reaction with nitrous oxide:

$$SiCl_2 + 2NO_2 \rightarrow SiO_2 + 2N_2 + HCl$$

Tetraethyl orthosilicate (TEOS). By far, most silicon dioxide is deposited from $Si(OC_2H_5)$ sources. The source is known as tetraethyl orthosilicate or TEOS. TEOS history goes back to the 1960s. Early systems relied on the simple pyrolysis of the TEOS in the 750°C range. Current depositions are based on the hot-wall LPCVD systems established in the 1970s, with temperatures of 400°C and above. TEOS sources used with plasma-assistance (PECVD or PETEOS) allowed deposition at temperatures below 400°C.[23] This process faces limits on conformal coverage of high-aspect ratio patterns in 0.5-μm

Figure 12.38 Step coverage. (*a*) good step coverage; (*b*) nonconformal coverage.

devices (Fig. 12.38). Step coverage is improved by the addition of ozone (O_3) to the gas stream.[24]

Another option is the reaction of silane with nitrous oxide in an argon plasma:

$$SiH_4 + 4N_2O \rightarrow SiO_2 + 4N_2 + 2H_2O$$

Doped silicon dioxide. Silicon dioxide layers are doped to improve their protective characteristics and flow properties, or for use as dopant sources. The earliest dopant used with deposited oxides was phosphorus. The phosphorus source is phosphine (PH_3) gas added to the deposition gas stream. The resultant glass is called phosphorus silicate glass or PSG. Within the glass the phosphorus is in the form of phosphorus pentoxide (P_2O_5), making the glass a dual compound or, more correctly, a binary glass.

The role of the phosphorus is threefold. The added dopant increases the moisture-barrier property of the glass. Mobile ionic contaminants become attached to the phosphorus and are prevented from traveling into the wafer surface. This action is called *gettering*. The third result is an increase of the flow characteristics (Fig. 12.39) which aid the planarization of the glass surface after a heating step in the 1000°C range. The phosphorus content is limited to about 8 weight by percent. Above this level the glass becomes hydrascopic and attracts moisture. The moisture can react with the phosphorus, form phosphoric acid, and attack underlying metal lines.

Boron is often added to the glass from a diborane (B_2H_6) source. The purpose of the boron is to also aid the flow characteristics

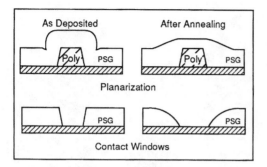

Figure 12.39 Planarization of surface by flowing glass.

(Fig. 12.39). The resultant glass is called a borosilicate glass (BSG). The boron and phosphorus are often used together in the glass. The result is referred to as BPSG (borophosphorus silicate glass).

Silicon nitride.   Silicon nitride is a replacement for silicon dioxide uses, especially for top layer protection. Silicon nitride is harder (which provides better scratch protection), is a better moisture and sodium barrier (without doping), has a higher dielectric strength, and resists oxidation. The latter property has led to its use in the local oxidation of silicon (LOCOS) for isolation purposes. Figure 12.40 illustrates the process, where patterned islands of silicon nitride prevent oxidation under the islands. After thermal oxidation and removal of the nitride, there are wafer surface regions ready for device formation, separated by isolating regions of oxide. A disadvan-

Figure 12.40 LOCOS process.

| Silane | $3SiH_4 + 4NH_3 \rightarrow Si_3N_4 + 12H_2$ |
| Dichlorosilane | $3SiCl_2H_2 + 4NHCl + 6H_2 + 6HCl$ |

Figure 12.41   Silicon nitride deposition reactions.

tage of silicon nitride is that it does not flow as easily as silicon oxide and is more difficult to etch. The etch restriction has been overcome with the development of plasma etch processes.

An early limit on the use of silicon nitride protective films was lack of a low-temperature deposition process. In APCVD systems, a temperature of 700 to 900°C is required for the deposition of silicon nitride from silane or dichlorosilane (Fig. 12.41). The result is a film with the composition Si_3N_4. The reactions also take place in LPCVD reactors but at a temperature low enough for deposition over an aluminum metallization layer. The advent of PECVD has opened up the use of different source chemistries. One use is silane reacted with ammonia (NH_3) or nitrogen in the presence of an argon plasma.

Conductors

The traditional metal conductors are aluminum and aluminum alloys deposited by evaporation or sputtering techniques (see Chap. 13). Perhaps the deposited conductor most often used is doped polysilicon. Polysilicon is used in silicon-gate MOS transistors. The addition of doping elements enables these transistors to function as surface conductors. High-density circuits have driven the increased use of deposited refractory metals alone and in conjunction with other conductor materials. The history and use of these deposited metals are explained in the next chapter.

Key Concepts and Terms

Amorphous	MBE
APCVD	MOCVD
Barrel system	Pancake system
BPSG	PECVD
BSG	Polysilicon
CVD	PSG
Deposited Si_3N_4	Selective epitaxy
Deposited SiO_2	SOI
Epitaxial	SOS
Induction heating	VPE
LPCVD	

Review Questions

1. Sketch and name the major subsystems of a basic CVD system.
2. Describe the differences between APCVD, LPCVD, and PECVD.
3. Define an epitaxial film.
4. Why is the deposition temperature of a silicon dioxide passivation layer limited to 450°C?
5. List an advantage of a horizontal vertical-flow CVD reactor.
6. Write the reaction equation for the deposition of silicon from silicon tetrachloride.
7. Describe the differences among MBE, VPE, and MOCVD systems.
8. What wafer surface condition is necessary for the deposition of a polysilicon layer?
9. Describe the advantages of plasma-assisted depositions compared to APCVD systems.
10. Why, and with what, are deposited silicon dioxide layers doped?

References

1. J. Hayes and P. Van Zant, *CVD Today Seminar Manual, Semiconductor Services,* San Jose, CA, 1985, p. 9.
2. S. Wolf and R. Tauber, *Silicon Processing for the VLSI Era,* Lattice Press, Sunset Beach, CA, 1986, p. 165.
3. M. L. Hammond, "Epitaxial Silicon Reactor Technology: A Review," *Solid State Technology,* May 1988, p. 160.
4. Ibid.
5. J. Hayes and P. Van Zant, *CVD Today Seminar Manual,* Semiconductor Services, San Jose, CA, 1985, p. 13.
6. S. Wolf and R. Tauber, *Silicon Processing for the VLSI Era,* Lattice Press, Sunset Beach, CA, 1986, p. 165.
7. B. Meyerson, H. Kaiser, and S. Schultz, "Extending Silicon's Horizon through UHV/CVD," *Semiconductor International,* Cahners Publishing, March 1994, p. 73.
8. R. Williams, *Gallium Arsenide Processing Techniques,* Artech House, Dedham, MA, 1984, p. 44.
9. S. Wolf and R. Tauber, *Silicon Processing for the VLSI Era,* Lattice Press, Sunset Beach, CA, 1986, p. 157.
10. P. Burggraaf, "The Growing Importance of MOCVD," *Semiconductor International,* November 1986, p. 47.
11. P. Burggraaf, "The Status of MOCVD Technology," *Semiconductor International,* Cahners Publishing, July 1993, p. 81.
12. P. Singer, "Molecular Beam Epitaxy," *Semiconductor International,* October 1986, p. 42.
13. P. Burggraaf, "The Growing Importance of MOCVD," *Semiconductor International,* November 1986, p. 48.
14. P. Burggraaf, "The Status of MOCVD Technology," *Semiconductor International,* Cahners Publishing, July 1993, p. 81.
15. A. Thompson, R. Stall, and B. Droll, "Advances in Epitaxial Deposition Technology," *Semiconductor International,* Cahners Publishing, July 1994, p. 173.
16. M. L. Hammond, "Epitaxial Silicon Reactor Technology: a Review," *Solid State Technology,* May 1988, p. 159.

17. R. P. Roberge et al., "Gaseous Impurity Effects in Silicon Epitaxy," *Semiconductor International*, January 1988, p. 81.
18. S. M. Sze, *VLSI Technology*, McGraw-Hill, New York, 1983, p. 119.
19. M. Venkatesan and I. Beinglass, "Single-Wafer Deposition of Polycrystalline Silicon," *Solid State Technology*, Penwell Publishing, March 1993, p. 49.
20. L. Jastrzebski, "Silicon CVD for SOI: Principles and Possible Applications," *Solid State Technology*, September 1984, p. 239.
21. K. Yallup, "SOI Provides Total Dielectric Isolation," *Semiconductor International*, Cahners Publishing, July 1993, p. 134.
22. P. Singer, "Directions in Dielectrics in CMOS and DRAMs," *Semiconductor International*, Cahners Publishing, April 1994, p. 57.
23. B. L. Chin and E. P. van de Ven, "Plasma TEOS Process for Interlayer Dielectric Applications," *Solid State Technology*, April 1988, p. 119.
24. K. Maeda and S. Fisher, "CVD TEOS/O$_3$: Development History and Applications," *Solid State Technology*, Penwell Publishing, June 1993, p. 83.

13

Metallization

Overview

A device or circuit is built in three major steps of the wafer fabrication process. In the first step, the active and passive parts are fabricated in and on the wafer surface. The last step comprises the series of substeps that are used to cover the completed chip surface with a protective or passivation layer. The step in between consists of the processes that put one or more layers of conducting metal on the wafer surface and a patterning process that leaves the circuit components electrically connected. These processes are collectively called metallization. In this chapter, the materials, specifications, and methods used to complete the metallization step are presented along with other uses of metals in chip manufacturing. Vacuum pumps, used in CVD, evaporation, ion implant, and sputtering systems, are also explained at the end of the chapter.

Objectives

Upon completion of this chapter, you should be able to:

1. List the requirements of a material for use as a chip surface conductor.

2. Draw cross sections of single and multilayer metal schemes.

3. Describe the purpose and operation of a thin-film fuse.

4. Make a list of materials used in the metallization of semiconductor devices and identify their specific uses.

5. Draw and identify the parts of a vacuum evaporator.

6. Describe the principle of sputtering.

7. Draw and identify the parts of a sputtering system.

8. Describe the principle and operation of oil diffusion, turbomolecular, and cryogenic high-vacuum pumps.

Introduction

The most common and familiar use of metal films in semiconductor technology is for surface wiring. The materials, methods, and processes of "wiring" the component parts together is generally referred to as *metallization* or the metallization process. Metallization includes all of the steps in the metallization sequence described in Chap. 5.

In the MSI era, metallization was relatively straightforward (Fig. 13.1), requiring only a single-level metal process. Small holes, called *contact holes* or *contacts,* are etched through the surface layers, to the device or circuit component parts. Following contact masking, a thin layer (10,000 to 15,000 Å) of the conducting metal is deposited by vacuum evaporation, sputtering, or CVD techniques over the entire wafer. The unwanted portions of this layer are removed by a conventional photomasking and etch procedure or by lift-off. This step leaves the surface covered with thin lines of the metal that are called *leads, metal lines,* or *interconnects.* Generally, a heat-treatment step called *alloying* is performed after metal patterning to ensure good electrical contact between the metal and the wafer surface.

1.
**Wafer with
Doped Regions**

2.
**Patterning:
Contact
Mask**

3.
**Layering: Conducting
Layer**

4.
**Patterning: Metal
Mask**

Figure 13.1 Metallization sequence.

In the role of surface conductor, a metal system must meet the following criteria:

- Good electrical current-carrying capability (current density)
- Good adhesion to the top surface of the wafer (usually SiO_2)
- Ease of patterning
- Good electrical contact with the wafer material
- High purity
- Corrosion resistance
- Long-term stability
- Deposition capability in uniform void-and hillock-free films
- Uniform grain structure

Conductor Metals

Aluminum

Prior to the development of VLSI-level circuits, the primary metallization material was pure aluminum. The reasons for this primacy and aluminum's limitations are instructive to the understanding of metallization systems in general. From an electrical conduction standpoint, aluminum is less conductive than copper or gold. Some early metallization schemes used gold. But its high contact resistance with silicon required a platinum interlayer and a top layer of molybdenum to overcome gold's softness. Copper also has a high contact resistance with silicon and raises havoc with device performance if it gets into the device areas. Aluminum emerged as the preferred metal because it avoids the problems just mentioned. It has a low-enough resistivity (2.7 $\mu\Omega$ cm),[1] and good current-carrying density. It has superior adhesion to silicon dioxide, is available in high purity, has a naturally low contact resistance with silicon, and is relatively easy to pattern with conventional photoresist processes. Aluminum sources are purchased at 5 to 6 "nines" of purity (99.999 to 99.9999 percent).

Today's metallization processes have evolved from the simple one-level, pure aluminum process. All of the changes have been driven by increased circuit density, increased number of surface layers, and miniaturization (scaling) of individual components.

Aluminum-silicon alloys

Shallow junctions in the wafer surface presented one of the first problems with the use of pure aluminum leads. The problem has to do with the need to bake aluminum-silicon interfaces to stabilize the electrical contact. This type of contact is called *ohmic* because the voltage-current characteristics behave according to Ohm's law. Unfortunately, aluminum and silicon dissolve into each other and at 577°C there exists a eutectic formation point. A eutectic formation occurs when two materials heated in contact with each other melt at temperatures much lower than their individual melting temperatures. Eutectic formations occur over a temperature range, and the aluminum-silicon eutectic starts to form at about 450°C, the temperature necessary for good electrical contact. The problem is that the alloy formation can melt down into the silicon wafer. If the surface has shallow junctions, the alloy region can extend completely through the junction, shorting it out (Fig. 13.2).

Two solutions to this problem are employed. One is a barrier metal layer (see "Barrier metals") that separates the aluminum and silicon and prevents the eutectic alloy from forming. The second is an alloy of aluminum with 1 to 2 percent silicon. During the contact-heating step, the aluminum alloys more with the silicon in the alloy and less with the silicon from the wafer. This process is not 100 percent effective and some alloying between the aluminum and wafer always occurs.

Aluminum-copper alloys

Aluminum suffers a problem called *electromigration*. The problem occurs when long skinny leads of aluminum are carrying high currents

Excessive Alloy

Aluminum / Silicon
    melted into Wafer

Aluminum with Si

Barrier Metals

Figure 13.2 Eutectic alloying of aluminum and silicon contacts.

over long distances, as is the situation on VLSI and ULSI circuits. The current sets up an electric field in the lead that decreases from the input side to the output. Also, heat generated by the flowing current sets up a thermal gradient along the lead. The aluminum in the lead becomes mobile and diffuses within itself along the direction of the two gradients. The first effect is a thinning of the lead. In the extreme the lead can become completely separated. Unfortunately, this event usually takes place after the circuit is in operation in the field, causing a failure of the chip. Prevention or moderation of electromigration is achieved by depositing an alloy of aluminum and 0.5 to 4 percent[2] copper or an alloy of aluminum and 0.1 to 0.5 percent titanium. Aluminum alloys containing both copper and silicon are often deposited on the wafer to resolve both alloying and electromigration problems.

Drawbacks of aluminum alloys are increased complexity for the deposition equipment and process, different etch rates, and an increase in film resistivity compared to the pure aluminum. The amount of the increase varies with the alloy composition and heat treatments, but can be as much as 25 to 30 percent.[3]

Barrier metals

A method of preventing the eutectic alloying of silicon and aluminum metallization is by a barrier layer. Both titanium-tungsten (TiW) and titanium nitride (TiN) layers are used. TiW is sputter-deposited onto the wafer and into the open contacts before the aluminum or aluminum alloy deposition takes place. The TiW deposited on the field oxide is removed from the surface during the aluminum etch step. Sometimes a first layer of platinum silicide is formed on the exposed silicon before the TiW is deposited.

Titanium nitride layers can be placed on the wafer by any of the deposition techniques: evaporation, sputtering, or CVD. It can also be formed by the thermal nitridation of a titanium layer at temperatures greater than 600°C in an N_2 or NH_3 atmosphere.[4] CVD titanium nitride layers have good step coverage and can fill submicron contacts. A layer of titanium is required under TiN films to provide a high-conductivity intermediate layer with silicon substrates.

Refractory metals and refractory metal silicides

Although the limitations of electromigration and eutectic alloying have been made manageable by aluminum alloys and barrier metals, the issue of contact resistance may prove to be the final limit on aluminum metallization. The overall effectiveness of a metal system is governed by the resistivity, length, thickness, and total contact resistance of *all* the metal-wafer interconnects. In a simple aluminum system there are two contacts: the silicon–aluminum interconnect and the aluminum interconnect–bonding wire. The number of connections becomes very large in an ULSI circuit with multilevel metal layers, barrier layers, plug fills, polysilicon gates and conductors, and other intermediate conductive layers. The cumulative effect of all the individual contact resistances can dominate the conductivity of the metal system.

Contact resistance is influenced by the materials, the substrate doping, and the contact dimensions. The smaller the contact size, the higher the resistance. Unfortunately, ULSI chips have smaller contact openings, and large gate array chip surfaces can be as much as 80 percent contact area.[5] These two factors make the contact resistance the dominant factor in ULSI metal system performance. Aluminum-silicon contact resistance, along with the alloying problem, have led to the investigation of other metals for VLSI and ULSI metallization. Polysilicon has a lower contact resistance than aluminum and is in use in MOS circuits (Fig. 13.3).

Refractory metals and their silicides offer lower contact resistance. The refractory metals of interest are titanium (Ti), tungsten (W), tantalum (Ta), and molybdenum (Mo). Their silicides form when they are alloyed on a silicon surface ($TiSi_2$, WSi_2, $TaSi_2$, and $MoSi_2$). The refractory metals were first proposed for metallization in the 1950s, but they stayed in the background due to a lack of a reliable deposition method. That situation has changed with the development of LPCVD and sputtering processes.

All modern circuit designs, especially MOS circuits, use refractory metals or their silicides as intermediate, barrier, or conducting layers. The lower resistivities and lower contact resistances (Fig. 13.4) make them attractive for conducting films, but impurities and deposition

Metallization
and Metal Mask

Figure 13.3 Silicon gate electrode extended for metallization lead.

Figure 13.4 Effect of contact resistance on RC time constant.

uniformity problems make them less attractive for MOS gate electrodes. The solution to the problem has been the polyside and salicide gate structures, which are combinations of a silicon gate topped by a silicide. The details of this structure are explained in Chap. 16.

A popular use of refractory metals is the filling of via holes in multilevel metal structures. The process is called *plug filling* and the filled via is called a *plug* (Fig. 13.5). The vias are filled by either selective tungsten deposition through surface holes onto the first layer metal or by CVD techniques.[6] Of the available refractory metals, tungsten finds a lot of use in aluminum-silicon barriers, MOS gate interconnects, and via plugs.

Doped polysilicon

The advent of silicon-gate MOS technology made the use of deposited polysilicon lines on the chip a natural consideration for conduction leads. For use as a conductor, the polysilicon has to be doped to increase its conductivity. Generally, the preferred dopant is phosphorus due to its high solid solubility in silicon. Doping is by either diffusion, ion implantation, or in situ doping during an LPCVD process. Each of the methods produces a different doping result. The differences relate to the doping temperature's effect on the grain structure. The lower the temperature, the greater the amount of dopant trapped in the polygrain structure, where it is unavailable for conduction. This is the situation with ion implantation. Diffusion doping results in the lowest film sheet resistivity. In situ CVD doping has the lowest dopant carrier mobility due to grain boundary trapping.

Doped polysilicon has the advantage of a good ohmic contact with the wafer silicon and can be oxidized to form an insulating layer. Polysilicon oxides are of a lower quality than thermal oxides grown on single-crystal silicon because of the nonuniformity of the oxide grown on the rougher polysilicon surface.

While polysilicon has a low contract resistance with silicon, it still exhibits too high a resistance to the metal material(s). Creating a multimetal stack of the polysilicon and a silicide (such as titanium silicide) is one solution. These stacks are called polycide structures (see Chap. 16).

Metal film stacks

Advanced circuit designs are requiring the use of metal film *sandwiches* or *stacks*. These stacks are only one layer thick, but contain two or more materials that are needed to make the system work. One example is a platinum-gold-molybdenum conduction system. Early device construction used some gold leads, but gold makes a poor direct electrical contact with silicon. Solution of this problem led to the use of a platinum layer between the gold and silicon. The platinum is heated to form a platinum silicide (platinum and silicon) that provides the necessary intermediate layer. The platinum silicide also prevents gold from getting into the devices, because gold, in many silicon devices, causes leakage currents that will destroy the device functioning. The platinum silicide also acts as an adhesion layer since gold does not readily stick to silicon dioxide. Another problem with gold is its softness; gold requires an overlayer of molybdenum for protection. Hence, the platinum-gold-molybdenum stack was created.

Multilevel metal schemes

The processes outlined in the previous sections are for single-layer metal systems. Increasing chip density has placed more components on the wafer surface, which in turn has *decreased* the area available for surface wiring. The answer to this dilemma has been dual- and triple-level metallization schemes (Fig. 13.5). The multilevel schemes

:: Via Plugs   M1 = First Metal   M2 = Second Metal

Figure 13.5 Cross section of typical planarized two-level metal VLSI structure showing range of via depths after planarization. (*Courtesy of Solid State Technology*)

start with a standard metallization process which leaves the surface components partially wired together. Next comes a layer of some dielectric material, called an *intermetallic dielectric layer* (IDL or IMD). This dielectric may be a deposited oxide, silicon nitride, or a polyimide film. This layer receives a masking step that etches new contact holes, called *vias* or *plugs,* down to the first-level metal. Normally this IMD layer is also a planarizing layer. As such it is generally thick, making plug filling a challenge. Tungsten (W) has become the plug filler of choice. This sequence is repeated until the desired number of layers are in place.

Multilevel metal systems are simple in concept but present one of semiconductor technology's biggest challenges. A circuit with a multilevel metal design will be dense and have a number of etched surface layers, leaving a topography of different step heights and a mix of surfaces. This type of surface requires planarization techniques (see Chap. 10) to minimize thinning of deposited conduction layers over the steps and allow precise imaging. Often the IMD will be globally planarized with a chemical-mechanical polishing process (CMP). Planarization techniques shown in Fig. 13.5 are SOG and flowed TEOS.

A multilevel metal system is more costly, of lower yield, and requires greater attention to planarization of the wafer surface and intermediate layers to create good current-carrying leads.

Copper

Faster devices and circuits require low resistance metallization. However, the longer and reduced cross sections required for larger chips increase the resistance of the metal wiring system. So far, barrier metal schemes, stacks, and refractory metals have been employed to reduce metal system resistance. Although aluminum provides a workable resistance, it is difficult to deposit in high aspect ratio using plugs. Using other metals for plug filling adds additional resistance to the circuit. Additional reductions needed for 0.25-μm and smaller devices has renewed interest in copper as a conducting metal. Copper is a better conductor than aluminum. Early research indicates that copper can fill plugs with good step coverage, is resistant to electromigration, and can be deposited at low temperatures.[7] Deposition can be by CVD, sputtering, electroless plating, and electrolytic plating. Plasma-enhanced CVD (PECVD) is emerging as an early leader. Drawbacks include etching problems, vulnerability to scratching and corrosion, and the need for barrier metals to keep the copper out of the silicon.

The slowing of circuit signals comes from the combination of the metal resistance (R) and capacitance (C) and is called the *RC constant* of the system. Research is underway on copper metallization with new low-capacitance (lower dielectric strength or low k) dielectrics, such as cobalt silicide ($CoSi_2$). Switching to a copper metal system and a low k dielectric could lower the RC constant by 400 percent.[8]

Metal Film Uses

MOS gate and capacitor electrodes

Most semiconductor devices operate as an electrical current passes through all or a part of the device. Capacitors are an exception. These devices (see Chap. 16) require two conductive layers, called *electrodes,* separated by a dielectric. In most designs the top electrode is a section of the conductor metal system.

MOS transistors are a capacitor structure and the top electrode, called a *gate,* is a critical structure in MOS circuits.

Fuses

The development of thin-film fuse technology allowed creation of the programmable read-only memory (PROM) circuit. The fuse allows field programming of data in the memory section of the chip. In this role the fuse is not a protective device, as in most electrical circuits, but is included specifically to be "blown" or disabled.

In the memory section of the chip, called the *array,* are a number of memory cells, each with a fuse between the cell and the main metallization system. The array is essentially a blank blackboard (Fig. 13.6). Information can be coded into the array in digital form (on/off) by having some of the cells operating and others not operating. Nonoperating cells can be created by blowing the fuse, thus removing

Figure 13.6 Schematic of fused memory array.

⚡ Fuse   ☐ Memory Cell

it from the circuit. The same system is used to program logic arrays. The fuse is blown by the heat generated by a high-voltage current (blowing current) passing through the narrow neck of the fuse, heating the material to the melting point and leaving it disabled (Fig. 13.7). Once the fuse is blown, the associated memory cell is permanently removed (electrically) from the circuit.

Films of nichrome, titanium-tungsten, and polysilicon are used for fuse materials. The deposited films are very thin, in the 200- to 500-Å range. Surface conditions and deposition uniformity are very important factors influencing the blowing current of the fuse. The heat generated in the neck is dependent on the current level and the amount of fuse material in the neck region. A nonplanarized surface or nonuniform film can have a great effect on the total amount of fuse material in the narrow neck and change the blowing current level.

Another fuse scheme employs a thin film of polysilicon or oxide in the contact hole. The fuse is blown when a high current is passed through the layer and destroys it by heating.

Backside plating

Gold is sometimes evaporated onto the entire back of the wafer just prior to wafer sort. The gold functions as a solder in certain packaging processes (see Chap. 18).

Deposition Methods

Metallization techniques, like other fabrication processes, have undergone improvements and evolution in response to new circuit

Blown Fuse

Figure 13.7 Thin film fuses.

requirements and new materials. The mainstay of metal deposition up to the mid-1970s was vacuum evaporation. Aluminum, gold, and the fuse metals were all deposited by this technique. The needs of depositing multimetal systems and alloys, along with the need for better step coverage, led to the introduction of sputtering as the standard deposition technique for VLSI circuit fabrication. Refractory metal use has added the third technique, CVD, to the arsenal of the metallization engineer. Some of the basic issues of the deposition of metals, by all techniques, are discussed in the following section on vacuum evaporation.

Vacuum evaporation

Vacuum evaporation is used for the deposition of metals on discrete devices and circuits of lower integration levels. It is also used for the deposition of gold to the back of a wafer for die adhesion into a package. Vacuum evaporation takes place inside an evacuated chamber (Fig. 13.8). The chamber can be a quartz bell jar or a stainless steel enclosure. Inside the chamber is a mechanism to evaporate the metal source, wafer holders, a shutter, thickness and rate monitors, and heaters. The chamber is connected to a vacuum pump(s) (see "Vacuum Pumps").

Since aluminum is the most critical of the materials evaporated, we shall focus on its deposition. The vacuum is required for a number of reasons. First is a chemical consideration. If any air (oxygen) molecules were in the chamber when the high-energy aluminum atoms were coating the wafer, they would form aluminum trioxide (Al_2O_3), a

Figure 13.8 Vacuum evaporator.

dielectric which, if incorporated into the deposited film, would compromise aluminum's role as a conductor. A second requirement for vacuum deposition is uniform coating. When the pressure is sufficiently reduced, the mean free path of the coating atoms is increased to exceed the dimensions of the chamber. This situation ensures that depositing atoms will strike the wafers before hitting each other and causing nonuniform depositions. The vacuum required for successful evaporation of aluminum is from 5×10^{-5} to 1×10^{-9} torr of pressure. (See Chap. 2 for a discussion on vacuum and pressure.)

Evaporation sources. Before describing the various methods of causing a metal to evaporate, a review of basic evaporation theory is in order. Most of us are familiar with the evaporation of a liquid from a beaker. This happens because there is sufficient energy (heat) in the liquid to cause the molecules to escape into the atmosphere. Over time some of them stay in the atmosphere; we call this evaporation. The same process of evaporation can be made to occur in solid metals. The requirement is to heat the metal to a liquid state so that the atoms or molecules evaporate into the surrounding atmosphere. Three methods are used to evaporate metals in a vacuum system. They are

- Filaments
- Electron beam
- Flash hot plate

Filament evaporation is the simplest of the three methods. It is used for noncritical evaporations such as backside gold layers. The material, in wire form, is wrapped around a coiled tungsten (or other metal that is able to withstand high temperatures) wire. A high current is passed through the tungsten wire where it heats the deposition metal to a liquid and evaporates into the chamber, coating the wafers. Another method uses a flat filament with a dimple to hold pieces of the deposition material.

Filament evaporation is not very controllable due to temperature variations along the filament. Another drawback is that contaminants in the source materials or on the filament are evaporated onto the wafers.

Alloys are difficult to deposit by this method. Each element has a different evaporation rate at a given temperature. When an alloy, such as nichrome, is evaporated, the nickel and the chromium each evaporate at different rates. The composition of the film on the wafer will be different than the composition in the starting material.

Figure 13.9 Electron gun evaporation source.

The need for evaporation control and low contamination led to the development of the electron beam evaporation source (Fig. 13.9) for aluminum. The system is called an *E-beam gun* or just *E-gun*. This evaporation source consists of a water-cooled copper crucible with a center cavity to hold the aluminum. At the side of the crucible is a high-temperature filament. A high current is passed through the filament, which, in turn, "boils" off electrons. The negative electrons are bent 180° by a magnet so that the electron beam strikes the center of the charge in the cavity. The high-energy electrons create a pool of liquid aluminum in the center of the charge. Aluminum evaporates from the pool into the chamber and deposits on the wafers in holders at the top of the chamber. The water cooling maintains the outer edges of the charge in a solid state, thus preventing contaminants from the copper crucible from evaporating. E-gun evaporation is relatively controlled for an elemental source such as aluminum or gold. It is less useful for the deposition of alloys due to the same limitation as with filament evaporation sources, that is, the different evaporation rates of different elements.

A more controlled evaporation of alloys is accomplished by including a second or third gun in the chamber (*coevaporation*). With multiple guns, the power to each is adjusted to achieve evaporation rates such that the alloy film on the wafer is the correct composition. Film thickness is controlled by shutters and by rate and thickness monitors. Closed automatic shutters in the path of the evaporating material allow the evaporation to reach a steady rate before depositing on the wafers. The shutter also allows rapid shutoff of the deposition. In-chamber monitors, located near or above the wafer holders, feed back information to the E-gun power supply which controls the evaporation rate. Rate control is important for consistent and uniform film structure.

Figure 13.10 Flash evaporation source.

The hot plate, or *flash,* system (Fig. 13.10) was developed to resolve some of the problems with the evaporation of alloys with E-gun sources. This source consists of a hot plate held at a temperature well above the melting point of the particular alloy. A thin wire of the alloy is fed automatically onto the hot plate's surface. Upon contact the tip of the wire melts and the material "flashes" into a vapor and coats the wafers in the chamber. Since all of the elements are "flashed" simultaneously, the composition of the film on the wafer is very close to the composition of the wire.

A major goal of any metal deposition system is good step coverage (Fig. 13.11). This is a challenge for vacuum evaporators because the

Good

Thin at Step

Figure 13.11 Step coverage.

Step Shadowed

source is essentially a point source. The problem comes when material from a point source is shadowed by steps on the wafer surface. The result can be that one side of the openings in the surface oxide can be too thin or have a void.

Several methods are used to ensure good step coverage. Primary is a planetary domed wafer holder (Fig. 13.12) suspended over the source. The planetary holders are designed so that when rotated at a high speed, the wafer is positioned at many angles to the source, allowing the even deposition of the film.

Quartz heaters in the chamber aid step coverage by maintaining evaporant mobility on the wafer for a short period of time. While cooling down, the atoms "fill in" the steps by a capillary action. Typical temperatures are about 400°C and cannot exceed the 450°C limit for aluminum on silicon. The third approach to step coverage is to round the edge of the contact holes by etching (called *slope etch*) in the masking area (Fig. 13.13). The rounded corners reduce the size of the "shadow," allowing easier access into the contact area. This method is not very effective for small contact holes with high aspect ratios.

Sputter deposition (PVD)

Sputter deposition (sputtering) is another old process adapted to modern semiconductor needs. It is a process first formulated in 1852 by Sir William Robert Grove.[9] Sputtering is a process that (in general) can deposit any material on any substrate. It is widely used to coat costume jewelry and put optical coatings on lenses and glasses.

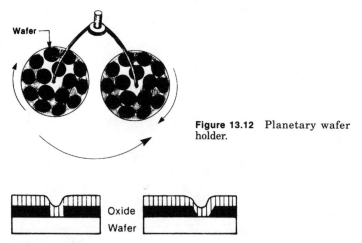

Figure 13.12   Planetary wafer holder.

Figure 13.13   Slope etch step coverage.

Oxide
Wafer

Discussion of the benefits of sputtering to the semiconductor industry is best left until the principles and methods of sputtering are covered. Sputtering, like evaporation, takes place in a vacuum. However, it is a physical, not a chemical, process (evaporation is a chemical process), and is referred to as physical vapor deposition (PVD).

Inside the vacuum chamber is a solid slab, called a *target,* of the desired film material (Fig. 13.14). The target is electrically grounded. Argon gas is introduced into the chamber and is ionized to a positive charge. The positively charged argon atoms are attracted to the grounded target and accelerate toward it. During the acceleration they gain momentum, which is force, and strike the target. At the target a phenomenon called *momentum transfer* takes place. Just as a cue ball transfers its energy to the other balls on a pool table, causing them to scatter, the argon ions strike the slab of film material, causing its atoms to scatter (Fig. 13.15). The argon atoms "knock off"

Figure 13.14 Principle of sputtering.

Figure 13.15 Typical sputtering equipment.

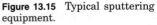

atoms and molecules from the target into the chamber. This is the sputtering activity. The sputtered atoms or molecules scatter in the chamber with some coming to rest on the wafer. A principal feature of a sputtering process is that the target material is deposited on the wafer without chemical or compositional change.

There are several advantages of sputtering over vacuum evaporation. One is the aforementioned conservation of target material composition. A direct benefit of this feature is the deposition of alloys and dielectrics. The problem of evaporating alloys was described in the preceding section. In sputtering, an aluminum and 2 percent copper target material yields an aluminum and 2 percent copper film on the wafers.

Step coverage is also improved with sputtering. Whereas evaporation proceeds from a point source, sputtering is a planar source. There is material being sputtered from every point on the target with material arriving at the wafer holder with a wide range of angles to coat the wafer surface. Step coverage is further improved by rotating the wafer holder and by heating the wafer.

Adhesion of the sputtered film to the wafer surface is also improved over evaporation processes. The higher energy of the arriving atoms makes for a better adhesion, and the plasma environment inside the chamber has a "scrubbing" action of the wafer surface that enhances adhesion. Adhesion and surface cleanliness can be increased by grounding the wafer holder and sputtering the wafer surface for a brief time prior to the deposition. In this mode the sputter system is functioning as an ion-etch (sputter-etch, reverse-sputter) machine, as described in Chap. 10.

Perhaps the greatest contribution of sputtering is the control of film characteristics available by the balancing of the sputtering parameters of pressure, deposition rate, and target material. Sandwiches of material can be sputtered in one process with multiple target arrangements.

Clean and dry argon (or neon) is required to maintain film composition characteristics, and low moisture is required to prevent unwanted oxidation of the deposited film. The chamber is loaded with the wafers and the pressure is reduced by pumps (pumped down) to the 1×10^{-9} torr range. The argon is introduced and ionized. Control of the argon amount entering the chamber is critical due to its effect of raising the pressure in the chamber. With the argon and sputtered material in the chamber, the pressure rises to a level of about 10^{-3} torr. Chamber pressure is a critical parameter in the deposition rate of the system. After liberating material from the target, the argon ions, the sputtered material, gas atoms, and electrons generated by the sputtering process form a plasma region in front of the target.

The plasma region is evident by its purple glow. The plasma region is separated from the target by a darkened region, known as the *dark space.*

Four sputtering methods are used in semiconductor applications. They are

- Diode [direct current (dc)]
- Diode [radio frequency (RF)]
- Triode
- Magnetron

The first two methods, called *diode sputtering,* are simple in concept. The target is connected to a negative potential with a positively charged anode present in the chamber. The negatively charged target ejects electrons, which accelerate toward the anode. Along the way they collide with the argon gas atoms, ionizing them. The positively ionized argon atoms then accelerate to the target, initiating the sputtering process. The ionized argon (+) and the target (−) form a diode.

A secondary effect of the ionization process is the impact of the electrons on the gas atoms, resulting in the plasma region visible as the glowing purplish region just in front of the target. Dark spaces exist just in front and to the sides of the target. Sputtering efficiency is enhanced when the plasma is confined to the region between the target and wafers. This condition is enhanced by placing "dark space" shields to the sides of the target. The shields prevent target material from being sputtered from the sides, material that will never deposit on the wafers.

Another problem arises from the out-gassing of contamination from the chamber walls while the chamber is under vacuum. This condition is called a *virtual leak,* as opposed to an actual leak of atmosphere into the system. Besides compromising the pressure level in the chamber, the contamination can be incorporated into the deposited film. This latter problem is addressed by placing a small negative bias (charge) on the wafer holder. The bias creates ions on the wafer surface and has the effect of dislodging stray out-gassed atoms from the growing film. Direct-current diode sputtering is used primarily to deposit metals.

Improved sputtering is gained by connecting the target to the negative side of a radio-frequency (RF) generator. The ionization of the gas takes place near the target surface without requiring a conductive target. Radio-frequency sputtering is necessary to sputter nonconducting materials (dielectrics) and is used also for conductors. Biasing is also used with the radio-frequency sputtering to achieve a cleaning effect at the wafer surface. Radio-frequency biasing offers the advantage of etching and cleaning of the exposed wafer surface. Etching and cleaning are achieved by putting the wafer holder at a different field potential than the argon, causing the argon atoms to impinge directly on the wafer. This procedure is called *sputter etch, reverse sputter,* or *ion milling*. The process removes contamination and a small layer from the wafer. Removal of contamination improves electrical contact between the exposed wafer regions and the film and improves adhesion of the film to the rest of the wafer surface.

In diode sputtering, a number of processes occur at or near the wafer surface. Upon impact of the argon atom, a number of electrons are created. These electrons cause heating of the substrate (as high as 350°C), which in turn can cause uneven film deposition. The electrons also create a radiation environment that can damage sensitive devices.

The heating produced with diode sputtering causes a serious problem with the deposition of aluminum. The heating causes residual oxygen in the target and in the chamber to combine with the aluminum to form aluminum oxide. The aluminum oxide is a dielectric and can compromise the conductive property of the deposited aluminum. More serious, a layer of aluminum oxide can form on the target surface and the impinging argon atoms do not (in diode sputtering) have enough energy to break through the layer. In effect, the target becomes sealed and the sputtering stops.

Triode sputtering avoids some of the problems of diode sputtering. The electrons necessary to ionize the argon are created by a separate high-current filament. In designs where the filament is located outside the deposition chamber, the wafers are protected from radiation damage. Films deposited by a triode method are more dense.

Another problem with diode sputtering is the electrons that escape into the chamber and do not contribute to the establishment of the plasma necessary for deposition. The situation is resolved in *magnetron* sputtering systems, which use magnets behind and around the target (Fig. 13.16). The magnets capture and/or confine the electrons to the front of the target. Magnetron systems are more efficient for increased deposition rates. The resulting ion current (density of ionized argon atoms hitting the target) is increased by an order of magnitude over conventional diode sputtering systems. Another effect is a lower pressure required in the chamber, which contributes to a cleaner deposited film. Magnetron sputtering leaves a lower target temperature, which makes it a favorite for the sputtering of aluminum and aluminum alloys.

Magnetic Field   Target (cathode)

AR⁺

Plasma Region
(Glow)

Wafers   Sputtered Atoms

Figure 13.16   Magnetron sputtering.

Production-level sputtering systems come in a variety of designs. Chambers are either batch systems or single-wafer in-line designs. Most production machines have load-lock capabilities. A load lock is an antechamber where a partial vacuum is created so that the deposition chamber can be maintained at vacuum. The advantage of a load lock is a higher production rate. Production machines are usually dedicated to one or two target materials, while development machines have a wider range of capability.

CVD metallization

Advances in LPCVD offer the deposition department a third choice for metal depositions. LPCVD offers the advantages of not requiring expensive and maintenance-intensive high-vacuum pumps, conformal step coverage, and high production rates. Perhaps the most often deposited CVD refractory metal film is tungsten (W).

Tungsten is used in a variety of structures, including contact barriers, MOS gate interconnects, and via plugs. The filling of via holes is a key to effective multimetal systems (see Fig. 13.5). The dielectric layer is relatively thick and the via holes have to be relatively thin (i.e., a high aspect ratio). These two factors make for a difficult continuous metal deposition to fill the vias without thinning of the metal in the via. Selective CVD-deposited tungsten plugs fill the entire via and present a planar surface for a subsequent conducting metal layer. For use as a barrier metal, tungsten can be deposited selectively by the silicon reduction of the gas tungsten hexafluoride (WF_6) by the reaction

$$2WF_6 + 3Si \rightarrow 2W + 3SiF_4$$

Tungsten can also be deposited selectively over aluminum and other materials from WF_6. The processes are called *substrate reduction*. Tungsten is also deposited from WF_6 by hydrogen reduction; the reaction is

$$WF_6 + 3H_2 \rightarrow W + 6HF$$

All of the depositions are performed in LPCVD systems at temperatures of about 300°C, which makes the processes compatible with aluminum metallization.

The depositions of tungsten silicide and titanium silicide proceed by the reactions

$$WF_6 + 2SiH_4 \rightarrow WSi_2 + 6HF + H_2$$

$$TiCl_4 + 2SiH_4 \rightarrow TiSi_2 + 4HCI + 2H_2$$

Vacuum Pumps

LPCVD, ion implantation, evaporation, and sputtering processes all take place in reduced-pressure (vacuum) chambers (see Chap. 2). Vacuum chambers provide process conditions free of contaminating gases. In the deposition processes, the vacuum increases the mean free path of the depositing atoms and molecules, which in turn results in more uniform and controllable deposited films. LPCVD takes place in the pressure range down to 10^{-3} torr (medium range) while the other processes take place at pressure ranges down to 10^{-9} torr (high to ultrahigh range). Medium range is reached with mechanical vacuum pumps. These same pumps are used to initially reduce the pressure in the high-vacuum process chambers. In this role they are called *roughing pumps*. Additionally, mechanical vacuum pumps are used on the outlet end of high-vacuum pumping systems to assist in the removal of gas molecules from the pump to the exhaust system.

After the rough vacuum is established, a high-vacuum pump takes over to establish the final vacuum. The pumps used for this purpose are *oil diffusion, cryogenic, ion,* or *turbomolecular.* Whatever the type, all pumps are constructed of materials that will not *outgas* into the system and compromise the vacuum. Materials used are typically type 304 stainless steel, oxygen-free high-conductivity copper (OFHC), Kovar, nickel, titanium, borosilicate glasses, ceramics, tungsten, gold, and some low-vapor-pressure elastomers.[10] Pumps used to evacuate corrosive and toxic gases or reaction by-products must have corrosion-free inside surfaces. Also, care must be taken in servicing pumps with these types of applications.

Pumps are selected and used based on a number of criteria, including

- Vacuum range required
- Gases to be pumped (lighter gases like hydrogen are more difficult to pump)
- Pumping speed
- Overall throughput
- Ability to handle impulsive loads (periodic outgassing)
- Ability to pump corrosive gases
- Service and maintenance requirements
- Downtime
- Cost

Recall from Chap. 2 that pressure in a system results from the activity of gas atoms or molecules in an enclosure striking the chamber walls with some force. Reduction of the pressure in a system requires the removal of the gas in the chamber. This is generally accomplished by the pump establishing a lower pressure, first within itself, which allows gas material in the process chamber to flow into the pump where it is removed entirely from the system. At very low pressures, there is not much material in the chamber and continued pressure reduction requires that the system be leak-free and not add to the pressure by its own outgassing. Some systems require traps to prevent material from the pump from *back-streaming* into the chamber. Cold traps are explained in the section on oil diffusion pumps.

Mechanical pumps

Mechanical oil rotary vacuum pumps have a basic design that can be traced back to the 1640s when Galileo and Torricelli were investigating the theory that air had weight. They are called *blower* type pumps. Air is removed from the pump by an eccentric rotating vane in a cavity (Fig. 13.17). As the vane rotates, it compresses and sweeps out the gas in front of it in the cavity, simultaneously leaving a reduced-pressure region behind it. The "pushed" material exits through a valve, while another valve opens to the cavity from the chamber and allows material in the chamber to flow into the cavity. As the vane rotates, more and more material is removed from the cavity, thus reducing the pressure in the chamber.

A critical part of the pump is the efficiency of the exit valve. An ineffective valve that leaks atmosphere back into the system will limit the

Oil Immersion
Outlet Valve

Inlet

Vane

Cavity

Figure 13.17 Mechanical rotary oil pump.

ultimate pressure level of the pump. This style of pump uses an oil-immersed valve that prevents back-streaming of the atmosphere.

Oil-filled pumps met their match with the advent of LPCVD and etch application systems. These processes produce a lot of physical residues that contaminate the oil quickly. *Oil-sealed, oilless, scroll, and roots* (called roots blowers) are all push type pumps that have replaced oil-mechanical pumps.[11]

Oil diffusion pumps

The oil diffusion pump (Fig. 13.18) was the mainstay of most semiconductor vacuum processes. The pump requires the services of an oil mechanical pump to first reduce the pressure in the chamber to the 10^{-3}-torr level. Either the same or a second mechanical pump is required at the outlet end. High vacuum, in the 10^{-8}-torr range (under production conditions), is achieved by a clever momentum transfer system. A low-vapor-pressure, hydrocarbon-based oil is heated in the base of the pump, where it rises up a structure called the *stack*. At the top of the stack is a series of downward-facing baffles. The hot oil molecules, which have gained speed and energy from the boiling, exit the stack in a downward direction. Outside the stack, they collide with gas from the chamber, causing them to be propelled toward the bottom of the pump where they are removed by a mechanical pump. The oil molecules return to the heated reservoir.

Two problems with oil diffusion pumps are the migration of oil molecules back up into the chamber and the inability of the pump to handle water vapor from the chamber. Both problems are resolved by the use of a *cold trap* between the pump and the chamber. A cold trap is similar in design to a liquid source bubbler. The fluid in the trap is liquid nitrogen, which reduces the temperature to $-196°C$. At this temperature, oil, contaminants, and water vapor molecules are frozen

Gas Molecules (Uncompressed)

To cold trap
and vacuum system.

Umbrella

coolant
(Air or Water)

Gas
Molecules
(Compressed)

To
Mechanical
Force Pump

Jet Stack

Pump Oil

Electric Heater

Figure 13.18 Oil diffusion pump.

to the inside walls of the trap and do not add to the pressure in the system.

Cryogenic pumps

Even with cold-trap technology, some processes cannot stand contamination from hydrocarbon oils such as are used in oil diffusion pumps. This situation has led to the use of cryogenic (cryo) pumps. A cryogenic pump (Fig. 13.19) uses the fact that gas molecules will "freeze" out on cold surfaces. The cold trap described in the preceding section and the frost that collects on the insides of refrigerators are examples of cryogenic activity.

Cryogenic pumps are designed with a central finned stack. At the low temperatures, gas from the chamber collects on the vanes, removing material from the system, which, in turn, reduces the pressure. The central stack, called an *expander,* is cooled as a compressor releases the liquid helium or nitrogen into it from the bottom. Cooling is by a phenomenon called adiabatic expansion. This is the same phenomenon that causes a pressurized can to cool when the nozzle is opened and the gas expands rapidly into the lower-pressure atmosphere. The top of the expander is at a higher temperature than the

To Chamber

First
Stage

Second
Stages

HC
or H$_2$
Compressor

Figure 13.19 Cryogenic pump.

bottom, resulting in different gas molecules freezing out at different levels. Cryogenic pumps can efficiently pump water vapor which turbomolecular pumps cannot. (See "Turbomolecular Pumps" below.)

Cryopumps operate without cold traps or mechanical roughing pumps. Due to their action of trapping gas material on the fins, they are of a type known as *capture* pumps rather than the displacement-type pumps that move the chamber material to the atmosphere. The total capture nature of the pump and its oilless operation drastically reduce the possibility of contamination. However, when the pump is brought to room temperature, by mistake or for maintenance, the frozen gases are released and care must be exercised to vent any toxic or flammable gases trapped on the vanes. The buildup of gases also affects the pumping speed. The pump's speed must be monitored and the system cleaned when the speed falls. Cryopumps are simpler to operate and maintain, since no cold-trap liquids or messy oils are needed. Additionally, cryopumps can handle bursts of outgassing from the process chamber and feature a fast pumping speed.

Ion pumps

Another capture type pump is the ion pump, also known as a *sputter ion pump* or a *getter pump* (Fig. 13.20). An ion pump operates in a manner similar to an ionization section in an ion implanter or sputter machine; only in this application, the atoms and molecules come from the chamber. A portion of those that drift into the ionization chamber are ionized to a positive charge by bombardment with electrons and attracted to a titanium cathode (negative potential). Upon collision with the titanium, some of the titanium is sputtered away and travels into the pump. The titanium atoms are chemically active enough to combine with other gases in the pump, which also accumulate on the

Figure 13.20 Principle of ion vacuum pump.

pump walls. Again, material is removed from the chamber which reduces the pressure. Ion pumps are capable of pressures down to 10^{-11} torr, which is the ultrahigh vacuum range.

Turbomolecular pumps

Turbomolecular pumps are similar in design to a jet turbine engine. A series of blades (Fig. 13.21) with openings are mounted and rotated at very high speeds (24,000 to 36,000 rpm[12]) on a central shaft. Gas molecules from the chamber encounter the first blade and gain momentum from the collision with the rotating blade. The momentum direction is downward to the next blade, where the same thing happens. The net result is a removal of gas from the chamber. The use of a

Figure 13.21 Turbomolecular pump.

Metal	Use			Deposition Method		
	Conductor	Fuse	Other	Evaporation	Sputtering	CVD
Aluminum	X			X	X	
Aluminum/Silicon	X			X	X	
Aluminum/Copper	X			X	X	
Aluminum/Copper/Silicon	X			X	X	
Gold/Molybdenum/Platinum	X			X	X	
Titanium/Platinum/Gold	X			X	X	
Titanium/Tungsten		X	Barrier		X	
Nichrome		X		X		
Gold			Die Attach/ Lifetime Control	X		
Doped Silicon	X				X	X
Molybdenum	X				X	X
Platinum Silicide	X		Barrier		X	X
Titanium Silicide	X				X	X
Tungsten Silicide	X				X	X

Figure 13.22 Metallization overview.

momentum transfer makes the pumping principle the same as an oil diffusion pump.

Major advantages of turbomolecular pumps are no back-streaming from oils, no need to recharge, high reliability, and pressure reduction into the high-vacuum range. Drawbacks are a slower pumping speed compared to oil diffusion and cryogenic pumps and vibration and wear due to the high rotational speeds.

An addition to turbo pumps is a *drag* type pump. Molecules are bounced off a rotating drum or disk rather than vanes or stators.[13] These *combination* pumps can exhaust at high pressures. Use of turbo pumps with corrosive gas processes requires coating the rotors and stators and/or heating the pump to keep the gases from forming solid particles that deposit on the pump parts.

Figure 13.22 is an overview of the metals used, their uses, and deposition methods.

Key Concepts and Terms

Barrier metals	Fuses
Conducting metals	Ion pumps
Contact resistance	Mechanical oil rotary pumps
Cryogenic pumps	Oil diffusion pumps
CVD metal deposition	Refractory metal
Doped polysilicon	Refractory metal silicides
E-gun evaporation source	Sputter deposition (PVD)
Electromigration	Turbomolecular pumps
Filament evaporation source	Vacuum evaporation

Review Questions

1. Discuss the reasons aluminum metallization is used rather than gold and copper.

2. Explain electromigration and how it is controlled.

3. Explain aluminum and silicon eutectic alloying and two processes used to prevent it.

4. Why are refractory metals and their silicides favored for VLSI circuit metallization?

5. What is a silicide and how is it formed?

6. Draw a diagram of a two-metal system, identifying all parts.

7. Why would there be two or more electron guns in an evaporator?

8. Draw a diagram of and label the parts of a sputter machine.

9. What are the advantages of a sputter process compared to an evaporator process?

10. List the four types of high-vacuum pumps used in semiconductor processes and their principles of operation.

11. Discuss the RC constant effect on circuits and the techniques used to reduce it.

References

1. S. Wolf and R. Tauber, *Silicon Processing for the VLSI Era,* Lattice Press, Sunset Beach, CA, 1986, p. 332.
2. P. Riley, S. Peng, and L. Fang, "Plasma Etching of Aluminum for ULSI Circuits," *Solid State Technology,* Penwell Publishing, February 1993, p. 47.
3. S. M. Sze, *VLSI Technology,* McGraw-Hill, New York, 1983, p. 347.
4. D. Pramanikm and V. Jain, "Barrier Metals for ULSI," *Solid State Technology,* Penwell Publishing, January 1993, p. 73.
5. Y. Pauleau, "Interconnect Materials for VLSI Circuits," *Solid State Technology,* February 1987, p. 61.
6. D. M. Brown, "CMOS Contacts and Interconnects," *Semiconductor International,* 1988, p. 110.
7. P. Singer, "New Interconnect Materials: Chasing the Promise of Faster Chips," *Semiconductor International,* November 1994, p. 53.
8. Ibid, p. 54.
9. A. J. Aronson, "Fundamentals of Sputtering," *Microelectronics Manufacturing and Testing,* January 1987, p. 22.
10. J. Ballingall, "State-of-the-Art Vacuum Technology," *Microelectronics Manufacturing and Testing,* October 1987, p. 1.
11. P. Singer, "Vacuum Pump Technology Leaps Ahead," *Semiconductor International,* Cahners Publishing, September 1993, p. 52.
12. S. Wolf and R. Tauber, *Silicon Processing for the VLSI Era,* Lattice Press, Sunset Beach, CA, 1986, p. 95.
13. P. Singer, "Vacuum Pump Technology Leaps Ahead," *Semiconductor International,* Cahners Publishing, September 1993, p. 53.

14

Wafer Test and Evaluation

Overview

The wafer-fabrication process requires a high degree of precision in process control, equipment operation, and material manufacture. One mistake can render the wafer completely useless. Throughout the process, various tests and measurements are made to determine both wafer and process quality. Increased wafer volumes and more critical processes are leading to more on-line measurements. Individual tests are described in this chapter. Statistical process control programs are addressed in Chap. 15.

Objectives

Upon completion of this chapter you should be able to:

1. Explain the difference between resistance, resistivity, and sheet resistance.
2. Draw a sketch of the parts and current flow in a four-point probe.
3. Compare the principles and uses of color interference, fringe counting, spectrophotometers, ellipsometers, and stylus for film thickness measurements.
4. Compare the principles and uses of groove and stain, SEM, and spreading resistance for junction depth measurements.
5. List the methods and advantages of microscopes and SEMs in the inspection of wafer surfaces.

6. Draw sketches of diodes in forward and reverse bias and their companion current-voltage curves.

7. Explain the effect of surface current leakage on a junction performance characteristic.

8. Draw sketches of a bipolar and MOS transistor in operation and their companion current-voltage characteristics.

9. List the process steps for a capacitance-voltage measurement and the principle of contamination detection.

10. Describe the principle and use of atomic force microscopes.

Introduction

Characterization of processes and circuit parameters is required for production-line control and product stability. Good characterization can warn of a process about to go out of control and device characterization is essential to analyze circuit performance and conformance to customer specifications. Consequently, every process step has a rigid set of controlled, operational parameters (temperature, time, etc.). And after every significant process step there is an evaluation of the result on the wafer or a test wafer. *Test wafers* are blank wafers or wafer pieces that are included in the process step for postprocess measurements. Many of the tests are destructive and cannot be performed on the device wafers or the actual components in the chip. In the preceding process chapters, the important parameters for each process were identified (film thickness, resistivity, cleanliness, etc.). Here, the basic theory, applicability, and range of sensitivity of test methods are examined.

Process control and improvement involve groups of measurements, some direct some indirect. One group includes electrical measurements of test wafers and the actual devices. They measure the direct effect of some of the processes, such as ion implantation. Device performance measurements usually include several processes and the results are used to infer individual process parameter control. Another group directly measures physical parameters such as layer thickness, width, composition, and the like. This group includes defect detection. A third group measures contamination in and on the wafer.

Not surprisingly, test and measurement methods have changed along with more complex levels of integration and smaller image sizes. ULSI technology is ushering in nanometer- and angstrom-level inquiry, called the *nanoanalysis era*.[1] And the price of in-line testing is going up. Larger wafers and denser circuits require more tests to properly characterize processes. High-volume processing requires real-time testing and analysis to guard against scrapping volumes of production wafers. Measurement systems for ULSI circuits usually include on-board statistical analysis and database management capabilities.

Wafer Electrical Measurements

Resistance and resistivity

The object of the fabrication process is to form, in and on the wafer surface, solid-state electrical components (transistors, diodes, capacitors, and resistors) that are wired together to form the circuit. Each of the components must meet individual electrical performance specifications if the entire circuit is to function. Throughout the process, electrical measurements are performed to judge the process and predict electrical device performance.

Resistivity measurements

The addition of dopants to the wafer, both during crystal growth and during the doping processes, alters the electrical characteristics of the wafer. The altered parameter is its resistivity, which is a measure of a material's specific "resistance" to the flow of electrons (Fig. 14.1). Whereas the resistivity of a given material is a constant, the resistance of a specific volume of the same material is a function of its dimensions and resistivity. This relationship parallels that of density and weight. For example, the density of steel is a constant, whereas the weight of a particular piece depends on its size.

The units of resistance (R) are ohms (Ω) and the units of resistivity (ρ) are ohm-centimeters ($\Omega \cdot cm$). Since adding dopants to a wafer will

$$R = \rho \frac{L}{A} = \rho \frac{L}{WXD}$$

Figure 14.1 Relationship of resistance to resistivity and dimensions.

alter its resistivity, measurement of resistivity is actually an indirect measure of the amount of dopants added.

Four-point probe

The parameters of resistance, voltage, and current are governed by Ohm's law. The three parameters are related mathematically in the following way:

$$R = V/I = (\rho)L/A = (\rho)L/(W{\times}D)$$

where R = resistance
V = voltage
I = current
ρ = resistivity of sample
L = length of sample
A = cross-sectional area of sample
W = width of sample
D = depth of sample

Theoretically, the resistivity of a wafer can be measured with a multimeter (Fig. 14.2) by measuring the voltage at a constant current through a sample of known dimensions and calculating the resistivity. However, the resistance between the probes and the wafer material is too great to accurately measure the resistivity of semiconductors with their relatively low quantity of dopants.

Ohms

Figure 14.2 Multimeter.

The four-point probe is the instrument used to measure resistivity on wafers and crystals. It is an instrument with four thin, in-line probes, connected to a power supply and voltmeter. A four-point probe consists of four thin metal probes arrayed in a line. The two outside probes are connected to a power supply and the inside probes are connected to a voltage meter. During operation, the current passes between the outer probes and the voltage drop (change) is measured between the inner probes (Fig. 14.3). The relationship of the current and voltage values is dependent on the resistance of the space between the probes and the resistivity of the material. The four-point probe cancels out the effects of probe-wafer contact resistance on the measurement.

Using a four-point probe, the voltage and current are related to the resistivity by the relationship:

$$\rho = 2\pi \, sV/I$$

where s is the distance between probes when s is less than the wafer diameter and less than the film thickness.

Sheet resistance

The four-point probe measurement just described is used to measure the resistivity of wafers and crystals. It is also used to measure the resistivity of thin layers of dopants added into the wafer surface by the dopant processes. When a four-point probe measurement is made on a thin layer of added dopants, the current is confined in the layer (Fig. 14.3). A thin layer is defined as a layer thinner than the probe spacing (distance between probes).

The electrical quantity measured on a thin layer is called sheet resistance R_s. This quantity has the units of ohms per square ($\Omega/\square$). The concept of ohms per square can be understood by considering the resistance of two squares of the same thin material of equal thickness (Fig. 14.4). Since the resistivity of ρ is the same for each piece and

Figure 14.3 Four-point probe measurement of a thin layer.

Figure 14.4   Resistance of a "square."

$T_1 = T_2$, the sheet resistance is the same for each piece. Or, the resistance of the thin sheet is a constant for any square of the same material.

The formula relating sheet resistance to the voltage and current is

$$R_s = 4.53 \; V/I$$

where 4.53 is a constant that arises from the probe spacing. Some companies elect to drop the constant 4.53 from the formula and just measure the V/I of a wafer as in Fig. 14.5.

Four-point probe thickness measurement

The thickness of uniform conducting layers on an insulating layer can be determined using a four-point probe. For thin films the formula is

Figure 14.5   Voltage/current (V/I) versus thickness of aluminum.

$$T = \rho_s/R$$

where T = layer thickness
ρ_s = resistivity
R_s = sheet resistance

Since the resistivity is a constant for pure materials such as aluminum (Fig. 14.5), the sheet resistance measurement is actually a measurement of the film thickness. This formula does not calculate the thickness of a doped layer since the dopants are not evenly distributed throughout the layer.

Concentration and Depth Profile

The distribution of dopant atoms in the wafer (see Chap. 11) is a major influence on the electrical operation of a device. The distribution (or dopant concentration profile) is determined by several techniques, one of which is the spreading resistance. After doping, a test wafer sample is prepared by the bevel technique. After the junction is exposed by the beveling, a series of electrical two-point probe measurements are made sequentially down the bevel (Fig. 14.6). At each point, the vertical drop of the probes is recorded and a resistance

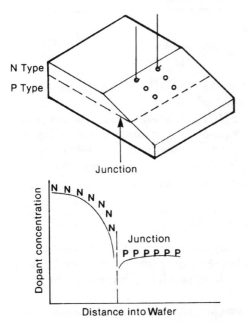

Figure 14.6 Spreading resistance.

measurement made. The resistance value at each point changes with the change in dopants at each level.

A computer is used to perform calculations that relate the depth and resistance values to the dopant concentration at each level. The computer uses the data to construct a dopant concentration profile for the sample. This measurement is usually made periodically off-line or when electrical device performance indicates that the dopant distribution may have changed.

Another method is to incrementally remove thin layers from the wafer surface with the anodic oxidation technique (See Chap. 7). After the oxide is grown, it is removed by etching and a four-point probe measurement is made on the new surface. The distance down into the wafer is a function of the oxide thickness and is related to the dopant concentration four-point probe sheet resistance values by a computer program.

Secondary ion mass spectrometry (SIMS)

SIMS is essentially a combination of the ion milling and secondary ion detection methods. Ions are directed at the sample surface, removing a thin layer. Secondary ions are generated from the removed material, which contains the wafer's material and the dopant atoms. The ions are collected and analyzed, providing a measure of the amount of dopant at each level, which in turn allows the construction of a dopant profile.[2]

Differential Hall Effect (DHE)

Like the other two profiling methods, DHE requires sequential removal of the doped layer down to the junction. As each layer is removed, the resistivity and the Hall coefficient (which relates to carrier mobility) are measured. Dopant concentration for each layer is calculated from the two measured parameters. A drawback to this method is the additional time it takes to prepare a specially shaped test structure (i.e., a van der Pauw structure).[3]

Device Electrical Measurements

During the process, it is necessary to make direct measurements of the actual device parameters. These measurements are usually made on special devices in the test die or on special structures in the scribe line areas.

A great deal of information about the process is available from these electrical tests. In this section, the basic tests and several of the more obvious and frequent device failures are explained. Often these tests are called *parametric testing,* referring to the measurement of the device electrical parameters as opposed to testing the overall function of the device or circuit, which occurs at wafer sort at the end of the fabrication process. A complete device and process troubleshooting guide is beyond the scope of this text.

Equipment

The basic equipment required to perform device electrical tests are a probe machine (Fig. 14.7) with the capability of positioning needlelike probes on the devices; a switch box to apply the correct voltage, current, and polarities to the device; and a method of displaying the results. If many devices are to be measured, the probes may be fixed on printed circuit-board-like structures called *probe cards.* These are the same cards used in the wafer sort process. Display may be by a curve tracer or by a special digital voltmeter. A curve tracer is a special-design oscilloscope set up to display voltage on the horizontal scale and current on the vertical scale. It shows the characteristic voltage and current traces for each type of measurement. Example displays are shown in the following sections for each test. In high-production situations, the tests are performed digitally and the results displayed on a readout. These measurement systems have the capability of storing and correlating data.

In advanced systems, the probe station may be automated to sequentially test several die. The switching can also be automated to allow the equipment to perform a series of tests in a predetermined sequence. Automated systems also include hard-copy printouts of the test results as well as analysis of the data. The individual tests will be explained, as performed on manual equipment. The shape and

Wafer

Figure 14.7 Device measurement equipment.

relationship of the traces displayed on the oscilloscope are very helpful for understanding device operation.

All the device measurements are made in basically the same way. A
voltage is applied to the component contact probe and the resultant
current flowing between the contacts is measured. The results are
displayed on the oscilloscope screen or display screen. The shape
of the trace is governed by the dimensions of the device, the doping,
and the presence of junctions. The current is modified by the resistance of the component or the presence of junctions. Properly working
components will display known traces. The fundamental relationship
is that of Ohm's law:

$$R = V/I$$

Resistors

Resistor measurements are made by contacting each end of the resistor and applying a voltage. Current passage between the probes will
be governed by the nature of the material or, in electrical terms, its
resistivity. During the test the applied voltage is varied from zero to
some higher value. On the screen the voltage value is displayed on
the x axis, with current measured on the y axis (Fig. 14.8).

Screen
Trace

V

Oscilloscope
Screen

Probed Resistor

Figure 14.8 Resistor.

The value of the resistance is calculated by dividing one of the voltage values by the corresponding current value. One might ask, "Why not determine the resistance by simply measuring the voltage and current values with a meter?" In other words, why display the values on the screen? The answer lies in the quality of information gained from the trace. A resistor's *V/I* relationship should be a linear one (straight line). Any deviation from linearity could indicate a process problem, such as high contact resistance or a leaking junction.

Diodes

Diodes function as switches in a circuit. This means that a diode can pass current in one direction (forward bias) and not in the other (reverse bias). Checking diode operation in either direction requires measuring the diode with the proper polarities, as shown.

As the voltage is increased in the forward direction, current immediately starts flowing across the junction and out of the diode (Fig. 14.9). The initial resistance to that flow comes from contact resistance and a small resistance at the junction. After the resistances are overcome, there occurs a "full" flow of current through the diode. A diode is designed to have this condition occur at some minimum voltage value. If the diode forward voltage value exceeds the design value, it is out of specification. This voltage is called the *forward voltage.*

In the reverse direction, the diode is designed to block current flow as long as the voltage stays below a specified value. In the reverse condition, a small current, called a *leakage current,* always flows across the junction (Fig. 14.10). Increasing the voltage increases the leakage current to a level where the junction breaks down, allowing "full" current flow. The value of the voltage where full flow begins is called the *breakdown voltage.* Circuits are designed to operate at a voltage level below the designed breakdown voltage of the diode to take advantage of the blocking nature of the junction. Lowered break-

Figure 14.9 Diode forward bias.

Figure 14.10 Diode reverse bias measurement.

down voltages are generally the result of processing mistakes or excessive contamination.

Exceeding the breakdown voltage does not usually permanently damage the junction. However, if the applied voltage is extremely high the diode (junction) can sustain permanent physical damage from a high current flow.

A second parameter determined during this test is the current at the breakdown voltage. A small amount normally occurs as described previously. Contamination and/or improper processing can result in additional leakage current.

Trace 1 in Fig. 14.11 shows a diode with a small amount of leakage. The amount of current increases as the voltage is increased. Eventually, the breakdown voltage is reached and the diode becomes fully conducting. In trace 2, gross leakage is demonstrated. The junction leaks current with every increase in voltage, and the problem is so severe that a breakdown level is never reached and the diode never operates as a current block.

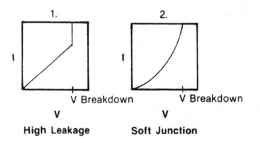

Figure 14.11 Junction leakage.

Oxide rupture (BV_{ox} or rupture voltage)

An electrical measurement, BV_{ox}, is used as a measure of oxide quality. The test structure used is the same as for capacitance-voltage (C/V) analysis. But in this case, the voltage is continually increased until the oxide is physically destroyed and current flows freely from the aluminum dot to the silicon. The maximum voltage that the oxide can withstand before breakdown is a function of its thickness, structural quality, and purity.

Bipolar transistors

Bipolar transistors, as explained in Chap. 16, are three-region, two-junction devices. Electrically, they can be thought of as two diodes back to back. Many tests are performed to characterize bipolar transistors. The individual junctions are characterized separately and the whole transistor operation measured. The junctions are probed for forward and reverse characteristics. The breakdown voltage (BV) tests are followed by the probing of the entire transistor.

Individual junction probe measurements are designated by the letters BV followed by subscripted lowercase letters that indicate the particular junction being probed. BV_{cbo} indicates the breakdown voltage measured between the collector and base regions. The "o" indicates that the emitter region is "open"—it has no voltage applied to it. BV_{ceo} is the breakdown measured between the collector and emitter.

The forward voltages of the two bipolar structure junctions are also measured. V_{be} is the forward voltage of the emitter-base junction and V_{bc} is the forward voltage of the base-collector junction. BV_{iso} probes the collector-isolation junction for leakage currents.

A principal electrical measurement of a bipolar transistor is the beta measurement (Fig. 14.12). This is a measurement of the amplification characteristic of the transistor. In a bipolar transistor, the current flows from the emitter to the collector, through the base (see Chap. 16). The base current is varied to change the resistance in the base region. The amount of current flowing out of the collector (from the emitter to base) is regulated by the base resistance.

Figure 14.12 NPN transistor beta measurement.

The amplification of the transistor is defined as the collector current divided by the base current. This number is known as *beta*. Thus a beta of 10 means that a 1-mA (milliampere) base current will give rise to a 10-mA collector current. The beta of a transistor is determined by junction depths, junction separations (base width), doping levels, concentration profiles, and a host of other process and design factors. Measurement of beta is done by a variation of the BV_{ceo} measurement. A BV_{ceo} measurement at a specific base current is performed. In this mode, the emitter base junction is forward-biased.

The collector characteristic of the transistor is displayed on the oscilloscope screen. The almost horizontal lines represent increasing base current values (IB_1, IB_2, etc.). With every increase in base current, a corresponding increase in collector current occurs. Calculation of the beta value takes place from the data displayed on the screen. Collector current is determined from the vertical axis (dotted line). The base current is calculated by multiplying the number of horizontal lines (steps) by the scale value for each step (from the oscilloscope).

MOS transistors

MOS circuits are also made up of resistors, diodes, capacitors, and transistors. The first three are measured by the same methods used to measure bipolar circuit components. Like the bipolar transistor, the MOS transistor is composed of three regions, in this case called the source, gate, and drain (see Fig. 14.13 and Chap. 16). Measurement of this type of transistor consists of determining the reverse and forward values of the source and drain junctions. The functioning of the gate is determined by the threshold voltage test.

In a MOS transistor, the source region is forward-biased. Because of the high resistivity of the gate region, the forward current does not reach the drain. A voltage applied to the gate at a specified level (threshold) will cause enough charges to appear in the gate region to form a conducting channel that allows the source current to reach the

Figure 14.13 Threshold voltage measurement.

drain region. Every MOS transistor is designed to operate at a specific threshold voltage. This value is measured using the capacitance-voltage technique. The gate voltage is continuously increased, while the capacitance of the gate structure is monitored.

A capacitor is a storage device. Initially, during the voltage-increase portion of the measurement, the capacitance does not change. At the threshold voltage level, the inversion layer forms and acts like a capacitor. Since two in-series capacitors have a combined lower capacitance than the sum of the two, the result is a drop to a combined lower capacitance. MOS transistors also exhibit amplification characteristics. The gain is defined as the source-drain current divided by the gate current. The source-drain characteristic for various gate currents is shown (Fig. 14.14).

Capacitance-voltage plotting

A variation of the threshold voltage test is used to test for the presence of mobile ionic contamination in the oxide. The test is performed on specially prepared test wafers. A thin oxide is grown on a "clean" silicon wafer. After oxide growth, aluminum dots are formed on the wafer by evaporation through a mask (Fig. 14.15). Dot evaporation is usually followed by an alloy step to ensure good electrical contact between the aluminum and oxide.

The "dotted" wafer is placed on a chuck and a probe is placed on the aluminum dot. This structure is actually a MOS capacitor. A voltage

$$\text{Gain} = \frac{\Delta I_{DS}}{\Delta V_{GS}}$$

Figure 14.14 Gain characteristic of a MOS transistor.

Wafer Oxide Growth Aluminum Dot Alloy

Figure 14.15 Preparation of C/V test wafer.

is applied to the dot and gradually increased as the capacitance of the structure is simultaneously measured. The results are printed out on an *x-y* plotter with capacitance on the *y* axis and voltage on the *x* axis (Fig. 14.16).

At a voltage level known as the *threshold voltage* (or inversion voltage) charge starts to build up at the silicon surface. The charges "invert" the conductivity type from N-type to P-type. The inverted layer has a capacitance of its own. Electrically, the structure now has two capacitors in series. The total capacitance value of the two is less than the sum of the two by the relationship

$$1/C_{\text{total}} = 1/C_{\text{oxide layer}} + 1/C_{\text{inversion layer}}$$

The trace on the *x-y* plotter drops vertically to the new capacitance level.

The second step in the process is to force the mobile positive ions in the oxide to the SiO$_2$-silicon interface. This is done by simultaneously heating the wafer to the 200 to 300°C level and placing a positive 50-V bias on the structure (Fig. 14.17). The elevated temperature increases the mobility of the ions and the positive bias "repels" them to the oxide-silicon interface.

The last step in the process is a repetition of the initial *C/V* plot. However, as the voltage increases, inversion does not start at the same level as in the initial test (Fig. 14.18). The positive charges at

Figure 14.16 *C/V* plotting—first test.

Figure 14.17 *C/V* plotting—ionic charge collection.

1. Original Plot
2. Second Plot
3. Voltage Shift
   or Drift

Figure 14.18 C/V replot.

the interface require additional negative voltage to "neutralize" them before inversion can happen. The result is a C/V plot identical to the original but displaced to the right. The additional voltage required to complete the plot is known as the *drift* or *shift*.

The amount of the shift is proportional to the amount of mobile ionic contamination in the oxide, the oxide thickness, and the wafer doping. C/V analysis cannot distinguish the element (Na, K, Fe, etc.) that was in the oxide, only the amount. Neither can this test determine where the contamination came from. It may have come from the wafer surface, any cleaning step, the oxidation tube, the evaporation process, the alloy tube, or any other process(es) the wafer has been through.

C/V analysis is usually a part of the evaluation of any process changes that may contaminate a wafer, such as a new cleaning process. To make the evaluation, C/V wafers are divided into two groups. One group receives normal processing as detailed above. The second group goes through the proposed cleaning process, usually between the oxidation and aluminum steps of the test. The drift on the standardly processed wafers is compared with the drift of the experimental group. An increased drift on the experimental wafers would indicate that the proposed cleaning process actually contaminated the wafers with mobile ionic contamination.

Acceptable C/V drifts vary between 0.1 and 0.5 V, depending on the sensitivity of the device being made. C/V analysis has become a standard test in a fabrication area. The test is made after any process change, equipment maintenance, or cleaning that could have the potential of contaminating the wafers. The C/V plot provides a wealth of other information, such as flat-band voltage and surface states. The thickness of the gate oxide can also be calculated from the C/V test data.

Contactless C/V measurement. C/V monitoring as described in the preceding section requires rigorous test wafer preparation, which is time-consuming and expensive. Another way to determine unwanted

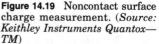

Figure 14.19 Noncontact surface charge measurement. (*Source: Keithley Instruments Quantox—TM*)

charges (drift) and the other MOS gate parameters is with a contact-less method called COS (Fig. 14.19). The MOS transistor method requires two electrodes separated by the gate oxide. A voltage on the top electrode (the metal) produces a charge buildup at the metal-oxide interface. A similar result can be obtained by using a corona source (the "C" of COS) to build a charge directly on the oxide surface. Increasing the charge puts the MOS transistor through the same paces as a metal electrode device, and the same information, that is, charge (drift), flat-band voltage, surface states, and oxide thickness, can be calculated. The data produced by this method correlate with the standard measurements.[4]

Device failure analysis–emission microscopy

When a semiconductor device is operating, it gives off certain emissions of visible light. When there are certain problems with the device, spots of visible light are given off at the trouble spot. For example, contamination-induced junction leakage will reveal itself as a light spot at the surface of the junction. Microscopes fitted with sensitive detectors and charge-coupled imaging devices (CCDs) can locate and produce an image of these trouble spots. This method is particularly useful when electrical measurements indicate a failure in a sector of a circuit but cannot pinpoint the exact device(s) causing the failure.[5]

Physical Measurement Methods

Product reliability and maintenance of yields requires on-line detection of defects, mistakes, and the like to allow removal of suspect material. Process control requires measuring the process results at each of the process steps and knowing the quantity, density, location, and nature of the various problems. These data come from a series of measurements and evaluations that vary with the sophistication of the circuit in terms of image size, sensitivity to contamination, and density.

Tests are performed on test wafers or directly on product wafers. Product wafer tests may call for revealing the subsurface structure by beveling, microsectioning, or using a focused ion beam (FIB) to remove portions of the circuit.

Layer Thickness Measurements

Color

Both silicon dioxide and silicon nitride layers exhibit different colors on the wafer. We know that while silicon dioxide is transparent (glass is silicon dioxide), an oxidized wafer has a color. The color seen is actually the result of an interference phenomenon, the same phenomenon that creates the colors of rainbows.

The silicon dioxide layer on a silicon wafer is actually a thin transparent film on a reflecting substrate. Some of the light rays impinging on the wafer surface reflect off the oxide surface while others pass through the transparent oxide and reflect off the mirrored wafer surface (Fig. 14.20). When the light rays exit the film, they combine with the surface-reflected ray, giving the surface an appearance of having a color. This phenomenon is the reason oxidized wafers change color as the angle of viewing is changed.

The exact color is a function of three factors. One, which is a property of the transparent film material, is the *index of refraction*. A second factor is the *viewing angle*. The third factor is the *thickness* of the film. The color of a thin transparent film becomes an indication of the thickness when the nature of the viewing light is specified (i.e., daylight, fluorescent), along with the viewing angle. The classic color versus thickness chart (Fig. 14.21) is a regular feature at oxidation and diffusion stations. Color alone is not an exact indication of thickness because of the consequences of the interference phenomenon.

As the film gets thicker, the colors change in a specific sequence and then repeat themselves. Each repetition of the color is called an *order*.

Figure 14.20 White light interference.

To determine the exact film thickness, a knowledge of the color order is necessary. A principal use of color charts is for process control.

Each oxidation or silicon nitride process is set up to produce a specified thickness. Naturally the thickness will vary from run to run. Operators quickly become sensitive to the wafer color. When a variation occurs, a quick check of the chart will indicate if the film thickness is out of specification. Rarely is a process so far off that the film thickness is a whole order (same color, different thickness) out of specification. The accuracy of color chart thickness determination is limited to the accurate perception of the colors (what exactly is red-orange?). A typical chart is accurate to $\pm$ 300 Å.

Fringes

When the order is not known, a fringe-counting technique can be used. When a test wafer edge is dipped in hydrofluoric acid for a few seconds the acid quickly eats through the oxide at an angle, leaving the oxide exposed to view (Fig. 14.22). When the wafer is viewed in white light, colored fringes are formed between the wafer surface and the top of the film. Thickness determination is made by first determining the order of the film thickness. It is easy to see the repeated sequence of colors. If three blue-red fringes exist, the thickness corresponds to the surface color in the third order on the color-thickness chart.

Film thickness, μm	Color* and Comments
Order I	
0.050	Tan
0.075	Brown
0.100	Dark Violet to Red Violet
0.125	Royal Blue
0.150	Light Blue to Metallic Blue
0.175	Metallic to Very Light Yellow
0.200	Light Gold or Yellow-Slightly Metallic
0.225	Gold with Slight Yellow Orange
0.250	Orange to Melon
0.275	Red Violet
0.300	Blue to Violet Blue
0.310	Blue
0.325	Blue to Blue Green
0.345	Light Green
0.50	Green to Yellow Green
Order II	
0.365	Yellow Green
0.375	Green Yellow
0.390	Yellow
0.412	Light Orange
0.426	Carnation Pink
0.443	Violet Red
0.465	Red Violet
0.476	Violet
0.480	Blue Violet
0.493	Blue
Order III	
0.502	Blue Green
0.520	Green (Broad)
*When viewed perpendicularly in daylight fluorescent light	

Figure 14.21 Silicon dioxide thickness color chart.

Figure 14.22 Color fringes.

A more accurate fringe-counting method uses monochromatic light as the viewing light. Monochromatic light consists of one color (wavelength), whereas white light is polychromatic (many wavelengths). The sample is prepared in the same way as it is for color fringe counting. However, in a microscope eyepiece, using monochromatic light, the fringes appear as alternating, evenly spaced black and white stripes (Fig. 14.23) with each fringe separation representing a specific vertical distance. Film thickness is determined by counting the number of fringes and multiplying by a correction factor. The correction factor is determined by the wavelength of the monochromatic light used. For sodium light the wavelength is 5890 Å.

Spectrophotometers and reflectrometry

Film-thickness interference and reflectance measurement techniques can be automated. To understand the method, let's review the interference effects. Light is actually a form of energy. The interference phenomenon can also be described in terms of energy. White light is really a bundle of rays (different colors), each with different energies. When the rays interfere through the transparent film, the result is a ray of one color, one wavelength, and one energy level. It is our eyes that interpret the energy as a color.

In a spectrophotometer, which is an automatic interference instrument, a photocell takes the place of the human eye. Monochromatic light in the ultraviolet range is reflected off the sample and analyzed by the photocell. To ensure accuracy, readings are made at different conditions. The conditions are changed by either using another monochromatic light (to change wavelength) or changing the angle of the wafer to the beam. Spectrophotometers specifically designed for use in semiconductor technology have onboard computers to alter the measurement conditions and calculate the film thickness. With visible and ultraviolet (UV) light sources these machines can measure films down to the 100-Å level.[6] These instruments are also used to

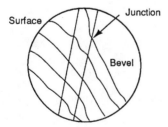

Figure 14.23 Junction depth determination by monochromatic fringes.

measure the index of refraction and silicon film thickness. Since silicon is opaque to ultraviolet light, an infrared source is used.

Ellipsometers

Ellipsometers are film-thickness instruments that use a laser light source and operate on a different principle than a spectrophotometer. The laser light source is polarized. Polarization is the creation of a wave with all the rays traveling in only one plane. Polarization can be imagined by considering looking into the beam of a flashlight. In an ordinary beam, light rays come to your eyes in many planes, like an arrow with many feathers. A polarized beam has all of the light in only one plane, or an arrow with only one feather (Fig. 14.24).

In the ellipsometer the polarized beam is directed to the oxide-covered wafer at an angle. The beam enters the transparent film and reflects off the reflective wafer surface. During its passage through the film, the angle of the beam plane is rotated. The amount of rotation of the beam is a function of the thickness and index of refraction of the film. A detector in the instrument measures the amount of rotation and an onboard computer calculates the thickness and index of refraction.

Ellipsometers are used to measure thin oxides (50 to 1200 Å) and the index of refraction of the film. Their accuracy in this range is unequaled by any of the other techniques. The technique is also used for multilayer thin-film stacks. Ellipsometer accuracy and range are enhanced by the addition of multiviewing angle capabilities, multiple wavelength sources, and reduction in beam spot size.[7]

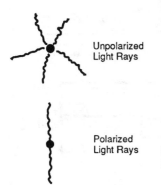

Unpolarized
Light Rays

Polarized
Light Rays

Figure 14.24 Unpolarized and polarized light.

Oxides thinner than 50 Å can be measured by the corona-oxide-semiconductor (COS) technique, described in the electrical measurement section.

Stylus

Some thin films such as aluminum cannot be measured by optical techniques. And in the case of aluminum and other very thin conductive films, the four-point probe thickness measurement is not sufficiently accurate. In these situations, a mechanical moving-stylus apparatus is used (Fig. 14.25). The method requires that a portion of the film be removed, creating a step on a test-wafer surface. This is normally done by a masking and etching step. The prepared sample is mounted and leveled on the pivoting stylus instrument stage.

After leveling, the measuring stylus is lowered gently onto one of the surfaces. The measurement is made as the stage is slowly moved under the stylus. As the stylus goes over the step, its physical position is changed. The stylus is linked to an inductor that generates an electrical signal in response to the stylus vertical position. This signal is amplified and fed into an x-y recorder.

While the leveled wafer is moving under the stylus, it does not move in the vertical direction and no change in signal is produced. The trace on the x-y plotter is a straight line. When the stylus reaches the surface step, it changes position, causing a change in the signal output. This change is evidenced by a change of pen position on the x-y chart trace. The change in position is relative to the step height, which is read directly from the calibrated x-y chart.

Figure 14.25 Step height measurement.

Junction Depth

A critical device parameter is the junction depth of the various doped regions. This parameter is measured after each of the doping steps. The measurement methods described are all performed off-line; that is, the test wafers or device wafers have to be taken to a measurement station or laboratory for the measurement.

Groove and stain

The traditional method of junction depth measurement is by the groove (or bevel) and stain technique. Grooving or beveling is a mechanical method of exposing the junction for viewing and measurement from the horizontal plane (Fig. 14.26). The extreme shallow depth of the junction requires either grooving or beveling of the wafer to expose the junction.

The junction itself is not visible to the naked eye. Two techniques, called *junction delineation,* are available to make it visible. Both techniques utilize the electrical differences between N-type and P-type regions. The first technique, the etch technique, starts with the placement of a drop of hydrofluoric acid and water mixture over the junction (Fig. 14.27). A heat lamp is directed onto the exposed junction. The heat and light cause holes or electrons to flow in each region. As a result of the flowing current, the etch rate of the $HF-H_2O$ mixture is higher on the N-type region, making it appear darker.

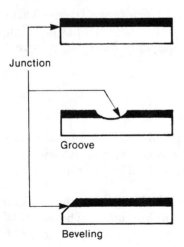

Junction

Groove

Beveling

Figure 14.26 Exposure of junction by groove or beveling.

Figure 14.27 Etching or staining of junction.

The second delineation technique is electrolytic staining. A mixture containing copper is dropped on the exposed junction. Again, the heat lamp is directed onto the junction. In effect a battery is formed, with the poles of the junction being the poles of the battery and the copper solution being the electrolytic connection. The current flowing in the liquid drop causes the copper to plate out on the N-type region side of the junction.

The final step, after exposure and delineation of the junction, is depth measurement. A number of methods may be used including optical interference and SEMs.

Optical interference methods create interference fringes (Fig. 14.28). Each fringe corresponds to a specific depth. This junction depth is determined by multiplying the number of fringes by a correction factor.

Scanning electron microscope (SEM) thickness measurement

The SEM technique (described next) can also be used to measure junction depths and film thickness. The wafer is cleaved with the break position over the junction. The exposed wafer junction is delineated by one of the methods described.

The exposed cross section is positioned to the SEM beam at right angles to the wafer surface and a photograph is taken. The depth is determined from the photograph and the scale factor of the SEM

Figure 14.28 Creation of fringes.

(Fig. 14.29). SEM and the groove and stain methods provide a visual look at the junction area and lateral diffusion that other methods do not.

Spreading resistance probe (SRP)

Spreading resistance is a technique also used for measuring the junction depth. As the probes pass through the junction, they sense the change in conductivity type (N or P). This information, when plotted on the profile curve, also gives the junction depth (Fig. 14.6).

Secondary ion mass spectrometry (SIMS)

A SIMS instrument bombards the wafer surface with a source that produces secondary ions from species on the surface. It can detect the dopant(s) in the junction region. After careful removal of the surface

Figure 14.29   SEM declination of device cross section.

down through the junction, junction depth can be determined by monitoring the dopant atoms as they disappear from the stream.

Scanning capacitance microscope (SCM)

One of the options with the atomic force microscope (AFM), see "General Surface Characterization") is a probe that measures capacitance. Since the capacitance of a doped layer changes as the junction is approached, the probe can be used for junction depth detection. Sample preparation is the same as that for spreading resistance measurements.

SCM offers the advantage of nanometer precision[8] and, in the near future, may be the concentration profile and junction depth measurement technique for submicron junctions.

Critical Dimensions (CD) and Line Width Measurements

The exact dimensions required of each component in the circuit are controlled and influenced by *all* processes. Vertical dimensions are set by the doping and layering processes. The horizontal surface dimensions are produced in the photomasking area. As part of that process, the critical dimensions are measured at both develop inspection and final inspection with microscopes.

Filar mechanical measuring eyepiece

The filar measuring eyepiece is a dimension-measuring instrument that is fitted to a microscope. The eyepiece features a movable hairline that requires calibration to an outside standard. A calculation is made to "correct" the filar measurements to the actual dimensions on the standard.

The pattern to be measured is focused in the field of view (Fig. 14.30). A hairline is positioned perpendicular to the measuring location path. Once oriented, the hairline is moved to the starting point of the pattern to be measured, and the value on the micrometer barrel is noted. The hairline is then moved in a smooth continuous motion to the other side of the pattern. The ending value is also noted. The actual width is calculated by subtracting the starting value from the ending value and multiplying the resulting number by the previously determined correction factor. This system is highly accurate and is used by the National Bureau of Standards. Accurate measurements require good operator techniques.

Filar systems are easily automated. The hairline movement mechanism can be motorized and the correction factor programmed into an onboard computer, resulting in a direct digital readout of the actual width. Operator fatigue is minimized by a video monitor rather than requiring the operator to view the wafer through a high-power microscope.

Figure 14.30 Manual filar width measurement.

Image shearing

An image-shearing attachment on a microscope is another method of critical-dimension measurement. A control on the unit allows the operator to optically separate (shear) the pattern into two images. To start the measurement, the two images are butted against each other (Fig. 14.31) then moved until the sheared images exactly overlap. The difference between the starting and ending values is the width of the pattern. Like the filar unit, an image-shearing unit must be calibrated to a standard. The procedure just described is called *single-image shearing* because the pattern is sheared once. Shearing the image twice is another technique. This is accomplished by following the same procedure used in single shearing but continuing the shear until the image is again doubled (Fig. 14.32). The pattern is actually sheared twice and the calculated width must be divided by 2.

This procedure can be automated for production work. The images are displayed on a cathode-ray tube and the shearing motion motorized. And like the filar automated measurement, a correction factor is programmed into the unit. The width of the pattern is read directly from the unit.

Reflectance

Both the filar and image-shearing techniques require some operator decision, which can be a source of error. A third type of dimension-measurement instrument is based on reflectance. Like the other two, the operator locates the pattern to be measured on a monitor. One edge of the pattern is positioned under a marker on the screen. The measurement is made automatically as a laser beam is swept along the direction of measurement and the reflection energy of the beam recorded (Fig. 14.33). When the beam comes to the edge of the pattern, it steps up (or down) to a new surface. The new surface causes a

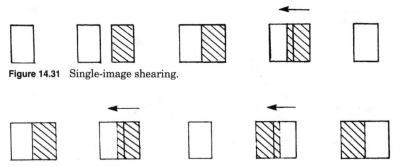

Figure 14.31   Single-image shearing.

Figure 14.32   Double-image shearing.

Figure 14.33   Reflectance CD measurement.

different reflectance of the beam, which is recorded by the detector. The width of the pattern is the difference between the starting point and the point where there is a change in reflectance. This value is read out automatically. Operator decision is limited to determining the starting point, resulting in a more constant accuracy.

The limitation of this method is with patterns that have a number of steps in them, such as metal steps. The instrument is designed to read to the first change in reflectance it senses. Another limitation is in measuring pattern edges that are sloped. The reflecting beams change gradually rather than abruptly, making a width determination difficult.

Some dimension-measuring microscopes use transmitted light for the measurement of mask and reticle critical dimensions. A light beam is passed from the bottom through the mask. The pattern to be measured is moved in front of the beam and a detector above the mask senses when the light is blocked by the opaque pattern. The horizontal distance is measured until the light again is detected by the sensor. The distance between the two is translated into the pattern dimension. A summary of the various surface-inspection techniques and their use in a fabrication process is tabulated in Fig. 14.34.

Scanning electron microscope (SEM)

The SEM inspection tool described in the following section is also used for very accurate measurement of line widths.

Contamination and Defect Detection

Detection of contamination and visual defects is essential to high yields and process control. Particulate contamination is detected primarily by visual techniques including high-intensity lights, microscopes, SEMs, and automatic machines. Chemical contamination is detected and identified by both Auger and ESCA techniques. Mobile ionic contamination in the wafer is detected by capacitance-voltage

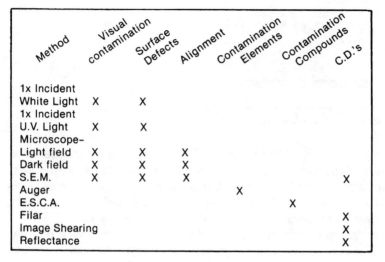

Method	Visual contamination	Surface Defects	Alignment	Contamination Elements	Contamination Compounds	C.D.'s
1x Incident White Light	X	X				
1x Incident U.V. Light	X	X				
Microscope– Light field	X	X	X			
Dark field	X	X	X			
S.E.M.	X	X	X			X
Auger				X		
E.S.C.A.					X	
Filar						X
Image Shearing						X
Reflectance						X

Figure 14.34 Overview of surface-inspection techniques.

plotting and by interpretation of the transistor and diode electrical tests. Many other sophisticated techniques can be used to make these inspections. The ones described here are those employed in a typical wafer-fabrication line.

1× Visual surface inspection techniques

The first line of defense is to look at the wafers, which is a magnifying power of 1, or, in microscope terminology, a 1× power. Operators quickly become sensitive to the way "normal" wafers look. Even minor changes in the surface appearance are picked up by the experienced eye.

1× **Collimated light.**   The resolving power of the naked eye (1×) can be assisted by using a high-intensity white light, like the beam of light from a slide projector (Fig. 14.35). Particulate contamination is highlighted in the light beam when the wafer surface is viewed at an angle. The effect is similar to the highlighting of dust in the air by light streaming through a window.

1× **Ultraviolet.**   In actuality, the eye cannot see ultraviolet light, but ultraviolet light from a mercury-vapor lamp emits blue, green, and even some red light. Since the ultraviolet is harmful to the retina, a filter is frequently placed over the light source to block out the ultraviolet. The primary benefit of the ultraviolet lights used in fabrication

Eye

Collimated light

Figure 14.35 Collimated light inspection.

areas is that they are very bright, which means the intensity of the scattered light is greater, therefore increasing the detection of surface contamination.

Microscopic techniques

Light-field microscope. The metallurgical microscope is the workhorse tool of surface inspection. The term *metallurgical* differentiates it from the standard microscopes found in biology labs. A biological microscope illuminates transparent samples by shining the light up through the sample. In a metallurgical microscope, the light is passed down to the nontransparent sample through the microscope objective (Fig. 14.36). The light reflects off the sample surface and is transmitted back up through the optics to the eyepieces. With white light illumination, the picture in the field of view exhibits the surface colors, which helps identify particular components on the wafer. Use of filters will change the surface colors.

A typical fabrication microscope is fitted with 10× or 15× eyepieces and a range of objectives from 10 to 100×. Increasing the total viewing power (eyepiece power times the objective power) reduces the field

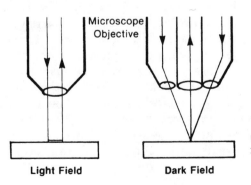

Microscope Objective

Light Field   **Dark Field**

Figure 14.36 Light and dark field inspection.

of view. This reduction requires more inspection time for the operator to look at the required sample inspection area on the wafer. The consequence is a slower inspection process. A tradeoff power level, when inspecting LSI and VLSI devices, is 200 to 300× magnification.

The industry typically uses a microscope inspection procedure requiring the operator to look at three to seven specific locations on the wafer. This procedure is easily automated with motorized stages. Most of the automated microscope inspection stations feature automatic wafer placement on the stage, automatic focusing, and automatic binning of the completed wafers. At the touch of a button, the operator can direct each wafer to a boat for passed wafers or into one of several boats for different reject categories. Obviously, a microscope inspection procedure is used to judge surface and layer quality and (in masking) pattern alignment.

Dark-field inspection. Dark-field illumination is achieved by fitting a metallurgical microscope with a special objective (Fig. 14.36). In this objective, the light is directed to the wafer surface through the outside of the objective body. It impinges on the surface at an angle and passes up through the center of the objective. The effect on the "picture" in the eyepieces is to render all flat surfaces black. Any surface irregularities, such as a step or pieces of contamination, appear as bright lines. Dark-field illumination is more sensitive than light-field to any surface irregularity. It has the drawback of limiting the operator's ability to discern the nature of the surface irregularity. A passable surface dimple may look the same as a rejectable piece of contamination. In practice, dark-field illumination is rarely used in on-line inspections but is reserved for diagnostic purposes.

Confocal microscopes. Distinguishing fine detail with a normal microscope is hampered by stray light rays interfering as they bounce off the various surfaces and from the different plane depths on a wafer surface. Confocal light sources minimize the scatter and provide greater detail by limiting the returning light rays to a narrower plane. This is done by passing an intense white light or laser beam through holes in a disk (positioned between the light source and the wafer) spinning at a high RPM. Only light rays from the inspection plane of interest bounce back through the holes. Confocal systems are capable of imaging submicron dimensions.[9] Laser beams used in a confocal system cause the sample to fluoresce (glow). The resultant image, after x-y scanning and computer processing, has resolution in the nanometer range and can display layers of registered translucent samples.

Other microscope techniques.   Optical technology is capable of providing many evaluation techniques beyond simple light- and dark-field viewing such as phase contrast and fluorescence microscopes. Each allows the viewer to determine additional visual information about the surface. Phase contrast brings out surface irregularities in the vertical plane and fluorescence-illuminated microscopes use ultraviolet illumination sources. In the ultraviolet light, organic residues (photoresist, cleaning chemicals) not easily visible in white light are brought into view. Their use and interpretation generally require technicians trained beyond the level of production operators.

Scanning electron microscope (SEM).   Conventional optical microscopes are limited in their ability to provide accurate information about the wafer surface. First, their resolving power is limited by their optical light source. The ability of a viewing system to distinguish detail is related to the wavelength of the light (radiation used). The shorter the wavelength, the smaller the detail that can be seen.

Depth of field is another viewing factor. It relates to the ability of the system to keep two planes in focus simultaneously. A conventional photograph with the subject in focus and the background out of focus has a background beyond the depth of field limit of the camera. In a microscope, the depth of field decreases as the power (magnification) of the system is increased. If the power is increased to see the surface "closer," the operator may not be able to see the top and bottom surfaces in focus. Constant refocusing results in loss of information and a slower inspection time.

Magnification is the third limiting factor of optical microscopes. An optical system with white light illumination is limited to about 1000× magnification with conventional objectives. The oil immersion technique pushes the limit up, but it is unacceptable because it is too slow, too messy, and a possible source of contamination to the wafer.

All three limitations are overcome by using a scanning electron microscope. The microscope varies from an optical one in many aspects. The "illumination" source is an electron beam scanned over the wafer or device surface. The impinging electrons cause electrons on the surface to be ejected. These secondary electrons are collected and translated into a picture of the surface (Fig. 14.37) on either a screen or a photograph.

SEM requires that the wafer and beam be in a vacuum. The electron beam has a much smaller wavelength than white light and allows the resolution of surface detail down to submicrometer levels. Depth of field problems do not exist; every plane on the surface is in focus. Magnification is similarly very high, with a practical upper

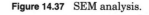

Figure 14.37 SEM analysis.

limit of 50,000×. A tilting wafer holder in an SEM allows the viewing of the surface at angles, which enhances the three-dimension perspective (Fig. 14.29). Surface details and features can be viewed at advantageous angles.

Some materials such as photoresist do not give off secondary electrons in response to e-beam bombardment. To inspect a photoresist layer in an SEM, the resist layer is covered with a thin layer of evaporated gold. The gold layer conforms to the topography of the photoresist layer. Under e-beam bombardment the gold gives off secondary electrons, thus resulting in an SEM picture that is an accurate reproduction of the underlying photoresist layer.

Transmission electron microscope (TEM). A standard SEM instrument has a resolution range of 20 to 30 Å,[10] which presents a barrier for inspection of ULSI devices. However, passing (transmitting) the electron beam through a thin sample increases the resolution to 2 Å. As attractive as the increased resolution may be, sample preparation is time-consuming and requires precision. TEM is best used as an off-line laboratory tool.

SEM-energy-dispersive x-ray spectroscopy (SEM-EDX). A spectrometer is simply an instrument that can analyze and compare spectrums. This principle is used to measure film thicknesses. In an SEM the energetic e-beam causes x rays to leave the surface. Fortunately the x-ray energies are reflective of the materials (elements) presence on the surface and can be analyzed for additional information. This is accomplished when an x-ray spectrometer is added to the SEM to detect dispersed x rays. This additional information is about the chemistry of the materials.

Automated in-line defect inspection systems

Laser scanning. The detection of ever smaller sized particles on the wafer surface has led to the use of laser beams as the detection illumination. Two advantages accrue from the use of lasers. First is the obvious ability to detect smaller particle sizes (Fig. 14.38) due to the high intensity of the reflected light, which makes very small surface particles detectable (helium and neon lasers are the usual sources).

The second advantage is automation. Laser inspection equipment is easy to automate so that the inspection is automated from cassette to cassette and the quantity and size of the contamination on the surface can be determined. The information is used to produce a map of the wafer surface showing the size, location, and density of surface contamination. This last factor is a big help in process control. Laser inspection of masks has become a standard production technique.

Image processing systems. Advances in image processing have allowed automatic defect and pattern distortion detection instruments. The machines mate image processing and computer technology. They have scanners with laser or optical light sources that move over the wafer surface. In one version, the computer is preprogrammed with the design pattern for the circuit. The scanner looks for added or missing parts from the expected pattern. Their location is recorded and surface maps printed out. Engineers can go back to the wafer or mask and determine the exact nature of the problem.

Another system compares adjacent die on the wafer or mask. A die is scanned and the pattern recorded in the computer memory. A second die is also scanned and any deviations between the two are recorded. This system will not detect any repeated pattern defect that occurs on

Surface View

Cross Section

Wafer with pinhole and surface irregularity in oxide layer

Silicon etch delineation of pinhole

After silicon etch

111   100

Figure 14.38 Laser detection of defects.

every die but will pick up random defects that have a very low probability of occurring in exactly the same spot on two adjacent die.

In both types of machines, the information from the surface is captured by charge-coupled device (CCD) cameras or photomultiplier
tubes.[11] For in-line inspection, a grave concern is calibration and
standardization. In addition to onboard verification of the electronics,
most systems use a standard wafer to verify machine operation.

General Surface Characterization

Atomic force microscope (AFM)

Several measurement techniques have multiple uses, for example,
SEMs. A new entry in the multiuse category is the *atomic force microscope* (AFM). It is surface profiler that scans a delicately counterbal-
anced probe over the surface (Fig. 14.39). Probe and surface separation is so small (about 2 Å) that atomic forces between the surface and
probe materials actually affect the probe. Sophisticated electronics
measure the relative position of the probe as it moves across the surface, thus compiling a three-dimensional map of the surface.[12] (Fig.
14.40)

Figure 14.39   Atomic force microscope (AFM). (*Semiconductor
International, August 1993, p. 66*)

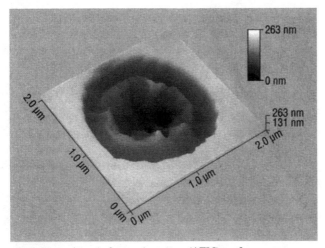

Figure 14.40 Atomic force microscope (AFM) surface map.

AFM sensitivity is in the 1-Å range and can be operated in contact or noncontact mode. The first AFM production instrument was introduced in 1990. Projections are for it to become an integral part of the inspection arsenal. By its nature, AFM can characterize grain sizes, detect particles, measure surface roughness, and provide critical dimension measurements—all in three dimensions. In one novel use, an AFM probe has been combined with an optical microscope objective.[13]

Scatterometry

The search for fast, accurate, and nondestructive surface inspection tools led to the *scatterometry* metrology. Accuracy in optical systems is limited by the wavelength of the light. Particles or surface features smaller than the wavelength used cannot be detected. However, a scattered beam can give information about surface features smaller than the wavelength. A scatterometry system has the wafer at the center of curvature of a screen (Fig. 14.41). An incident laser beam is scanned over the surface and is reflected and "scattered" from the surface onto the screen. A camera with a microprocessor captures the screen image to reconstruct the surface that produced the particular pattern on the screen. Like AFM, this technique has the potential of measuring grain sizes, contours, and CDs. It can also measure latent images in undeveloped photoresist and characterize phase shift masks.

Figure 14.41 Arrangement of a scatterometer. (*Source: Solid State Technology, March 1993, p. 30*)

Contamination Identification

Advances into the 0.18-μm era are requiring more detailed information of the nature of the various contaminants that end up on the wafer surface or in the deposited layers. Contamination types, forms, amounts, and other data are needed to maintain clean processes and products. In this section, the instruments used to collect these data are examined. In general, all of the techniques are based on a common phenomenon. Whenever a surface is excited with energy there will be energy given off. The energy given off will have characteristics reflecting the interaction of the incoming energy wave with the material(s) on the wafer surface.

Auger electron spectroscope (AES)

In a SEM, a range (spectrum) of secondary electrons is released by the impinging electron beam. One portion of this spectrum contains the electrons that are released from the top several nanometers of the surface. These electrons, known as *Auger electrons,* have energies characteristic of the element that emits them. Thus sodium and chlorine each give off different Auger electrons.

The collection and interpretation of Auger electrons allows the identification of the surface materials, including contamination. In operation the e-beam is scanned across the wafer. The ejected Auger electrons are analyzed for their energies (wavelengths) and printed out on an *x-y* plotter (Fig. 14.42). Energy peaks at specific wavelengths indicate the presence of specific elements on the surface.

Figure 14.42   Typical auger trace.

Scanning Auger microanalysis (SAM) is limited to the identification of elements. This technique cannot identify the chemical state of the element, what it may be combined with, or its quantity on the surface. Salt (NaCl) contamination is identified only as the presence of sodium and chlorine on the surface.

Electron spectroscope for chemical analysis (ESCA)

Solving surface contamination problems often requires a knowledge of the chemical state of the contaminant. An Auger detection of chlorine on the surface does not reveal whether the chlorine is present as hydrochloric acid or a trichlorobenzene. Knowledge of the form of the chlorine expedites locating and eliminating the process source of contamination.

The electron spectroscope for chemical analysis (ESCA) is an instrument used to determine surface chemistry. The instrument works on principles similar to the Auger technique. However, x rays instead of electrons are used for the bombarding radiation. Under bombardment, the surface gives off photoelectrons. Analysis of the photoelectron information leads to the determination of the chemical formula of the contamination. Unfortunately the ESCA x-ray beam is wider than most integrated circuit features. The beam diameter limits the technique to a macro surface analysis. By contrast, an Auger electron beam can zero in on specific bits of contamination.

Time of flight–secondary ion mass spectrometry (TOF-SIMS)

TOF-SIMS surface analysis can use a number of incident radiation sources, such as an Nd YAG laser, a Cs ion beam, or a Ga ion beam.[13]

For this method, the impinging energy pulses and ionizes surface material and produces secondary ions, which are accelerated to a mass spectrometer. In the spectrometer, each ion's time of flight from the surface is measured. The time of flight is indicative of the species on the surface.[14] A SIMS analysis can also determine the junction depth.

TOF-SIMS can sample material depths of only several tenths of a nanometer.[15] Because of this precision and the ability to characterize both organic and inorganic contaminants, TOF-SIMS is a mainstay of analytical laboratories.

Vapor phase decomposition–atomic absorption spectroscopy (VPD–AAS)

VPD–AAS is used to detect trace inorganic contaminants. The sample is exposed to vapor phase hydrogen fluoride gas, which chemically reduces (decomposes) the contaminant to its fluoride salt. Next, the reduced material is dissolved in water or an acid and analyzed for elements with standard atomic absorption spectroscopy.

Key Concepts and Terms

Contamination detection	Junction depths
Critical dimensions	Light-field microscope
C/V measurements	Measurements
Dark-field microscope	Resistance/resistivity
Device	SEM
Ellipsometer	Spectrophotometer
Four-point probe	Transistor measurement
Fringe counting	Visual inspections
Groove and stain	

Review Questions

1. Changing the size of a resistor will change which of the following?
   a. Resistance
   b. Resistivity
   c. Both

2. In a four-point probe the current is flowing in the inside probes. (true/false)

3. The color of an oxide film on silicon is an indication of its exact thickness. (true/false)

4. Which is the most accurate determination of a thin oxide layer (less than 1000 Å)?

    a. Color

    b. Fringes

    c. Ellipsometer

    d. Spectrophotometer

5. Explain why an oxidized silicon wafer changes color when rotated.

6. Name a method to delineate an N-P junction for the groove and stain technique.

7. A collimated light is used to detect sodium contamination. (true/false)

8. An SEM uses which magnifying power as compared with a light-field microscope?

    a. Higher

    b. Lower

    c. The same

9. Name two methods of measuring CDs.

10. If an oxide layer is contaminated with sodium, what is the effect on the C/V drift?

    a. C/V drift will be higher

    b. C/V drift will be lower

    c. C/V drift will remain the same

References

1. B. McDonald, "Analytical Needs," *Semiconductor International,* Cahners Publishing, January 1993, p. 36.
2. S. Felch, "A Comparison of Three Techniques for Profiling Ultrathin p⁺–n Junctions," *Solid State Technology,* January 1993, p. 45.
3. Ibid.
4. M. Peters, "COS Testing Combines Expanded Charge Monitoring Capabilities with Reduced Costs," *Keithley Instruments, Inc* (product description paper), 1995.
5. T. Adams, "IC Failure Analysis: Using Real-time Emission Microscopy," *Semiconductor International,* Cahners Publishing, July 1993, p. 148.
6. P. Burggraaf, "Thin Film Metrology: Headed for a New Plateau," *Semiconductor International,* Cahners Publishing, March 1994, p. 57.
7. Thin Film Measurements, Rudolph Engineering, product brochure, 1985.
8. R. McDonald, "How will we examine ICs in the year 2000?" *Semiconductor International,* Cahners Publishing, January 1994, p. 46.
9. *Product Description Bulletin,* Technical Instrument Company, San Francisco, CA, 1995.

10. S. Wolf and R. Tauber, *Silicon Processing for the VLSI Era,* Lattice Press, Sunset Beach, CA, 1986, p. 597.
11. P. Burggraaf, "Patterned Wafer Inspection Now Required!" *Semiconductor International,* Cahners Publishing, December 1994, p. 57.
12. L. Peters, "AFMs: What Will Their Role Be?" *Semiconductor International,* Cahners Publishing, August 1993, p. 62.
13. "Lithography News," *Semiconductor International,* December 1994, p. 24.
14. "Time-of-Flight Secondary Ion Mass Spectrometer" brochure, Charles Evans & Associates, Redwood City, CA, 1994.
15. P. Chu and R. Hockett, "New Ways to Characterize Thin Films," *Semiconductor International,* Cahners Publishing, June 1994, p. 143.

15

Manufacturing Technology

Overview

Now into its fifth decade, the semiconductor industry has advanced from revolutionary development to evolutionary advances with a focus on manufacturing. While product and process advances are still very much industry issues, greater emphasis is being placed on the "business" factors. Though high yields are a necessity, new processes and equipment are being measured by overall productivity. Strategies to increase the bottom line include maximizing equipment cost of ownership, automation, cost control, computer-automated manufacturing, computer-integrated manufacturing, and statistical process control.

Objectives

Upon completion of this chapter you should be able to:

1. List the major cost factors that influence fabrication costs.
2. Describe the intent and factors of cost of ownership (COO) models.
3. List the advantages of statistical process control.
4. Identify the parts and use of a control chart.
5. List and discuss the different levels of automation.
6. List the factors that enter into an evaluation of a particular piece of equipment.

7. Define WIP, JIT, and their importance to production costs.

8. Sketch a "product manager"-oriented organizational chart.

9. Define the terms CIM and CAM and their use in a manufacturing setting.

Fabrication and Factory Economics
Overview

The semiconductor industry started supplying commercial products in the late 1940s. The manufacturing lines were little more than laboratories, and the workers were primarily trained technologists. By the 1970s, total factory sales of integrated circuits alone were some $1 billion per year. The manufacturing scene had changed to clean rooms with highly specialized equipment attended by skilled production workers. A fabrication area of 2000 to 3000 square feet could be built for 2 to 3 million dollars.

The VLSI era dramatically changed the look and cost of wafer-fabrication areas. The size stayed about the same but the cost had shot up to some 100 million dollars. Most of the increase has been in the expense of building class 10 and 1 clean rooms and newer, even more specialized and automated equipment and process control systems.

By the mid-1990s the industry was shipping large memory and microprocessor devices and facing $1- to 2-billion fabrication facility costs. The price tag is not so surprising considering the factors driving the processes. Market forces are demanding higher capability and faster circuits. That means larger chips with more transistors (and other devices) in a given area with the devices getting smaller. Maintaining productivity and yields at this level requires larger wafers at higher costs. The equipment to process larger wafers with tighter tolerances becomes more expensive. Improvements are needed to maintain uniformity across a larger wafer (or more tools processing single wafers). Automatic handling equipment is required. Of course, all of the materials and the processing environment must get cleaner since the smaller, more closely packed devices are more sensitive to contamination. And don't forget that at these levels the number of tests and characterizations must increase to maintain product quality and process control in a fabrication line that is rapidly moving large volumes.

Of special note is the increase in process steps at the ULSI level. Wafers are best suited for semiconductor devices, while the non-semiconductor circuit parts (wiring, for example) can be stacked in layers above the wafer surface. Squeezing the devices closer together and making them smaller has introduced problems that must be solved by the addition of new process steps, such as planarizing techniques to overcome topography-generated imaging problems. The additional steps drive up the process and inventory expense.

All these cost increases have come as individual chip prices continue to erode through improved productivity and competition. To stay profitable, chip manufacturers have to continually improve their efficiency, yields, and control costs at all levels.

Despite all the changes in processes, costs, and markets the overall financial measurement of a fabrication area has remained the same: it is *the cost per functioning die shipped out of fabrication.* When extended to a complete merchant facility with assembly capabilities, the measure becomes *the cost per die shipped.*[1] In the world of the megachip, the cost per transistor is becoming an indicator parameter.[2]

The remainder of this chapter examines the various factors involved and strategies used to increase process and factory performance while controlling costs.

Wafer-Fabrication Costs

A number of factors contribute to the cost of producing a functioning die (Fig. 15.1). They are generally divided into fixed and variable categories. Fixed costs are those that exist whether or not any die are being made or shipped. Variable costs are those that depend on the volume of product being produced.

Fixed costs

- Overhead—administration and facilities
- Equipment

Figure 15.1 Fabrication cost factors.

Variable costs

- Materials—direct and indirect
- Labor
- Yield

Overhead

Overhead costs are all those incurred by the administrative and executive staff plus the cost of providing and maintaining a facility. A curious fact of company growth is that after a certain level the number of administrative personnel grows faster than the manufacturing workers. As companies grow, more information is generated internally and more information must be handled from customers and suppliers. To be effective, the information must be available to an ever-growing staff. The two needs result in more and more staff processing "information" rather than product. Also, decision making becomes more formalized (and costly) as more departments become stakeholders in the outcomes. Currently some 50 percent of the workforce of the industrialized economies is involved in information processing.[3] A primary overhead expense is the design activity. With expensive CAD systems and a large professional design team, the cost of circuit design is considerable. Design cost goes up for ASIC fabrications where a large number of individual circuits are designed and produced with a fast turnaround requirement.

The facility cost and maintenance of the facility is a major contributing cost. A fabrication area occupies only about 20 percent of the total facility area, yet requires the majority of the expense. Air conditioning, chemical storage and delivery, and the cost of the clean room are all major expenses. Fabrication clean-room costs for an ULSI facility are thousands of dollars per square foot. Actual floor costs are a factor of the cleanliness strategy chosen (see Chap. 4). A total clean room layout is more expensive than a hybrid minienvironment approach. But in the latter, there is more expense for the equipment.

Materials

Manufacturing materials are divided into direct and indirect. Direct materials are those that go directly in or on the chip. This includes the wafer materials, the added layer materials, and the packaging material costs. Indirect materials are the masks and reticles, chemicals, stationery supplies, and other materials that support the process but do not enter into the product.

Equipment

This cost is the equipment used directly in the fabrication of the devices and wafers. It shows up in the cost calculations as fixed overhead or as depreciation. Depreciation is the loss of value of the machines as they wear out or become obsolete.

Labor

Labor also has a direct and indirect component. Direct labor takes in those workers actually handling the wafers and equipment. Indirect labor is the support personnel such as supervisors, engineers, facility technicians, and office workers.

Yield

The overall fabrication yield (see Chap. 6) determines how the various costs affect the final die cost. If the die yield is low, the cost per die goes up. Not only are the fixed costs distributed over fewer die but the variable costs go up as more materials are required to get out the die. When the yield is calculated into the cost, the term used is *yielded die cost*. The cost of producing the wafers without considering die yield is the *unyielded die cost*.

Die costs are a function of the wafer size, die size, and wafer-sort yield. A $300 wafer-manufacturing cost for a 6-in-diameter wafer with 300 die will translate into an unyielded die cost of $1 each. If the die sort yield is 50 percent, the die cost rises to $2 each ($300 wafer cost divided by 150 functioning die).

Cost factor distribution

Typical contributions of each of the factors to the overall unyielded die cost are shown in the following table.[4] The percentages will vary with the wafer and feature size, degree of automation, and number of process steps.

Factor	Contribution, %
Overhead and utilities	8
Materials (total)	44
Equipment depreciation	25
Labor	
Direct	17
Indirect	6

Actual wafer and die costs

In any discussion about wafer cost factors, one is naturally curious about the actual costs. This number changes from line to line, with the complexity of the devices and with the position of the product in the maturity cycle (Fig. 15.2). The more mature the product and process, the higher the yields and the lower the equipment depreciation factor. Newer products suffer from operator learning curves, equipment shakedown times, and development of new processes. Wafer volume and type are major influences on cost. High-volume products, such as DRAMS, generally have the lowest per transistor or wafer cost due to high manufacturing efficiencies. ASIC wafers, on the other hand, have higher costs due to smaller production runs and the higher design and processing costs required for a varied product mix. The range of costs is from about $100 for simple, discrete ICs to several thousand dollars for advanced products.

Figures of merit

Aside from the financial cost figures, fabrication-area efficiencies are judged by several figures of merit. One is the number of wafers out per operator. This number, for a reasonably defined process, is about 16. A second figure is the number of wafers out per square foot of fabrication area. This number is about two wafers out per month per square foot of fabrication.

Yield improvements

While it is true that more attention is being given to traditional business factors, no less attention is paid to process and wafer yields. The

Figure 15.2 Wafer production loss versus product maturity.

Process	Change	Yield		Assembly	Overall
		Fab	Sort		
Base Line		0.80	0.50	0.95	0.38 (38%)
	Fab to 0.81	0.81	0.50	0.95	0.384 (38.4%)
	Fab to 0.85	0.85	0.50	0.95	0.404 (40.4%)

Figure 15.3   Yield impact of fabrication yield improvement.

effect of a yield improvement can be significant in terms of dollars. Consider the yield figures in Fig. 15.3. Line 2 shows that a wafer-fabrication yield improvement of five percentage points increases the overall yield from 38 to 40.4 percent, a 1.2 percent increase. If the fabrication area starts 10,000 wafers per month and there are 350 die per wafer and the selling price is $5 per die, the increased revenue is

$$10,000 \text{ wafer starts} \times 350 \text{ die per wafer} \times 0.012 \, (1.2\%) \times \$5$$
$$= \$210,000 \text{ per month}$$

Whether or not this amount of increased income is significant depends on the cost of effecting the improvement.

Manufacturing strategies

Minimizing the cost per wafer (or transistor) involves a host of factors, including the layout of the fabrication area, contamination control systems, equipment and automation levels, process control programs, and material management schemes. These are discussed in the following sections.

Contamination control systems

There are two major contamination control systems in use (see Chap. 4). Some manufacturers elect to build and maintain a total clean room environment with class 1 or better air quality. This approach involves the highest facility expense and requires rigorous maintenance and contamination control programs. Wafer isolation technology (WIT) or use of minienvironments is the other approach. It requires a large investment in clean boxes and equipment interfaces but the fabrication area can be at a lower class value.

Equipment

In this section we examine equipment issues. Over the years semiconductor processing has changed from a laboratory activity to full manufacturing using sophisticated and dedicated equipment. The development of semiconductor-specific equipment has allowed the advances into the ULSI era and the spread of the technology throughout the world. Up to the early 1970s, most of the equipment was basic and manual with process effectiveness dependent on process engineers monitoring and adjusting the processes—the so-called "processing tweaking." IBM coined the term *tool,* for semiconductor process machines and it has become part of the industry vocabulary. As the tools have become more sophisticated, the prices have risen to the point where they are a major expense in any new facility. An ASA–Dataquest calculation shows a worldwide chip business of $804 billion and an equipment–materials business of $77 billion.[5]

Equipment selection factors

A number of considerations (Fig. 15.4) go into the selection of a particular process tool. They can be roughly divided into two broad categories, performance and economic. Performance factors relate to the ability of the tool to produce the required results. Economic factors relate to price, costs, and support factors. Figure 15.4 shows how these factors have shifted as the industry has moved into the manufacturing era.

While the importance of performance has decreased as a selection criterion, nevertheless, any tool must have the basic capability to meet the process requirements. If the machine cannot routinely produce the right product, it is of no use. Other performance factors are repeatability, flexibility, upgradability, ease of operation and setup, and reliability

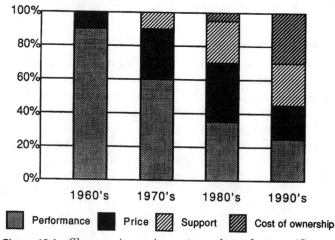

Figure 15.4 Changes in equipment purchase factors. (*Source: Semiconductor International, May 1993, p. 58*)

and downtime factors. Repeatability is the ability to produce the same result every time and over long periods of time. Flexibility relates to the ease with which a machine can be switched to run a variety of products and processes. This is an important factor for ASIC and other small fabrication lines. Upgradability is the ability of the machine to handle future process requirements, such as increasing wafer diameters. This factor is very important for fabrication lines doing ASIC work where staying abreast of the latest product technology is vital. In fabrication areas dedicated to long production runs of the same (or similar) products, the ability to upgrade is less important.

Ease of operation and setup factors address the issues of minimizing operator mistakes by good design and user-friendly controls. Setup issues relate to the amount of time required to bring a tool on-line (tests, calibrations, etc.) and the related loss of production time.

Most companies run equipment two to three shifts per day, sometimes six to seven days a week. Scheduled maintenance and unscheduled breakdowns stall production flow and run up expenses. These are collectively called downtime factors, and include *scheduled maintenance frequency and time, mean time to failure* (MTF), and *mean time to repair* (MTR). Given the expense of most machines, having backups is a costly luxury. It falls on the vendor to provide a machine that runs for long periods of time (MTF), can be repaired quickly (MTR) when failure occurs, and does not need frequent and lengthy routine maintenance. Cost also includes the price of materials needed for the machine process. Machines that waste materials also waste money.

Vendor support has emerged as a critical factor. The advancement of process technology and equipment sophistication has forced close cooperation and alliances between the chip manufacturers and their suppliers. The cost of machine development requires input from the using-process engineers, and the detailed nature of the machines requires on-time backup from the vendors. Management guru Tom Peters sees modern corporations as having "fuzzy edges." By this he means that more and more a corporation must rely on its vendors and customers for information and codevelopment of needed equipment and materials. This is certainly true of the semiconductor industry and is the model practiced by the very successful Japanese semiconductor industry.

Cost of Ownership

Shifting equipment purchase decisions from purely technical to business-related factors has driven the development of cost of ownership (COO) models. These models attempt to bring together all of the relevant factors driving the total cost of ownership of a tool, process, or facility over its expected lifetime. Beyond the initial equipment purchase price, tools differ in the amount of expensive floor space they occupy; the amount of power and materials they require to operate; the yield of in-specification wafers; and maintenance, repair, and failure rates. One COO formula was developed by Sematech to evaluate equipment purchases:[6]

$$C_w = \frac{\$F + \$V + \$Y}{L \times TPT \times Y_{TPT} \times U}$$

where C_w = cost of the finished wafer, based on the particular cost associated with the tool and process for the lifetime of the tool

$\$F$ = the fixed costs, which include the equipment purchase prices, facility costs, initial modifications, and so on.

$\$V$ = the material, labor, and process costs generated when the tool or process is operating

$\$Y$ = the cost of wafers scrapped due to tool-scrap and defect-induced loses

L = the lifetime of the tool in hours

TPT = the wafer throughput rate, in hours, as reduced from the maximum by maintenance requirements, setup, test-wafer monitoring, and so on

U = the tool utilization factors that reduce available process time from the maximum

Y_{TPT} = yield factor

Each of these equation terms are calculated from formulas that consider the subfactors for a particular process. The COO formula provides the method to determine tradeoffs for various tool factors. For example, a tool may have a high initial cost that is offset by lowered operating costs or a long time between failures. Or a tool may provide a high yield but require so much adjustment and calibration that additional machines would be required to meet production schedules.

Automation

As industries mature, the technology becomes stabilized and the market drives up demand. These two conditions are the precursors to process automation. Since 1940, automation of oil refineries has reduced the number of workers by a factor of 5.[7] Automation of semiconductor processes has been in progress since the 1970s when process tools were designed to accept wafers in cassettes. Since then, automation has marched along toward the fabled dream of the total "peopleless, lights-out" fabrication process. Automation stages start with the process tool and extend to the factory level.

Process automation

The first level of automation is of the process itself. Most semiconductor equipment, by definition, automates a part of the process. Photoresist spinners automatically dispense the primer and resist at the correct speeds and for the correct time. Automatic gas-flow controllers dispense gases to the tool in the right amount, at the right pressure, and for the right time. Process automation brings consistency to the process and the product by reducing reliance on operator skills, training, morale, and fatigue.

Most tools are controlled by a set of instructions programmed in an onboard computer. The program is called a *recipe*. The recipe is loaded into the machine by the operator or from a central host computer.

Wafer-loading automation

The next level of automation is the loading and unloading of the wafers. The industry has settled on the wafer cassette as the primary wafer holder and transfer vehicle. Cassettes are placed on the machine by various mechanisms. Elevators and/or wafer extractors or robots feed the wafers into the particular process chamber, spin chuck, etc. In some processes, such as process tubes, the entire cassette is placed in the process chamber. This level of automation is referred to as the "one button" operation. With one button the operator activates the loading system and the wafers are processed and returned to the cassette. At the end of the cycle, the machine sounds an alarm or light and the operator removes the cassette.

Some machines have buffer storage systems that maximize the machine efficiency by always having fresh wafers available for processing. The operator places the cassette(s) on the machine loader and pushes a start button, after which the machine takes over the processing.

At the 12-in-diameter-wafer level the cassette will probably give way to individual wafer holders or transporters.

Clustering

Combining two or more process steps in a single unit represents another level of automation. Generally this level is called *clustering*.

The industry has been clustering for a long time. Photoresist spinners were long ago combined with soft bake modules and other track equipment.

Recent clustering designs (Fig. 15.5) have been driven by both technical and economic forces. On the technical side, some clustered processes make a better product, such as a silicon wafer that is cleaned after etch and before metal deposition. Another process advantage is sequential deposition of different materials in the same chamber. In these cases, the deposition process is better because the wafer is not exposed to air between steps. Certainly any time a wafer loading or unloading step can be eliminated, both cleanliness and cost are favorably affected. For vacuum processes, time and cleanliness factors are enhanced when two or more processes can be performed with only one vacuum pump down. Two or more sequential processes performed in a cluster are called *integrated processes*.[8] On the economic side, some processes clustered for increased throughput are called *parallel processes*.

Despite the obvious attractions of clustering there are drawbacks. Clustering for critical process advantages is easy to justify. But a cluster of the same types of processes requires interlocks, electronics, and software more sophisticated than individual tools. A shutdown for maintenance or repair idles a larger part of the production capacity when clusters are used. Another barrier occurs when cluster modules cannot be provided by the same vendor. Customers prefer one responsible vendor and vendors are slow to pair up when responsibilities are clouded.

Figure 15.5 Three-chamber cluster tool.

Wafer-delivery automation

The third level of automation is when wafers are automatically brought to and loaded on and removed from the machines. Early delivery systems used traveling robotic carts that duplicated human delivery (Fig. 15.6). Called *automated guided vehicles* (AGVs), the carts travel along the aisles and dispense wafer cassettes when the machines need them. This approach has the advantage of being retrofitable to fabrication lines where the equipment is lined up in rows.

Another approach is the use of an overhead rail[9] (also called a gantry). The wafer cassettes arrive at the process tool area (Fig. 15.7) where a secondary system (usually a robot) removes them from the overhead rail and places them in the tool buffer. This system works best with machines that are grouped in bays rather than in the traditional linear layout.

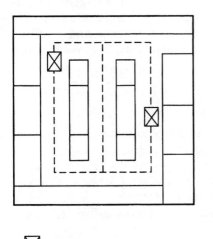

⊠ AGV

| AGV Path

Figure 15.6 Automated guided vehicle wafer delivery system.

Figure 15.7 Overhead rail wafer delivery layout.

Closed-loop control-system automation

The industry is just embarking on the last level of total automation, the closed-loop feedback system. There are two aspects. First, some tools have onboard sensors that measure critical process parameters. Through feedback electronic circuits, the tool adjusts itself to maintain process operation specifications. This stage of automation is difficult to achieve for many processes. It requires measuring sensors that operate in hostile environments, such as heated-tube furnaces and sputtering chambers. One area of progress is in the area of mask making where recognition systems are advanced enough to compare the manufactured plates with the design criteria.

Second, some tools have, or are connected to, automatic inspection or measuring subsystems. The subsystem measures important parameters in real time (as they are happening), compares them with a standard, and feeds the information back. This may trigger a shutdown of the out-of-specification tool or make process parameter adjustments. The development of closed-loop process and machine control is necessary before the goal of a "peopleless, lights-out" wafer fabrication plant can be realized.

Factory-Level Automation

The advent of higher levels of process and tool automation and inventory control systems requires higher levels of centralized control and information sharing. Most companies have computer-based management information systems (MIS) handling the paperwork and details of employment and finances. These systems are being expanded to the entire manufacturing environment in a process called *computer-integrated manufacturing* (CIM).

CIM is the computerization of all plant operations and the integration of those operations into one computer design, control, and distribution system. The processes involved in CIM are all related and interdependent, as illustrated in Fig. 15.8. The major activities of CIM are business functions, product design (mask and circuit), manufacturing planning (inventory, shop floor priorities, etc.), manufacturing control, and the fabrication processes. A complete CIM system is interactive at all levels. This means that each of the five functional areas inputs data to the system in real time and that the information is available to all who need it.

Computer-aided design (CAD) and computer-aided manufacturing (CAM) are two subsystems within the CIM system. The role of CAD has been discussed in mask and circuit design. CAM is the part of the

Figure 15.8   Provinces of factory computer control systems.

system that does the planning and control of the manufacturing operation. A CAM system includes a computer network and the automated process tools and material delivery systems.

In concept, a CIM system kicks into operation when a customer order is received. The computer logs the order and initiates the CAD system to start the design (if it is a custom order). It also triggers the ordering of needed materials in the right quantities and schedules their delivery times. This subsystem goes by the name *computer-aided process planning* (CAPP). Through the CAM program the individual process recipes are downloaded to the individual equipment computers. Once processing begins, the CAM system controls the number of wafers in process (WIP) and makes necessary priority decisions to meet shipping schedules. It also keeps track of equipment performance and schedules repairs and maintenance.

An important feature of the CAM system is yield monitoring and reporting. Important measuring systems are connected directly to the factory computer. If a poor yield problem appears, the system reports it to the engineering and facility staff and (if necessary) will reorder materials to make up the losses. At the end of the production run, the system calculates costs and yields and schedules shipment to the customer.

Other packages included in CIM systems are facilities monitoring, process modeling, and security systems. Facility monitoring might include power consumption and environmental factors inside the plant. Some facility CAM systems include monitoring the levels of liquids and gases in storage with automatic alerting and/or reorder. Monitoring and precise control of these factors can save appreciable amounts of money. Process modeling is a system of testing a particular design against the known process variations. The computer can run many variations, simulating the changes expected in fabrication. Good modeling can identify weak points in the design or process before wafers are committed to the line. Security monitoring may include the entering and exiting of employees (and/or intruders) and the securing of expensive finished product or materials. A security system may also include fire and other hazards control.

Fabrication Floor Layout

The traditional fabrication floor layout, consisting of equipment lined up in rows, has given way to the bay or tunnel system (Fig. 15.7). This system groups sections of the process in separate rooms to minimize contamination from personnel traffic and cross-contamination from other process areas.

Automation is forcing a rethinking of the most efficient way to move material while maintaining cleanliness. Robots have become the material movers of choice. Semiconductor processing has never lent itself to conveyor belt manufacturing lines. There are too many stations with different loading mechanisms and groupings of wafer lots. Also driving robot use is the increasing weight of wafer lots (because of larger diameters), especially in SMIF-type transfer modules.[10] In vacuum tools, robotic design and performance are challenging, especially if the process uses corrosive gases.

One equipment layout option is the process island. The various tools for a particular process segment are grouped around a single loading–unloading robot (Fig. 15.9). The tools may be single or in clusters.

Batch versus Single-Wafer Processing

For the first two decades of process development, the goal was to produce ever-larger batches of wafers. Simple economies of scale were behind this goal: to yield more products per process sequence. However, the arrival of larger-diameter wafers and VLSI- and ULSI-level circuitry has changed the picture. Larger wafers require larger process chambers, which in turn pushes the limits of uniformity. Many processes, particularly the deposition steps, are easier to con-

Figure 15.9 Robot wafer delivery in process island,

trol in a smaller (single-wafer) chamber. Fast-cycle-time fabrications (ASIC and small-volume devices) can use the flexibility of single-wafer processing to good advantage.[11]

Improved process control is offset by a loss of productivity as compared to batch processing. Larger-diameter wafers help somewhat (300-mm wafers have almost four times the area of a 150 mm wafer). Tool designs that allow one vacuum pump down for an entire batch while wafers are processed individually, along with other improvements, increase single-wafer process productivity.

On the downside, single-wafer processing requires higher levels of repeatability. Assuming same-size wafers, a single-wafer processor will have to perform 25 times compared to a one-batch 25-wafer batch process. Also, the equipment is more expensive. Each single wafer processor requires essentially the same components as a batch tool.

Statistical Process Control (SPC)

This text, for the most part, has presented the process technology with few mathematical formulas. The next few sections on statistical process control will break with that approach. SPC is a powerful and necessary tool to maintain process control and improve yields. SPC is based on statistics, which is applied using the language of mathematics. We will attempt to convey the background, use, and interpretation of commonly used statistical tools with illustrative charts and graphs. No formulas will be presented.

The first question to address is, "What is the object of process control?" The answer is simple: to produce a product that falls within its design and operational specifications and do it at a high enough rate to be profitable. Challenging this simple goal are processes that are influenced by a multitude of parameters, including the conditions already on the wafer.

Processes do not run in control without monitoring and adjustments. Process control provides the information to make the necessary changes. Statistical process control does it in a way that relies on established mathematical principles that govern the type of processes used to make chips. Process control techniques can vary from simple to very complex. The simplest (and most familiar) is the calculation of the average of a group of numbers (called a population). We also know that the extremes of a population (range) along with the average give us an idea of the distribution of the data within the population.

For example, the two groups in columns A and B (Fig. 15.10) have the same average. If the numbers were the sheet resistance of wafers from furnaces A and B, we could easily come to the decision that furnace A has the most in-control process because of its tighter distribution. This fact is illustrated when the data are plotted (Fig. 15.11). The plots are called *histograms* and visually display data distributions that a simple average calculation would not reveal. Average calculations and histograms are statistics in action. Histograms are usually the first step in determining whether or not a process is in control.

Their power comes from a mathematical distribution known as the gaussian distribution. Named after the famed mathematician Gauss, its origin is interesting. Gauss set out to reconcile the different star

Reading no.	Run A	Run B
1.	25	28
2.	24	25
3.	26	23
4.	23	26
5.	25	25
6.	26	22
7.	25	23
8.	24	25
9.	25	27
10.	26	25
Average	24.9	24.9
Range	26 − 23 = 3	28 − 22 = 6

Figure 15.10 Sheet resistance reading ($\Omega/\square$).

Figure 15.11 Frequency distribution of sheer resistances.

positions reported by different astronomers. His approach was to make all the necessary corrections to the observations, taking into account the time of the year and from where on the earth the observations were made. He expected that when all the corrections were made, there would be agreement from all the position calculations on the position of a particular star. After all, reason dictates that a star can occupy only one position at a time, and we should be able to determine that position.

However, the final data did not confirm his hypothesis. After all corrections were made, there were still a number of locations for each star. Fortunately Gauss did not scrap the project but went on to establish the basis for the field of statistics and distribution probabilities. If the math-aversion readers will hang on, they will find that the concept of probability is not so difficult. The way Gauss analyzed the data was to plot the various calculated locations for a star. He calculated the center point (average) and drew a circle that encompassed the star location the farthest from the center. He reasoned that there was a 100 percent probability (a probability of 1) that the real star location was within that circle. He also reasoned that the probability

of finding the star within the confines of smaller circles was less than
1. In fact, the smaller the circle, the lower the probability that the
star was within those boundaries.

It also turns out that processes that produce data, as in the example given, and most semiconductor processes follow the gaussian distribution. A good example is the height of blades of grass in a lawn. If
all the blade heights are measured and plotted on a histogram, the
distribution will be the familiar bell-shaped curve, also called a *normal curve* (Fig. 15.12). From a probability consideration, it predicts
that there is a higher probability that any given blade of grass will
have a height closer to the average (center values) and that there is a
lower probability that any given blade will be very short or very tall.
The mathematical conditions that result in the distribution include
human heights, IQ distributions, and (in most cases) semiconductor
process parameter distributions, such as sheet resistance.

A first step in process control is to make a histogram of the particular process parameter and determine if the distribution is a normal
distribution. If it is not, the chances are good that there is something
wrong in the process. However, there are now normally distributed
processes. If the distribution is a normal one, the next step is to compare the range of the distribution with the design limits for the particular parameter (Fig. 15.12). This comparison is made to determine if
the natural process distribution limits fall within the design limits. If
they do not, the process must be fixed or some percentage of the parameter readings (and the wafers) will always be out of specification.

Another useful statistical tool is the paredo chart. This is a form of
the histogram but with the x axis divided into unrelated sections
rather than a continuum of some parameter such as sheet resistance.

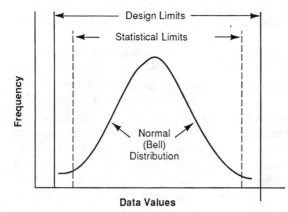

Figure 15.12 Comparison of normal distribution to design
limits.

Defect inspection results are a good candidate for paredo charts. On the x axis is a list of the defect categories. A mark or vertical column over each defect type indicates the frequency of occurrence of each of the defect types. The operator or process engineer can see from the chart which defects are occurring most often and what processes must be improved.

So far the statistical methods explained are the after-the-fact history of a process. A most powerful statistic of real-time process control is the X-R control chart (Fig. 15.13). The chart is constructed in two parts with the y axis representing the parameter values. The top graph has a horizontal line for the historic average of the parameter. On each side of the average are control limit lines that are calculated from the historic data.[12] A control limit represents the limits that the individual data values will range between when the process is in control. Also on the chart are the process or design limits which represent the extremes the individual data points may have, before being rejected. The bottom graph is constructed by calculating and plotting the amount that each data point varies from the average. When plotted, these values give further visual evidence of the amount of control in the process.

The value of the X-R bar control charts is their predictive powers. A process in control will produce data points that tend to vary in a regular pattern about the average (top of Fig. 15.13). The mathematics of a controlled process predicts this regular fluctuation. It also predicts *when* a process is going out of control *before the data points exceed the*

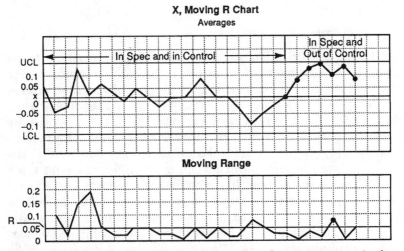

Figure 15.13 Moving R chart contains the averages of measurements x in the upper plot. The lower plot shows the moving range R which is descriptive of process stability.

control limits (bottom of Fig. 15.7). The data points in part B have shifted to the top of the control limit range. This is an unnatural pattern for an in-control process. When this situation occurs, the production operators, who maintain the charts as the data are produced, alert the proper personnel so the process can be brought back into control *before the data points exceed the control or design limits* and wafers have to be scrapped. A number of more sophisticated controls used in processing are beyond the scope of this text.

Another powerful statistical tool is multivariable experiment analysis. Most measured quality-control parameters (sheet resistance, line width, junction depth, etc.) are influenced by a number of variables in the process. Line width, for example, varies with the resist solution, film thickness, exposure radiation time and intensity, baking temperatures, and etch factors. Any one or all can contribute to an out-of-spec condition. Multivariable evaluations allow the engineer to run tests that separate and identify the contribution(s) of each of the individual variables.

Designing an SPC system for a process requires selection of the proper statistical tool and parameter or indicator population. Ideally, all the die on all the wafers from all the batches, day in and day out, should meet the specifications. In reality, picking the parameter population is not always an easy chore, because there are variations across the wafer, variations from wafer to wafer within a batch, and variations from tool to tool. Since every die cannot be measured, selecting the right sample point and sampling level becomes a demanding task.[13]

Inventory Control

A critical issue in fabrication cost control and yield is inventory levels and control. As the number of processing steps has increased, so have the length of the processing time and the number of wafers in process (WIP). The problem is that the company pays for the wafers when purchased but doesn't receive payment until the finished devices are shipped. This period can vary from two to eight months for a production line making similar circuits. Fabrication lines doing ASIC circuits have an even heavier burden when many different circuit types are going through the system. To get an idea of the burden, consider a CMOS-type process with 50 major steps and 4 substeps each for a total of 200 processes. The high cost of the equipment generally requires some buffer inventory to ensure that the machines are operating at maximum efficiency. If each buffer has 4 cassettes of 25 wafers each, the total inventory of WIP is 40,000 wafers. At a cost of

$100 per wafer (large diameter) the total inventory burden becomes $4 million.

Excessive WIP affects productivity by having the capability of hiding process and equipment problems.[14] With a lot of inventory at the stations, wafers can keep flowing out the back end while parts of the process are shut down. With a lower WIP these problems are readily apparent and force the fixing of problems. WIP also influences the overall fabrication yield. The collective experience of the industry is that the longer wafers are in the process the lower their wafer-sort yield.

Just-in-time inventory control (JIT)

Just-in-time inventory control is based on a philosophy with the objective of "Make only what is required, only as required."[15] The system is simple in concept. All buffer inventories are reduced to an absolute minimum, from the storeroom to the machine buffers. To work effectively, excellent vendor relationships must be established. The incoming materials have to be of the highest quality as JIT leaves little cushion for extensive incoming inspections and returns. Second, vendors are asked to maintain ready-to-ship materials at their facilities. In effect, the vendors are asked to hold the inventory previously held by the chip manufacturer, a situation that they are not always happy about. The chip manufacturers have to be very efficient in assessing their quantity, quality, and delivery needs, and must have a system that expedites material to the proper process tool and detects quality problems quickly.

JIT also has applications in the process flow. Some companies will only commit wafers to a process if the work can flow unimpeded through all of the substeps. This system gives higher yields and, if on a properly balanced line, increases throughput time, even though parts of the line are idle. This system is called *demand-pull*. Wafers "upstream" are worked on only when there is demand from downstream process stations about to run dry of wafers.

An effective JIT procedure can reduce the number of operators in the fabrication area since a good proportion of their time is spent sorting, staging, and delivering work to the process tools. Fabrication-area layouts can change from the traditional linear arrangement of process stations. A linear layout benefits a line making only a few different product types. Wafers are introduced into the front of the line and physically move on a first in–first out (FIFO) basis. Problems arise when particular lots of wafers must be moved quickly. Usually tagged as *hot lots*, they are given priority processing at each station.

Besides the control problems associated with keeping track of the lots, hot lots have the effect of making regular product sit in a queue waiting for the hot lots to clear. This layout is particularly cumbersome for ASIC lines where shipping dates and product types have constantly changing priorities.

JIT-CAM systems offer the advantage of knowing where all the lots are in the production cycle. The computer can stage the work to minimize disruption and keep a steady flow of product. Further, JIT-CAM systems teamed with automatic delivery systems can be more efficient when the process stations are grouped rather than strung out in a line. This concept, called *work cells,* is more efficient, especially when a cell can run a number of different products. When a machine is inoperable in a linear layout, the line comes to a halt and inventory builds up in front of the machine that is down.

Equipment and Material Standards

Given the many suppliers of materials and equipment and the very stringent technical needs of chip manufacturers, the need for standards is obvious. In many industries, different manufacturers attempt to establish their standards as the industry standards, as a way of maintaining a competitive edge. Fortunately, the semiconductor industry informally settled on many standards, for example, the use of standardized cassettes.

In 1973, Semiconductor Equipment and Materials International (SEMI) established a standards program. Since then, supplier and user personnel have come together to establish standards by a consensus process.

At the automation level, SEMI has published two communication protocols for equipment interfaces called *S*emiconductor *E*quipment and Materials International *C*ommunications *S*tandard (SECS I and SECS II). There is also a standard mechanical interface (SMIF) standard and a standard for vacuum-type cluster tools from the Modular Equipment Standards Committee MESC.[16]

Quality control and ISO 9000 certification

SPC and other product, process, and personnel quality programs fall under the general category of quality control. In the semiconductor industry, there are two broad quality segments, quality control (QC) and quality assurance (QA). QC generally refers to all of the techniques used to monitor and control the *process* and the quality of the wafers while they are in the process. QA refers to the efforts to ensure that the customer is receiving product that meets required specifications.

In 1987, the ISO 9000 system of quality requirements for organizations was introduced. Developed by the International Organization for Standardization (ISO), the requirements are more like guidelines, and are sometimes called an "umbrella standard"[17] rather than a strict cookbook set of standards.

The ISO 9000 guidelines and standards guide the development of and give a critique of a complete quality system for product, process and management. Individual companies decide on how best to comply and implement programs. With certification, customers and vendors know that there is a comprehensive quality program backing the product. The European Common market (EC) requires that all companies doing business in the EC have ISO 9000 certification. In the United States, the American National Standards Institute (ANSI) and American Society for Quality Control (ASQC) have developed equivalent standards. Assisting semiconductor suppliers is a program sponsored by the Semiconductor Equipment and Materials International (SEMI).[18]

Line Organization

Most fabrication areas are organized around the product-line concept, in which fabrication areas are built to accommodate products with similar processing needs. For example, there are bipolar lines, CMOS lines, and so on. This arrangement makes for more efficient processing since most of the machines are in use most of the time and the staff can gain experience in processing a few products.

The staff of these lines are also fairly self-contained. The primary responsibility falls to the fabrication or product manager (Fig. 15.14). Reporting to this individual are an engineering supervisor, a production manager (or general supervisor), a design department, and the equipment maintenance group. The production manager is responsible for producing the finished wafers to specification, to cost, and to schedule. The engineering group is responsible for the development of high-yield processes, documentation of the processes, and the daily sustaining of the line process. Both the production and engineering staffs are divided into groups focusing on a particular part of the process. This organization has the virtue of high focus on the fabrication area's primary goal of producing chips at a profitable level.

As the processes become more automated and arranged in process cells, small group organizational teams and responsibilities are emerging. A cell is attended by the operator(s), the equipment technicians, and the process engineer(s). These small groups make floor-level decisions with the information provided by the CIM system. However, few companies have formalized this arrangement with an

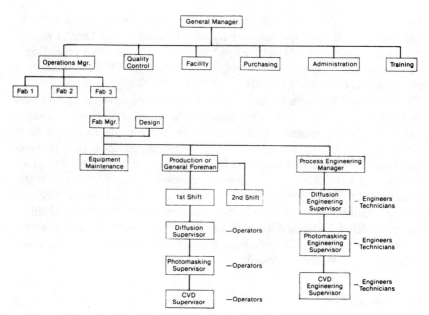

Figure 15.14   Typical semiconductor product line table of organization.

organization structure, and the teams tend to exist as cross-depart-ment cooperatives.

Key Concepts and Terms

Automation levels	Histogram
CAD	Inventory control
CAM	Just-in-time (JIT) inventory
CAPP	MIS
CIM	Normal distribution
Clustering	Process limits
Control chart	Product-line organization
Cost of ownership	Statistical process control
Data averages	Wafers out per operator
Design limits	Wafers out per square foot
Fabrication cost factors	WIP

Review Questions

1. List the major factors determining the manufacturing cost of a wafer.

2. Rank the factors according to their level of influence.

3. Describe how WIP impacts profit.

4. Discuss how statistical process control can enhance wafer yields.

5. List the four levels of automation in a fabrication process.

6. Describe the philosophy and operation of a JIT inventory system.

7. What is the difference between CIM and CAM?

8. What is a process cell and how does it compare with a traditional line layout?

9. List the production, process, and yield advantages of a totally automated fabrication area.

10. Draw an organizational chart of a product-line-managed fabrication area.

References

1. J. G. Harper and L. G. Bailey, "Flexible Material Handling Automation in Wafer Fabrication," *Solid State Technology,* July 1984, p. 94.
2. W. Maly, "Growth of Silicon Cost and Resulting Changes in Manufacturing Paradigms," *Semiconductor International,* Cahners Publishing, June 1994, p. 158.
3. S. M. Sze, *VLSI Technology,* McGraw-Hill, New York, 1983, p. 5.
4. J. G. Harper and L. G. Bailey, "Flexible Material Handling Automation in Wafer Fabrication," *Solid State Technology,* July 1984, p. 94.
5. J. Owens, Industry Strategy Symposium, *Semitech, Semiconductor Equipment and Materials Institute (SEMI),* October 1994.
6. P. Burggraaf, "Applying Cost Modeling to Stepper Lithography," *Semiconductor International,* Cahners Publishing, February 1994, p. 40.
7. S. Shinoda, "Total Automation in Wafer Fabrication," *Semiconductor International,* September 1986, p. 87.
8. P. Singer, "The Thinking Behind Today's Cluster Tools," *Semiconductor International,* August 1993, p. 46.
9. Ibid.
10. P. Singer, "The Thinking Behind Today's Cluster Tools," *Semiconductor International,* August 1993, p. 46.
11. M. Moslehi, "Single-Wafer Processing Tools for Agile Semiconductor Production," *Solid State Technology,* Penwell Publications, January 1994, p. 35.
12. D. M. Campbell and Z. Ardehale, "Process Control for Semiconductor Manufacturing," *Semiconductor International,* June 1984, p. 127.
13. W. Levinson, "Statistical Process Control in Microelectronics Manufacturing," *Semiconductor International,* Cahners Publishing, November 1994, p. 95.
14. K. Levy, "Productivity and Process Feedback," *Solid State Technology,* July 1984, p. 177.
15. Ibid.
16. P. Singer, "The Thinking Behind Today's Cluster Tools," *Semiconductor International,* August 1993, p. 46
17. E. Hnatek, "ISO 9000 in the Semiconductor Industry," *Semiconductor International,* July 1993, p. 88.
18. P. Dunn, "The Unexpected Benefits of ISO 9000," *Solid State Technology,* March 1994, p. 55.

16

Semiconductor Devices and Integrated Circuit Formation

Overview

Integrated circuits are composed of individual conductors, fuses, resistors, capacitors, diodes, and transistors. The operation and formation of each is explored. The formation of the major integrated circuits from the components is also presented in this chapter.

Objectives

Upon completion of this chapter you should be able to:

1. Sketch and identify the structural parts of the individual components of an integrated circuit.

2. Explain the role and different isolation structures used for integrated circuits.

3. Sketch and identify the operation of a bipolar and MOS transistor.

4. List the types and advantages of the different MOS gate structures.

5. Sketch and identify the parts of a Bi-MOS circuit.

Semiconductor-Device Formation

The previous chapters have focused on the individual processes used to make semiconductor devices (also referred to as components or circuit components) and integrated circuits. In this chapter the actual formation of the devices and circuits is explained. It is assumed that the reader has already read about (or is familiar with) the processes and has a good understanding of the electrical performance of the individual components as explained in Chap. 14.

There are literally thousands of different semiconductor device structures. They have been developed to achieve specific performances, either as discrete components or in integrated circuits. In this chapter the basic structures are explained along with the major variations used for specific applications. The circuits components are:

- Resistors
- Capacitors
- Diodes
- Transistors
- Fuses
- Conductors

Resistors

Resistors have the effect of limiting current flow. This is accomplished by the use of dielectric materials or high-resistivity portions of a semiconductor wafer surface. In semiconductor technology resistors are formed from isolated sections of the wafer surface, doped regions, and deposited thin films.

The value of a resistor (in ohms) is a function of the resistivity of the resistor material and its dimensions (Fig. 16.1). The relationship is

$$R = \rho L/A$$

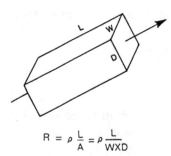

$$R = \rho \frac{L}{A} = \rho \frac{L}{W \times D}$$

Figure 16.1 Relationship of resistance to resistivity and dimensions.

where ρ = resistivity of material
   L = length of resistive region
   A = cross-sectional area of the resistive region

The area (A) becomes $W{\times}D$, where W = width of the resistor and D = depth of the resistive region. For doped resistors, the length and width are the surface pattern openings and the depth is the junction depth.

It should be obvious that every doped region is also a resistor and the basic resistor formula governs electrical flow. A conductor is simply a resistor with a low resistance. The conceptual importance of Ohm's law is that the electrical resistance of any region in the device or circuit is altered by any change in dimensions or change in the doping level.

Doped resistors.   Most of the resistors in integrated circuits are formed by a sequence of an oxidation, masking, and doping operation (Fig. 16.2). A pattern is opened in the surface oxide. Typical resistor shapes are dumbbells (Fig. 16.3) with the square ends serving as contact regions and the long skinny region in between serving the resistor function. The resistance of this region is calculated from the sheet resistance of the region and the number of squares ($\square$'s) contained in

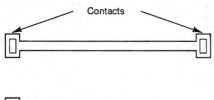

Figure 16.2   Diffused resistor formation.

Contacts

Figure 16.3   Resistor shapes.

the region. The number of squares is calculated by dividing the length by the width.

After doping and a subsequent reoxidation, contact holes are etched in the square ends to contact the resistor into the circuit. A resistor is a two-contact, no-junction device. The term *no-junction* means that the current flows between the contacts without crossing an N-P or P-N junction. (But the junction serves to confine the current flow in the resistive region.)

Resistors doped by ion implantation have more controlled values than those in diffused regions. Doped resistors can be formed during any of the doping steps performed during the fabrication process. A bipolar base mask will have the base pattern and a set of resistor patterns. In MOS circuits, resistors are formed during the source-drain doping step. The resistor has the same doping parameters (sheet resistance, depth, and dopant quantity) as the transistor part.

EPI resistors.   A resistor can be formed by isolating a section of an epitaxial region (Fig. 16.4). After surface oxidation and contact hole masking, what is left is a three-dimensional region functioning as a resistor.

Pinch resistors.   Ohm's law shows that the cross-sectional area of the resistor is a factor in its value (Fig. 16.5). One way to reduce the cross-sectional area (and increase the resistance) is to dope the resistor region and then do another doping of the opposite conductivity type. This occurs in bipolar processing when a resistor region is formed during the P-type base doping with a "pinched" cross section

Figure 16.4 Epitaxial layer resistor.

Figure 16.5 N-type resistor "pinched" with P-typed doped region.

formed from a subsequent N-type region formed along with the emitter doping.

Thin film resistors. Doped resistors don't always have the resistance control needed in some circuits and are poor performers in radiation environments. Radiation, such as found in space, generates unwanted holes and electrons that allow the current to leak across the confining junction. Resistors formed from deposited thin films of metal do not have this radiation problem.

The resistors (Fig. 16.6) are formed by either film deposit and masking sequences or by lift-off techniques. After resistor formation of the wafer surface, it is "wired" into the circuit by contact between the resistor ends and the leads of the conducting metal. Nichrome, titanium, and tungsten are typical resistor metals.

Capacitors

Oxide-silicon capacitors. Silicon planar technology is based on a silicon wafer with a grown silicon dioxide layer on top. The conducting metal leads lie on top of the oxide. This is a capacitor structure, in fact, a metal-oxide-metal (MOS) capacitor structure. However, the oxide thickness has to be thin enough (about 1500 Å[1]) for the structure to act as a capacitor. The top electrode is also called a *cell plate* and the bottom electrode is also called the *storage node*.

A capacitor is a device that stores charge. A battery is a capacitor. When a voltage is applied to the metal, a charge builds up in the wafer layer under the oxide (Fig. 16.7). The amount of charge is a function of the oxide thickness, the dielectric constant of the oxide, and the area, as defined in the metal top plate. Capacitors of this structure are called parallel-plate, monolithic, or MOS (after the *m*etal *o*xide *s*emiconductor materials of the sandwich).

Figure 16.6 Formation of a thin film resistor.

Figure 16.7 Monolithic capacitor.

In dense circuits, an oxide-nitride-oxide (ONO) dielectric sandwich is used. The combination film has a lower dielectric constant, allowing a capacitor area smaller than a conventional silicon dioxide capacitor.

Junction capacitors. A capacitor is formed at every junction in a device. When a voltage is applied across any junction, carriers on each side of it move away from the junction, leaving a depleted region (Fig. 16.8). This depleted region acts as a capacitor in the device or circuit. The value of this junction capacitance must be taken into account when the circuit is designed. Some circuits actually use junction capacitors as part of the circuit design. In some circuits, the natural junction capacitance has the effect of slowing up the circuit operation. This is due to the time required to "fill up," or charge, the depleted region before current flows. A finite time is also required for the various junction capacitors to discharge. Both of these times affect circuit switching and operational speeds.

Trench capacitors. Preservation of wafer surface area is always a design criterion. One of the problems with oxide-metal capacitors is their relatively large area. Trench (or buried) capacitors solve the problem by creating a capacitor in a trench etched vertically into the wafer surface (Fig. 16.9). The trenches are etched either isotropically with wet techniques or anisotropically with dry etch techniques. The trench sidewalls are oxidized (the dielectric material) and the center of the trench filled with deposited polysilicon. The final structure is "wired" from the surface, with the silicon and polysilicon serving as the two electrodes with the silicon dioxide dielectric between them.

Stacked capacitors. Another alternative to conserving surface area is to build stacked capacitors on the wafer surface. This effort has been

Figure 16.8 Depletion layer junction capacitor.

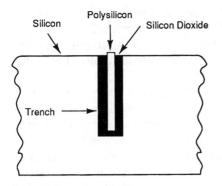

Figure 16.9 Trench capacitor.

driven by the need for small high-dielectric capacitors for dynamic random access memory (DRAM) circuits. The storage part of a DRAM cell is a capacitor. In DRAM cells, the bottom electrode is usually polysilicon or hemispherical grain polysilicon (HSG). Capacitor shapes can be planar, cylindrical, or with "fins."[2]

Capacitor dielectric materials under consideration include tantalum pentoxide (Ta_2O_5) and barium strontium titantate ($BaSrTiO_3$ or BST).

Top electrode materials may be titanium nitride (TiN), tantinum nitride (WN), platinum (Pt), polysilicon, or one of the other semiconductor conducting materials.

Diodes

Doped diodes. A diode is a two-region (two-contact) device separated by a junction. A diode either allows current to pass easily or acts as a current block. Which function it performs is determined by the voltage polarity, called *biasing* (Fig. 16.10). When the current voltage is the same as in the diode region the diode is in forward bias and the current flows easily. When the polarities are reversed the diode is

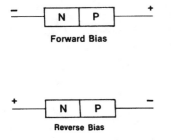

Figure 16.10 Forward and reverse biasing.

reverse-biased and the current is blocked. A reverse-biased diode can be forced into a conducting state by raising the current voltage until the junction goes into *breakdown*. This condition is temporary; when the voltage is reduced, the diode once again becomes a blocking device (see Chap. 14).

Diodes are used in circuits to steer the current around the circuit. By proper choice of the current polarities and diode polarities, the current is allowed to pass into some branches and is blocked out of others. A planar diode is formed from a doped region and two contacts on either side of the junction where it intersects the surface (Fig. 16.11). Diodes are usually formed along with transistor doping steps. Thus in bipolar circuits, there are base-collector diodes and emitter-base diodes. In MOS circuits, most of the diodes are formed with the source-drain doping step.

Schottky barrier diodes.   In 1938 (10 years before the invention of the transistor) W. Schottky[3] discovered that whenever a metal is in contact with a lightly doped semiconductor, a diode is formed (Fig. 16.12).

Figure 16.11 Formation of a P/N planar diode.

Figure 16.12 Schottky barrier diode.

This diode has a faster forward time (it responds faster) and operates with a lower voltage than a doped silicon junction diode. Metal contacts to highly doped regions (greater than 5×10^{17} atoms per cubic centimeter) are regular ohmic contacts. This is the situation for the majority of contacts in a silicon circuit. This Schottky diode effect is taken advantage of in some NPN bipolar transistors. The structure and effect are explained in the section on bipolar transistors.

Transistors

Transistor operational analogy. A transistor is a three-contact, three-part, two-junction device that performs as a switch or an amplifier. An often used analogy to explain the role of the parts and the operation of a transistor is the water flow system in Fig. 16.13. The flowing water represents current flow. In this system, the tank is the source of the water, the valve controls the flow, and the bucket collects the water. The system can be operated as a switch simply by turning the valve on and off. It can even be imagined in an amplifier role. Consider the valve as a high-mechanical-advantage miniature water wheel activated by a small external stream of water. A small trickle onto the valve wheel could open the valve to allow a large flow through the system. If the whole system was enclosed so that an observer saw only the trickle going in and a large flow coming out, he or she might conclude that the system was *amplifying* the water trickle.

Bipolar transistors

The same basic parts and functions are present in solid-state transistors. A bipolar transistor is shown both as a simple block arrangement and in a double doped planar form in Fig. 16.14. The current flows from the emitter region (tank) through the base (valve) into the collector (bucket). When there is no current to the base, the transistor is turned off. When it is on, the current flows. It only takes a small current to turn the base on enough to allow current flow through the

Figure 16.13 Water analogy of transistor operation.

Figure 16.14 Bipolar transistor operation.

whole transistor. The size of the base current regulates the larger amount of current flowing through the transistor (called the *collector current.*) There is an amplification of the base current to the collector current. The base current in effect changes the resistance of the base region. In fact, the term *transistor* comes from an early term for a bipolar transistor: *transfer resistor.* During operation both positive and negative currents flow in the base, hence the term *bipolar,* literally, two polarities.

The amplification property, called gain or beta, is numerically the result of dividing the collector current by the base current (see

Chap. 14). For efficiency, the emitter region has a higher doping concentration than the base, and the base doping is higher than the collector. A typical doping concentration versus distance plot is shown in Fig. 16.15.

Most bipolar circuits are designed with NPN transistors. NPN represents the respective conductivity types of the emitter, base, and collector. Some applications require PNP transistors, with many of them being formed laterally (Fig. 16.16). NPN transistors are more efficient because of the ease (higher mobility) of moving electrons in the N-type regions.

Bipolar transistors feature fast switching speeds. The speed is governed by a number of factors, of which the most important is the base width. Applying common sense, it stands to reason that the shorter the distance an electron or hole has to travel the less time it will take. Bipolar transistors can switch on and off in as little as a billionth of a second. To achieve this speed the transistor is maintained at an "on" state. This means that the base always has power applied, which is a downside of bipolar-based circuits. Another penalty for this necessary condition is a buildup of heat in the transistor. This heat will eventually affect circuit operation and is the reason for the cooling fans and air-conditioned rooms of earlier bipolar-based computers.

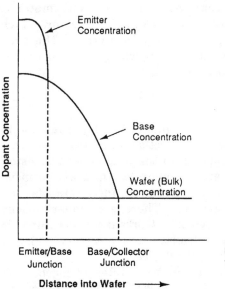

Figure 16.15 Dopant concentration profile of bipolar transistor.

Figure 16.16 Lateral and vertical PNP transistors.

Figure 16.17 Schottky barrier bipolar transistor.

Schottky barrier bipolar transistors. The Schottky barrier diode principle mentioned in a previous section is put to use in some bipolar transistors (Fig. 16.17). The construction requires that the base contact be extended into the collector region. When covered with metal, a Schottky diode is formed between the base and collector, which results in a faster-responding transistor. In a circuit the time required to turn the transistor on and off (switching speed) is critical when millions of transistors are operating.

Field-effect transistors (FETs)

Metal-gate MOSFET. As early as 1948, William Shockley noted another type of transistor operation, a field-effect-actuated transistor current flow. That effect was developed into the field-effect transistor, with the MOS structure now the most popular design. A MOS transistor, like a bipolar (Fig. 16.18), has three regions, three contacts, and two junctions but in a different structure. There is a similar analogy to the water system described previously. Current travels from the source region (tank), through the dielectric gate material (valve), and into the drain (bucket) before exiting the device.

A MOSFET gate controls current flow by a different mechanism than a bipolar base. The MOS structure shown in Fig. 16.18 is a simple metal-gate type capacitor and operates in the same way as a

Figure 16.18   MOS transistor operation.

capacitor. When a voltage (the gate voltage) is applied to the gate through the gate metal, a *field effect* takes place in the surface of the semiconductor. The effect is either a buildup of charge or a depletion of charges in the wafer surface under the top plate. Which event occurs depends on the doping conductivity type in the wafer under the gate and the polarity of the gate voltage.

The buildup or depletion of charge creates a channel under the gate which connects the source and drain. The surface of the semiconductor is said to be *inverted*. The source is biased with a voltage and the drain is grounded relative to the source. In this condition, a current starts to flow as the inverted surface creates an electrically connecting channel. The source and drain are essentially shorted together. Applying more voltage to the gate increases the size of the channel, allowing more current to flow through the transistor (see Chap. 14). By controlling the gate voltage, a MOS transistor can be used as a switch (on/off) or as an amplifier. However, MOS transistors are voltage amplifiers, unlike the current amplification of bipolar transistors.

If the source and drain are N type formed in a P-type wafer, the channel must be of N type for conduction to occur. This type of MOS transistor is called N channel. MOS transistors with P-type sources and drains are P channel. Most high-performance MOS circuits are built around N-channel transistors due to the higher mobility of electrons in silicon. The mobility makes N-channel transistors faster and they consume less power than P-channel circuits. They often are referred to as NMOS transistors. Figure 16.19 shows the major steps in the formation of an N-channel metal-gate MOS transistor.

Silicon gate MOS.   A certain amount of voltage must be applied to the gate metal before the channel forms. This voltage is called the *threshold voltage* or V_t. The value of the threshold voltage is an important and critical circuit parameter. A lower V_t means fewer power supplies and faster circuits.

A primary parameter that determines the threshold voltage is the *work function* between the gate material and the doping level in the

Starting Wafer

Field Oxidation

Source/Drain Mask and Doping

Gate Mask and Gate Oxide

Contact Mask and Metallization

Figure 16.19 N-channel metal gate MOS process.

Not Shown: Passivation Layer

semiconductor. The work function can be thought of as a kind of electrical compatibility. The lower the work function, the lower the threshold voltage, the lower the power required to run the circuit, etc.

Deposited doped polysilicon has a lower work function than aluminum as a MOS gate material and has become the standard gate electrode material for MOS transistors. The formation of the transistor is shown in Fig. 16.20. The polysilicon is heavily doped N type to reduce its resistance. Thus doped it serves as the gate electrode and as a circuit conduction line. A polysilicon gate can withstand subsequent high-temperature processing without degradation.

An additional benefit of the silicon-gate process is the self-aligned gate. In the metal-gate process sequence a hole for the gate oxide

Figure 16.20 Silicon gate MOS process steps.

must be patterned between the source and drain. To ensure that the source and drain are bridged by the gate, overlap for alignment tolerances must be allowed. This results in some overlap of the gate into the source and/or drain. The overlap becomes a region of unwanted capacitance. In the silicon-gate process the gate is formed first and acts as a mask to locate the source and drain. Thus, whatever the gate placement, the source and drain *self-align* to it.

Other factors, besides the gate metal, affecting the gate threshold voltage and operation are:

- Gate oxide thickness
- Gate material (dielectric constant)
- Source-drain separation (channel length)
- Gate doping level
- Sidewall capacitance of the doped source and drain regions

In the section on capacitors, the development of high-dielectric constant materials for higher performance in a smaller area were discussed. These same factors apply to MOS gate materials.

The thinner the gate oxide, the faster the device and the lower the threshold voltage. Current gate oxide thicknesses for production devices are in the 100- to 300-Å range. The technology roadmap projects that as the industry enters the 0.35- and 0.18-μm design rule era, gate thicknesses will fall into the 90- to 60-Å range.[4] In this range, maintaining thickness control, pinhole-free gates, and small area gates has driven the development of new gate materials. They are the same materials discussed in the section on capacitors. The new materials are oxide-nitride sandwiches (ONO), oxide tantalum pentoxide (OTa_2O_3), and amorphous silicon film stacks.

Channel lengths also affect the speed. Channel lengths have continually shrunk to the submicrometer range. In self-aligned structures, the channel length is established by the gate width. The gate doping level influences the threshold voltage by modifying the work function difference between the gate metal and surface. Ion implantation, with its ability to dope through thin oxides, is often used to set the gate doping level. Sidewall capacitance of the doped source and drain also serves to slow up the operation of the device, as enough charge must be built up to overcome the junction capacitance.

Polycide-gate MOS. MOS development was impeded by the inability of the industry to grow noncontaminated and thin oxides in the 1960s. Contamination, especially the mobile ionic variety, interferes with the field effect, making very unreliable gates. In fact, clean gate oxides and the oxide-silicon interface are so well understood and effective that gate oxide replacement is proceeding very cautiously.

Consequently, the quest for a higher-efficiency gate has led to the polycide structure (Fig. 16.21). The gate sandwich retains the thin

Contact Barrier   SiO$_2$ Aluminum-Based
Layer   Metal

N   N

Metal Silicide

Gate
Oxide   Polysilicon

Figure 16.21 Polycide gate structure.

oxide on the wafer surface topped by a layer of polysilicon. The polysilicon provides a low work function (and lower threshold) gate and the reliable polysilicon-oxide interface is preserved. The new layer is a refractory metal silicide on top of the polysilicon. The silicide makes a low contact resistance with the polysilicon (as compared with aluminum) and reduces the overall sheet resistance of the polyside sandwich.

Salicide-gate MOS. The self-aligned process that uses the polycide-gate structure is called a *salicide* gate. Its formation is illustrated in Fig. 16.22. The process combines the best features of a polysilicon gate with self-alignment. The source and drain are lightly diffused around the polysilicon gate. Then a layer of silicon dioxide is deposited and anisotropically etched to form spacers on the side of the gate. These spacers act as ion implantation masks for a subsequent heavier doping of the source and drain. The more lightly doped "finger" under the gate is called a lightly doped drain extension (LDD) and is necessary for channel lengths less than 2 μm.[5] After ion implantation, the refractory metal is deposited, and the silicide is formed by reaction with the underlying polysilicon layer by an alloy step. The final step is the removal of the unreacted refractory metal from the wafer surface.

The LDD process illustrated puts lightly doped fingers in both the source and drain areas. There are processes designed to create asymmetrical LDD structures that place the finger only in the drain region.[6]

V-groove MOS (VMOS). Invented in the late 1970s, this MOS transistor structure offered the promise of higher-density circuits since the transistors were formed on the sidewalls of a V-groove etched into the surface (Fig. 16.23). The formation of the transistors on the sidewalls

Figure 16.22 Salicide-gate structure.

Channel Width

Figure 16.23   VMOS structure.

eliminated device problems associated with surface damage and contamination. VMOS is a difficult structure to fabricate owing to the inconsistencies in patterning the components on the nonplanar groove sides. While not finding a home as an integrated circuit technique, VMOS has proved a good structure for discrete power MOS transistors.

Diffused MOS (DMOS).   DMOS refers to a diffused MOS structure (Fig. 16.24). The channel length is established by two diffusions through the same opening. As the second diffusion is taking place the first

Figure 16.24   DMOS structure.

moves laterally to the sides. The second diffusion functions as the source and the bulk semiconducting material of the wafer functions as the drain. The difference between the two diffusion widths is the channel length of the transistor. DMOS structures feature narrow and well-controlled gate widths.

Memory MOS (MMOS).   MMOS is a structure that provides a more or less permanent storage of the charge in the gate region. The storage is provided by a thin layer of silicon nitride between the wafer and the gate oxide (Fig. 16.25). When the gate is charged to store data, the silicon nitride layer traps and retains the data. This type of transistor is used in nonvolatile circuits where protection against memory loss is important (see Chap. 17).

Junction field-effect transistors (JFET).   A junction field-effect transistor (Fig. 16.26) is similar in construction to a MOSFET but has a junction formed under the gate. During operation the current flows *under* the diffused region from the source to the drain. As the gate voltage is increased, a region depleted of charge (the depletion region) spreads under the junction toward the N-type and P-type interface. The depleted region will not support current flow and has the effect of restricting the current flow as it increases in depth.

A JFET operates opposite from a MOS transistor. In the MOS version, increasing the gate voltage *increases* the current flow. In a JFET, increasing the gate voltage *decreases* the current flow. JFETs are standard gallium arsenide devices. The transistor is formed in an N-type GaAs layer that is on top of a semiinsulating GaAs wafer (Fig. 16.27).

Figure 16.25   MMOS structure.

Figure 16.26   Junction field-effect transistor.

Figure 16.27   GaAs JFET.

Figure 16.28   GaAs MESFET.

Metal semiconductor field-effect transistor (MESFET).   The MESFET is the basic GaAs transistor structure (Fig. 16.28). The MESFET operates in the same manner as the JFET but the gate metal is deposited directly onto the N-type GaAs layer.

Fuses and conductors

The formation of fuses and surface wiring conductors were detailed in Chap. 14. In very dense circuits, precious surface area is preserved by the use of subsurface *underpass conductors*. These are created from heavily doped regions formed under surface wiring leads.

Creating fuses and conductors involves depositing and patterning multiple metal layers. Key technologies are planarization, step coverage, and plug filling. Development of low-contact resistance metal film and film sandwiches will be required in the 0.5-μm design rule era.

Integrated-Circuit Formation

Integrated circuits contain all the components described in the previous sections. The components are formed in specific sequences. The process designer will attempt to form as many component parts as possible with each doping step, for example, at the source/drain doping step circuit resistors are also formed.

Circuits are designated by the transistor type. A bipolar circuit means that the circuitry is based on bipolar transistors. MOS circuits are based on one of the MOS transistor structures. For the first 30 years of the semiconductor industry, bipolar transistors and bipolar circuits were the structures of choice. Bipolar transistors had fast speed (switching times), control of leakage currents, and a long history of process development. These qualities fit nicely into the logic, amplifying, and switching circuits that were the first offerings of the industry. These circuits handled the computational requirements of the growing computer industry. The internal memory functions of the early computers were handled by core memories. These memories were limited in capacity and slow. Much of the information needed was stored outside the computer on tape, disks, or punch cards. While bipolar memory circuits were available, they could not compete economically with core memories.

MOS transistor circuits held the promise of fast, economical, solid-state memory, but earlier metal-gate MOS circuits suffered from high leakage currents and poor parameter control. Even so, the built-in advantages of MOS transistors drove the development of MOS memory circuits. The advantages are smaller dimensions, which allow denser circuits, and relatively faster switching speeds. Yields tend to be higher on smaller-dimensioned circuits since a given defect density will affect fewer transistors and components.

Perhaps the biggest density factor advantage of MOS components is the smaller area required for isolation of adjacent components. The various isolation schemes used are discussed in the next sections. Another advantage is low-power operation. First, MOS transistors sit in the circuit in the "off" mode, not soaking up power or generating heat like bipolar transistors, which must be "on" all the time to be in a "ready" state. Second, MOS transistors, being voltage-controlled devices, require a lower power to operate. CMOS circuits use an integrated circuit design that reduces power requirements even further.

An initial advantage of MOS circuits was fewer processing steps and smaller die sizes, which made for lower processing costs and higher yields. These advantages have disappeared as MOS circuits have evolved to VLSI and ULSI size with the additional steps required to fabricate CMOS circuits. In general, the faster switching speed of bipolar circuits has made them favored for logic circuits. MOS circuits, with their smaller component dimensions and lower power requirements, have been incorporated into memory circuits. By the 1980s these traditional uses blurred, with bipolar memories and MOS logic circuits readily available. These topics are addressed further in Chap. 17.

Bipolar circuit formation

Junction isolation. The bipolar transistor structure and basic performance have been illustrated. However, combining transistors with the other devices to form a circuit requires additional structures. Basic structures include a component isolation and a low-resistance collector contact. If two transistors or other devices are fabricated next to each other, they will not work because they are electrically shorted together (Fig. 16.29). An early challenge to bipolar IC designers was a way to *isolate* the various circuit components. This need led to the epitaxial (epi) layer bipolar structure (Fig. 16.30).

The process starts with a P-type wafer into which an N-type diffusion is made (the flow diagram does not show the oxidation and

No Isolation

Figure 16.29 Adjacent bipolar transistors with common collectors.

Figure 16.30 NPN bipolar process.

masking steps required to create the diffused layer). After the diffusion step, an N-type epitaxial layer is deposited, leaving the N-type diffused region "buried" under the epitaxial layer. The N-type region is known as a *buried layer* or as the subcollector of the transistor. Its function is to provide a lower-resistance path for the collector current as it flows out of the base region on its way to the surface collector contact.

After the deposition of the epitaxial layer, it is oxidized and a hole opened up on each side of the buried layer. A P-type doping step is performed deep enough to reach the P-type wafer surface. The doping step divides (*isolates*) the epitaxial layer into N-type islands, each surrounded on the sides (the doped regions) and the bottom (the P-type wafer) by P-type doped regions. Components formed on the surface of each of the islands are electrically isolated from each other (Fig. 16.31). The electrical isolation occurs because the N-P junction is "wired" into the circuit to function in the reverse-bias mode; that is, no current crosses the junction. This scheme is called *junction isolation* or *doped junction isolation*.

Dielectric isolation.   In high-radiation environments, such as outer space or in the vicinity of atomic weapons, doped junctions produce holes and electrons that compromise the junction functions. Besides causing circuit component failure, the radiation swamps out the isolation protection of the doped regions. Dielectric isolation schemes provide the necessary electrical isolation and radiation protection.

1. EPI and Collector	4. Collector Contact	7. Metalization
2. Isolation	5. Base	8. Passivation Layer
3. Surface Oxide	6. Emitter	9. Buried Layer

Figure 16.31 Bipolar structures.

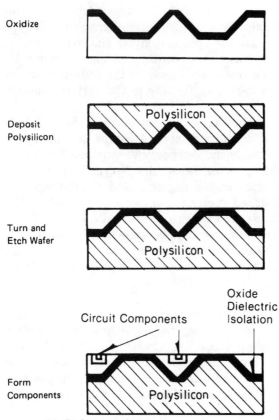

Oxidize

Deposit
Polysilicon

Polysilicon

Turn and
Etch Wafer

Polysilicon

Oxide
Dielectric
Isolation

Circuit Components

Form
Components

Polysilicon

Figure 16.32 Dielectric isolation.

The process (Fig. 16.32) starts with the etching of pockets or trenches in a wafer surface. The etching can be either an isotropic wet etch or an anisotropic dry etch. Isotropic etch profiles follow the orientation structure of the wafers. Dry etch processes allow the shaping of the trenches. One goal of the etch step is to minimize the area of the pocket on the wafer surface. Wide pockets limit the packing density of the circuit.

After etch, the pocket sides are oxidized and backfilled with deposited polysilicon. Next the wafer is turned over and the silicon of the wafer is lapped until the oxide layer is reached. These steps leave a wafer surface containing oxide-isolated pockets of the original single silicon material. The circuit components are fabricated in the silicon pockets, with each pocket being isolated on three sides by the layer of silicon dioxide. The dielectric property of the silicon dioxide prevents leakage currents in both normal and radiation environments.

Localized oxidation of silicon (LOCOS). Junction isolation and dielectric isolation take up valuable surface area. Dielectric isolation also requires extra processing steps. A popular alternative is LOCOS (Fig. 16.33). In this process, a layer of silicon nitride is deposited and etched on the wafer surface. Active devices will be formed in the area defined by the silicon nitride layer. In the partially recessed version an oxidation follows. Oxygen will not penetrate the silicon nitride, so the oxide will grow only on the exposed silicon surface. The silicon for the silicon dioxide comes from the wafer surface, and since silicon dioxide is less dense than silicon, the oxide layer forms slightly above the original silicon surface. It is *partially recessed* relative to the wafer surface. After oxidation, the silicon nitride is removed, leaving the area free for the formation of circuit components.

A variation of the LOCOS isolation process is illustrated in Fig. 16.34*b*. In this process, the silicon surface is etched *before* oxidation. By calculating the proper removal amount, the subsequent oxidized layer is *fully recessed* to the original surface.

Start

Deposit and
Pattern Silicon
Nitride Layer

Grow
Thermal
Oxide

Remove
Nitride

Figure 16.33 LOCOS process.

Figure 16.34 (*a*) Conventional bipolar transistor; (*b*) LOCOS isolated bipolar transistor. (*Source: VLSI Fabrication Principles by Ghandhi*)

Collector contact. Note that in the cross section of a circuit integrated bipolar transistor (Fig. 16.31) there is a doped region under the collector contact. This doped region is put into the surface along with the emitter N-type doping. The emitter is usually designated N+ to indicate that it is highly doped. The N+ region under the collector contact is present to create a lower resistance between aluminum metallization and the silicon of the collector.

MOS integrated circuit formation

MOS LOCOS isolation. While MOS transistors are somewhat self-isolated (because they don't share common electrical parts), there is electrical leakage between devices, especially as the spacing becomes very close. Isolation is necessary to block leakage currents. Isolation structures are generically called *channel stops*.

LOCOS is the preferred isolation technique. However, there are several problems to overcome to make a LOCOS structure effective in advanced circuits.[7] One problem is the *bird's beak* spur that grows under the edge of the blocking silicon nitride layer (Fig. 16.35). The

Figure 16.35 Bird's beak growth. (*a*) No pre-etch; (*b*) 1000 angstrom pre-etch; (*c*) 2000 angstrom pre-etch. (*Source: VLSI Fabrication Principles by Ghandhi*)

beak takes up surface area, effectively enlarging the circuit. At a performance level, it induces stress damage in the silicon during the oxidation step. The stress comes from the mismatch in thermal expansion properties between silicon nitride and silicon. A solution for the stress problem is to grow a thin oxide layer under the silicon nitride film. This is called a *pad oxide.*

Minimizing the bird's beak and reducing the stress in the active device region have spurred a number of variations on the LOCOS process. One, called SWAMI, was developed by Hewlett Packard (Fig. 16.36).[8] The process starts the same as a standard LOCOS. After the nitride and pad oxide steps, grooves (or trenches) are etched with an orientation-sensitive etchant. On <100> oriented material, the grooved sidewalls are at a 60° angle, which reduces stress in the silicon. Next another stress-relief oxide (SRO) layer is grown and covered by a layer of silicon nitride, which provides conformal coverage. A LPCVD silicon dioxide completes the sandwich before etching (c). This oxide protects the silicon nitride layer from removal. Finally the

Figure 16.36 SWAMI process. (*From SST, June 1993, p. 97*)

field oxide (FOX) is grown. The length of the nitride layer governs the size of the bird's beak encroachment. Removal of the original nitride and/or SRO and the second nitride leaves a somewhat planarized surface for device formation. LOCOS isolation schemes usually include an ion-implanted layer between active regions to provide further channel stop capabilities.

Trench isolation.   Trench isolation is also used for MOS circuits (Fig. 16.37). The procedure is the same as forming trench capacitors.

CMOS.   Complementary MOS (CMOS) is a MOS circuit formed with both N-channel and P-channel transistors. CMOS has become the standard circuit for many applications. It is the CMOS circuit that has made possible digital watches and hand-held calculators. It allows circuits on one chip that would require several chips using N-channel and P-channel only circuits. CMOS circuits also use lower amounts of power than comparable circuits.

CMOS structures (Fig. 16.38) are formed by first fabricating an N-channel MOS transistor in a deep P-type well formed in the wafer surface. After N-channel transistor formation, a P-channel transistor is fabricated. The transistor structures are silicon gate or other advanced structures. CMOS processing uses the most advanced tech-

Figure 16.37 MOS trench isolation.

N Channel   P Channel

1. Wafer   4. Gate
2. P Well   5. Gate Oxide
3. Source/Drain   6. Metallization

Figure 16.38 CMOS structure.

niques since smaller, more densely packed, and higher-quality components all increase the advantages inherent in the CMOS design.

An isolation problem particular to CMOS structures is latch up.
Figure 16.39 shows a cross section of a portion of a CMOS chip. The
side-by-side MOS transistors form a lateral bipolar (NPN) transistor.
During circuit operation, the lateral bipolar transistor can function as
an unwanted amplifier, increasing its output to a point where it causes the memory cell to go into a state where it cannot be switched. This
is *latch up*. In this state the cell cannot be addressed for its information. One solution to preventing latch up is a low-resistivity epitaxial
layer that shunts (shorts) the emitter of the bipolar transistor so that
it does not "turn on," preventing the latch up.

Another solution is the use of LOCOS or other structures. One is a
hardened-well design, using a buried layer which effectively breaks
up the lateral bipolar transistor (Fig. 16.40).

Formation of the deep P well by diffusion techniques requires a
large thermal budget that contributes to lateral diffusion problems
and stress growth. Ion-implanted wells have become common.
Additionally, with ion implantation, the wells can be dopant graded in

Silicon Gate

N   N
P

Figure 16.39 Cross section of silicon gate MOS transistor.

Figure 16.40 CMOS well structure Silicon on Insulator (SOI) with buried layer and trench isolation. (*Source: Semiconductor International, July 1993, p. 34*)

the vertical direction to maximize performance. Called *retrograde wells,* they are formed by high-energy implants (MeV) such that the bottom of the well has a higher dopant concentration. Implanted wells offer the advantage of fewer steps compared to conventional well formation[9] (Fig.16.41). Also, since the N and P wells are formed in the same surface, planarity is preserved. In conventional well processing, the two well surfaces are offset and can cause depth of focus problems.

Bi-MOS

The unique advantages of bipolar and CMOS transistors and their respective circuits come together in Bi-MOS (or Bi-CMOS) circuits. The circuits (Fig. 16.42) have in them bipolar, P-channel, and N-channel transistors along with memory cells (see Chap. 17). The low-power advantage of CMOS is used in logic and memory sections of the circuit while the high-speed performance of bipolar circuitry is used for signal processing.[10] These circuits represent great challenges to the processing area with their large die size, small component size, and large number of processing steps. Figure 16.43 lists the number and type of implants required for a typical Bi-MOS structure.

Silicon on insulator (SOI) isolation

There has been renewed interest in forming CMOS and Bi-CMOS circuits in thin epitaxial silicon films deposited over an insulating sub-

Retrograde Well   **Conventional Well**

Figure 16.41 Implanted and conventional wells. (*Source: Semiconductor International, June 1993, p. 84*)

Figure 16.42 Bi-MOS structure.

strate (see Chap. 12). Elimination of a conductive substrate minimizes leakage and latch-up problems.[11]

Superconductors

Much interest has been generated by developments in superconductivity materials. Superconductivity is a phenomenon that occurs in certain materials when they are cooled close to absolute zero (−273°C). In ordinary metals, current is passed along by the flow of electrons. In a rest (nonconducting) state, the electrons exist in orbits

BIPOLAR   CMOS

N$^+$ Buried Layer
P$^+$ Buried Layer
Epitaxial Layer
                          N-Well
                          P-Well
                          Isolation Implant
                          LOCOS
                          Capacitor
                          Threshold Implant
                          Gate Oxidation
Collector
Base
Emitter   NMOS Source/Drain
                          PMOS Source/Drain
                          Passivation

Figure 16.43 Bi-MOS implant steps.

around the nuclei of the atoms. To become conducting they must gain energy to overcome the internal resistance of the particular material. The energy must be continuously supplied to maintain the current flow.

In a superconducting material electrons exist in a "conducting state" and can support an electrical current with little or no additional energy input. The prospect of a resistanceless material has the potential of revolutionizing electronic devices.

Semiconductor researchers have investigated the superconducting effect for years. In 1962, B. D. Josephson described the effect (named after him): when a thin oxide separates two superconducting materials, electrons will pass through the oxide with zero resistance. The structure is called a *Josephson junction*. This effect (called *tunneling*) is very complicated and requires quantum physics concepts to understand (well beyond the scope of this text!). The result is that the oxide has the functional aspects of a gate and Josephson junctions can perform basic switching, logic, and memory functions.[12] While IBM and Bell Labs (now AT&T) have development programs in superconductivity devices, no practical devices are currently available.

Physical Semiconductor Devices

Semiconductor junctions have the property that almost anything done to them produces a change in their functioning. The anything can be a physical strain, exposure to light, radiation, or heating. Miniature devices can be constructed using wafer-fabrication techniques that either produce an electrical signal in response to some physical input or produce a mechanical result from a signal. These devices are called *microelectromechanical systems* (MEMS).[13] MEMS devices made with wafer-fabrication techniques include motors, motion sensors (as in air-bag sensors), spectrometers, and miniature scanning tunneling microscopes.

Strain gauges

A device that takes advantage of a junction reaction to physical pressure is the strain gauge. The gauge is made by back-etching through the wafer until only a very thin membrane is left. A junction and supporting circuitry are formed in the membrane. When the membrane is deflected by some force, as in a weight scale or pressurized gas line, the semiconductor circuit produces an output proportional to the deflection. The output of the circuit is correlated with the amount of pressure on the membrane and displayed on the appropriate meter.

Batteries

Thin-film, rechargeable batteries have been made from lithium and vadium pentoxide layers. They could be formed on a package and provide backup for CMOS devices in the event of power failures.[14]

Light-emitting diodes

One effect when certain compound junctions are reversed-biased is the production of photons. Photons are a form of radiation that humans see as light. The devices are called *light-emitting diodes* (LEDs). These are the displays that are used in consumer electronics equipment and automobile displays.

The devices (Fig. 16.44) are made on gallium-arsenic-phosphide (GaAsP) wafers that are covered with thousands of diodes, wired so that they can be turned on or off individually. Groups of diodes are turned on in groups to form letters and numbers. GaAsP material

Figure 16.44   LED structure.

produces the familiar red displays, with other colors being produced when dopants are added to the wafer.

Optoelectronics

Not only can semiconductor junctions emit light, they can respond to it. An integrated optoelectronic connection has been sought for years for local area networks (LANs). In this use the opto devices (sensitive to laser and other lights) are connected to a fiber-optic network or wave guide. Like LEDs, the receivers are mostly III–V-based ICs. An onboard conventional IC processes the data and sends them through an output LED or laser.[15]

Solar cells

Solar cells are devices that respond to light, also. This property is taken advantage of in the solar cell (Fig. 16.45). The cell is composed of diodes formed in a thin layer of semiconducting material such as amorphous silicon. When sunlight strikes the junction region, a current passes across it. The current is captured by onboard circuitry.

Temperature sensing

Semiconductor junctions are temperature-sensitive. Heating a device will produce more current across the junction. This effect is taken advantage of in a variety of devices such as solid-state medical thermometers and industrial control units.

(a)

(b)

Figure 16.45 Light-sensitive semiconductor structures. (*a*) Photodiode; (*b*) solar cell.

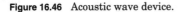

Figure 16.46 Acoustic wave device.

Acoustic wave devices

Acoustic wave devices (Fig. 16.46) are nonsilicon solid-state components used in microwave communication systems. They serve the
function of converting microwaves into electrical impulses. A compound material such as $Be_{12}GeO_{20}$ has the property of reacting to the
wave and setting up an electrical response in a solid-state circuit
formed on the chip by normal semiconductor processes.

Bubble memories

Bubble memories are nonsemiconductor, solid-state devices used for
memory storage (Fig. 16.47). The material is an iron alloy deposited
on a substrate. The alloy is capable of forming tiny magnetic *bubbles*
in response to an electrical pulse from a semiconductor-type metal
system formed on the surface. The information to be stored is coded
in long lines (strings) in the alloy surface as magnetic bubbles that
are oriented either north or south (up or down). The information is
retrieved by moving the strings of bubbles under a surface sensor

Figure 16.47 Bubble memory
device.

that records the polarity of the bubble and sends the information to other circuitry that reconstructs it into the required word form.

Flat panel displays

Flat panel displays (FPDs) are most often seen as laptop computer screens. Their future is the replacement of cathode ray tubes (CRTs) because they can provide a thin and more rugged option. However, improvements in clarity, brightness, and cost are necessary before they become regular replacements for CRTs.[16]

The majority of FPDs are the color-active matrix liquid-crystal displays (AMLCD). Both color and noncolor LCD displays are made using microchip-fabrication technology. A display is a sandwich of two glass plates with a liquid crystal material in between (Fig. 16.48). On the bottom plate is an array of simple MOS devices with onboard driver circuitry. The transistors are formed by standard process steps, deposition, patterning, and the like. The top panel is a color filter (for the color display) formed by multiple photoresist steps. In between is sandwiched the liquid crystal.

The display shown on the screen is first formed as a pattern in the MOS transistors (on/off conditions).[17] This pattern is replicated by the liquid crystal, which can change states in response to an electrical signal. For the color displays, the top filter transmits the proper color. Black and white displays simply transmit the on/off condition of the LCD.

Field emission displays (FEDs)

In CRTs and FPDs, the information required to put a pattern on the screen has to come from electronic signals (such as a computer) and be transformed into a physical entity. In conventional CRTs, the entity is a stream of electrons from an electron gun. An alternative device to produce electrons is a FED (Fig. 16.49). Formed from semiconductor materials using wafer-fabrication techniques, the end result is an etched microtip. When a voltage is placed between the cathode and gate, electrons are emitted. FPDs made with FEDs have the promise of brighter displays, a wider viewing angle, and improved speed.[18]

COLOR FILTER FORMATION

GLASS CLEANING

COLOR FILTER PATTERNING

ITO DEPOSITION

ORIENTATION FILM APPLICATION & RUBBING

THIN FILM TRANSISTOR FORMATION

GLASS CLEANING

THIN FILM DEPOSITION

DEVICE PATTERNING

ORIENTATION FILM APPLICATION & RUBBING

ASSEMBLY

SEPARATION

LIQUID CRYSTAL INJECTION

Figure 16.48 Active matric liquid crystal (AMLCD) display process flow. (*Source: O'Mera Associates*)

Vacuum deposited microtip

Gate

Insulator

Cathode

Glass substrate

Figure 16.49 Cross section of field emission device.

Key Concepts and Terms

Acoustic wave	Junction isolation
Bipolar transistors	LED
Bubble memory	MESFET
CMOS	MMOS
Dielectric isolation	Oxide capacitors
DMOS	Polycide gate
Doped diodes	Salicide
Doped resistors	Salicide gate
Epitaxial resistors	Schottky barrier diodes
FETMOS transistors	Strain gauge
Isoplanar and LOCOS isolation	Thin film resistors
JFET transistors	VMOS
Junction capacitors	

Review Questions

1. Sketch and name the three parts and functions of a bipolar and MOS transistor.
2. Explain the function of a diode, resistor, and capacitor.
3. What is the operational advantage of a Schottky barrier diode?
4. Sketch and name the layers in a metal, polyside, and salicide MOS gate structure.
5. What is the structural difference between MOSFET, JFET, and MESFET gate structures?
6. Describe why component isolation is required in integrated circuits.
7. What are the differences and advantages of junction and dielectric isolation schemes?
8. Sketch the process steps used to create a LOCOS isolation scheme.
9. Name three ways that a semiconductor junction in operation can be influenced.
10. How does a CMOS structure differ from an NMOS structure?

References

1. H. R. Camenzind, *Electronic Integrated Systems Design,* Van Nostrand Reinhold, Princeton, NJ, p. 85.

2. P. Singer, "Gearing Up for Gigabits," *Semiconductor International*, November 1994, p. 34.
3. H. R. Camenzind, *Electronic Integrated Systems Design,* Van Nostrand Reinhold, Princeton, NJ, p. 141.
4. P. Singer, "Directions in Dielectrics in CMOS and DRAMs," *Semiconductor International,* April 1994, p. 56.
5. Y. Pauleau, "Interconnect Materials for VLSI Circuits," *Solid State Technology,* April 1987, p. 157.
6. "Industry News," *Semiconductor International,* Cahners Publishing, April 1994, p. 16.
7. S. K. Ghandhi, *VLSI Fabrication Principles,* John Wiley and Sons, New York, 1994, p. 717.
8. S. Wolf, "A Review of IC Isolation Technologies—Part 8," *Solid State Technology,* Penwell Publishing, June 1993, p. 97.
9. L. Peters, "High Hopes for High Energy Ion Implantation," *Semiconductor International,* Cahners Publishing, June 1993, p. 84.
10. C. B. Yarling, "M. I. Current, Ion Implantation for the Challenges of ULSI and 200 mm Wafer Production," *Microelectronic Manufacturing and Testing,* March 1988, p. 15.
11. K. Yallup, "SOI Provide Total Dielectric Isolation," *Semiconductor International,* Cahners Publishing, July 1993, p. 134.
12. P. W. Anderson, "Electronic and Superconductors," in E. Ante'bi (ed.), *The Electronic Epoch,* Van Nostrand Reinhold, Princeton, NJ, p. 66.
13. K. Gabriel, "Engineering Microscopic Machines," *Scientific American,* September 1995, p. 150.
14. J. Bates, "Rechargeable Thin-Film Lithium Microbatteries," *Solid State Technology,* Penwell Publishing, July 1993, p. 59.
15. P. Singer, "The Optoelectronics Industry: Has It Seen the Light?" *Semiconductor International,* Cahners Publishing, July 1993, p. 70.
16. P. Singer, "Flat Panel Display: An Interesting Test Case for the U.S.," *Semiconductor International,* Cahners Publishing, July 1994, p. 79.
17. Ibid.
18. K. Derbyshire, "Beyond AMLCDs: Field Emission Displays?" *Solid State Technology,* Penwell Publishing, November 1994, p. 55.

17

Integrated Circuit Types

Overview

The solid-state semiconductor components wired into integrated circuits perform many different functions in electronic instruments and machines. In this chapter the major circuit families and their functions are explained.

Objectives

Upon completion of this chapter you should be able to:

1. Explain the concept of binary numbering.
2. List the three major integrated circuit functions.
3. Compare the basis of analog and digital logic circuits.
4. List the user and production advantages of logic gate arrays and PAL circuits.
5. Explain the two major memory circuit types.
6. Make a list of the four nonvolatile memory circuits.
7. Compare the performance and cost factors of DRAM and SRAM memory circuits.

Introduction

The majority of units produced by the semiconductor industry are integrated circuits. Countless numbers and types of circuits can be created using the processes described in this text. A circuit catalog from a major integrated circuit (IC) producer like National Semiconductor or Motorola is as large as a New York City phone book. IBM estimates that its internal circuit catalog lists over 50,000 separate circuits!

Becoming familiar with integrated circuits is not as awesome a task as these high numbers might imply. The burden is eased by the fact that most circuits fall into three primary families: logic, memory, and microprocessor (logic and memory) (Fig. 17.1). Within each circuit family, there are a few principal designs and functions. The multiplicity of circuits comes from the many parameter variations required for specific uses.

The major functional circuit categories and their circuit designs are explained in this chapter. In the last section, we look at the future in IC circuitry from the perspective of the industry today. What the circuits will actually be like in 2010 can only be imagined, just as in 1950, no one predicted the megabit RAM or the microprocessor.

Circuit Basics

The question of how an integrated circuit actually works is the subject of other texts, but all circuits are based on the processing of data in binary notation. The binary system is a way of representing any number with just two digits—for example, a zero and a one. It is actually an accounting system that keeps track of the place and value of the components of a number. Numbers can be expressed as the sums of numbers. For example:

$$1 = 1 + 0$$

$$3 = 2 + 1$$

$$7 = 4 + 2 + 1$$

$$10 = 8 + 2$$

Another way to express numbers is as powers of their factors. Yet another way is to express numbers as powers of their roots.

Figure 17.1   IC circuit functions.

$$1 = 2^0$$
$$2 = 2^1$$
$$4 = 2^2$$
$$8 = 2^3$$

Figure 17.2 Powers of 2.

The basis of binary notation is the powers of the number 2. Figure 17.2 shows the numbers 1, 2, 4, and 8 expressed as powers of 2. We could also represent those numbers just as the power of 2. Thus 1 becomes 0, 2 becomes 1, 4 becomes 2, and 8 becomes 3.

Now for the clever part. Any number can be expressed as the sum of some numbers that are powers of 2. The number 25 is the sum of $16(2^4)+8(2^3)+1(2^0)$. In the number 25, there is one 2^4, one 2^3, zero 2^2, and one 2^0. Or the number 25 can be represented by the code 1101, each of the digits representing the presence or absence of a particular power of 2. The chart in Fig. 17.3 lists some numbers in binary notation.

Translating numbers into binary notation is easily accomplished by establishing a grid with each column representing a power of 2. The actual number is represented by a string of zeros or ones that indicate the presence or absence of the particular powers of 2 that make up that number.

Binary notation has been known for centuries. Buckminster Fuller, in his book *Synergistics,* has an amusing account of the use of binary coding by the ancient Phoenicians to keep track of cargo amounts. He claims that the Phoenician sailors were considered stupid because they could not count in the system of the day, when actually they were accurately keeping track of large amounts of cargo with only two "numbers"; they were counting in binary notation.

With binary notation only two numbers are necessary, a one or a zero. In the discussion above, binary notation was represented by the

Standard	32	16	8	4	2	1
Power of 2	2^5	2^4	2^3	2^2	2^1	2^0

Number

Standard	= Binary Number					
1	0	0	0	0	0	1
7	0	0	0	1	1	1
18	0	1	0	0	1	0
33	1	0	0	0	0	1

Figure 17.3 Binary notation.

**Power
of 2**

	2^3	2^2	2^1	2^0
	8	4	2	1
7 =	0	1	1	1

7 =

7 =

7 = ■   □   □   □

Figure 17.4 Binary representation of the number 7.

7 = Off   On   On   On

numbers zero and one. In the physical world, binary numbers can be represented by any system that has two conditions. Figure 17.4 shows several different ways to code the number 7. The last row could represent binary coding by the off-on states of a transistor or memory cell.

Inside a circuit numbers are coded, stored, and manipulated in their binary code. The numbers can be stored and manipulated because capacitors can be charged to have a charge or not have a charge and transistors can be either on or off. The smallest piece of information in a circuit is called a "binary digit" or "bit."

The binary coding system is simple. The problem of how the coded numbers could be added, subtracted, and multiplied was solved by George Boole, a nineteenth-century mathematician. He developed a logic system capable of handling numbers in binary notation. Until the development of computer logic, his Boolean logic (or Boolean algebra) was an academic curiosity.

Chips and computers are designed to handle a specific size of binary number or word. An 8-bit machine manipulates numbers with 8 binary bits at a time. A 32-bit machine can handle a number composed of 32 binary bits. The more bits a machine can handle at one time the faster and more powerfully it processes data. Every 8 bits is known as a byte. Thus, an 8-megabyte storage capacity can hold 64 million bits of information.

Integrated-Circuit Types

A solid-state integrated circuit is composed of a number of separate functional areas. Each chip, regardless of the circuit function, has an input and encode section where the incoming signals are "coded" into a form that the circuit can understand. Most of the circuit area contains the circuitry required to perform the circuit function, either memory or logic. After the data are manipulated by the circuit, the result goes to a decode section where it is changed back into a form that is usable by the machine's output mechanism. The circuit's output section sends the data to the outside world.

Although this is an overly simplified explanation of a circuit, it illustrates the fact that the interior of a chip is composed of definite separate functional areas. In many circuits, these areas perform the same functions as the main parts of a computer. Circuit types fall into three broad categories: logic, memory, and a combination of logic and memory (microprocessors).

Logic circuits perform a specified logical operation on the incoming data. For example, pushing the plus (+) key on a calculator instructs the logic portion of the chip to add the numbers presented to it. An onboard automobile computer goes through a logical operation to direct the signal from a sensor indicating an open door to light up the correct warning light on the dashboard.

Memory circuits are designed to store and give back information in the same form in which it was entered. Pushing the (pi) key on a calculator activates the memory part of the circuit where the value of pi (3.14) is stored. The value 3.14 is displayed on the screen. Every time that key is activated, that value is displayed.

A third type of circuit function combines both logic and memory in a circuit called a microprocessor. In 1972, Intel Corporation introduced the first practical microprocessor. Microprocessors can be programmed to perform many different circuit functions. To accomplish this, they contain logic and memory circuitry as well as the necessary encode, decode, input, and output sections. It was the microprocessor that allowed the design of powerful personal computers, digital watches, and one-chip calculators and the transfer of so many business machines to solid-state electronics, from phone systems to vending machines.

The microprocessor has been dubbed "a computer on a chip." While it contains all the functional areas of a computer, it is not truly a complete computer. Even simple computers require vast amounts of memory capacity, which microprocessors do not have. Within many personal computers, microprocessors function as the central processing unit (CPU). Additional memory chips have to be included in order to make the computer of practical use.

Actually, every integrated circuit contains both logic capability and memory sections. For example, the logic circuitry of a calculator must have certain constants stored in a memory section in order to perform calculations. And memory circuits must have some logic functions to direct the flow of electrons and holes to the right parts of the circuit for storage.

Logic circuits

Analog and digital logic circuits.   Logic circuits fall into two main categories: analog and digital (Fig. 17.5). Analog logic circuits were the earliest logic circuits developed. An analog circuit has an output that is proportional to the input. Digital circuits, on the other hand, feature a predetermined output in response to a variety of inputs. A wall light dimmer is an analog device. Turning the control varies the voltage to the dimmer, which in turn varies the brightness of the light. A standard on-off light switch is a digital device. Only two brightness conditions are possible: on or off. Most audio circuits are of the analog type. Changing the level setting of the volume produces a proportional change in the sound coming out of the speaker.

Analog logic circuit types.   Analog circuits were the first type designed in integrated form. The home computer hobby kits of the 1950s were the analog type. These simple circuits were based on Ohm's law ($R = V/I$). The circuit contains a resistance meter and a means for generating a current and measuring a voltage. The three quantities are related by Ohm's law. Any other three variables similarly related can be represented by the resistance, voltage, and current. Varying one changes the other two. The circuit thus becomes a computer for solving any equation of the form $A = B/C$.

The accuracy of analog circuits is dependent on the precision of the relationship between the input and the output. In the simple computer illustrated, accuracy is dependent on the precision of the components in the circuit, the clarity of the meters for setting the input and reading the output, and the immunity of the circuit to outside "noise." Unless the circuit contains a section to regulate incoming voltage levels, a change in the line voltage would alter the output, and hence the accuracy.

Both simple and complex analog circuits are vulnerable to variations in the incoming signal and to internal noise. Analog circuits are

Figure 17.5 Logic circuit types.

also dependent on precise control of the resistor values. Unfortunately, diffused resistors cannot be fabricated with a resistance variation from design value better than 3 to 5 percent, which is unacceptable for many applications. Greater resistance precision is gained by the use of matched resistor pairs, in which the effective resistance in the circuit is the difference between two resistors. This difference can be more tightly controlled than one resistor alone.

Ion implantation also provides the analog circuit designer with a tool for producing resistors with a higher degree of control. Many analog circuits feature thin film resistors to achieve the required precision. The growth and popularity of digital circuits is based on their ability to produce a set output. If a 5-V signal is absolutely required to operate a device, this can be achieved by designing a digital circuit that produces a 5-V signal every time, regardless of the input variation or internal electrical noise.

Digital circuits, however, do not respond as fast as linear circuits. The term used in electronics is *real-time response*. In some applications, such as airplane controls, real-time response is mandatory. Recent development in digital circuit speed is speeding the encroachment of digital circuitry into this traditional use of analog circuits. A major advantage of digital circuits over analog circuits is in general-purpose computers. Analog circuits are more difficult to design to respond to a general range of problems. All modern general-purpose computers are based on digital circuits.

The most popular use for analog circuits is in amplifiers. They are designed in a variety of configurations, for many different applications. All have the same basic principle—the incoming signal or pulse is amplified. Audio circuits require amplification of a weak signal from the record tone arm or other input in order to produce the level required to operate a speaker.

The real-time aspect of analog circuits also makes them real-world circuits. Wherever there is a real-world measurement such as temperature or movement, analog circuits are used. Even when most of the circuitry in a system are digital, analog circuits are often part of the interface with the outside world.

Most analog amplifier circuits are of the differential operational type. These circuits produce an output voltage amplified from and proportional to the difference of two input signals. Bipolar technology is favored for these circuits because bipolar circuits are modular electric current devices and are better suited to the applications required of analog circuits.

The output signal of an analog device can have a "one-to-one" relationship with the input signal. These circuits are called *linear*. If the input is changed, the output changes in a linear manner. So many analog circuits are of linear design that the two terms are often used interchangeably. However, there are nonlinear circuits, such as those featuring a logarithmic relationship between the input and output.

Logic circuitry is built around the logic gate. A gate controls and directs passage through a barrier. The size and design of a gate influence the amount of passage allowed. For example, a room with many "in" doors and only one "out" door is a gate. Many people can enter the room but their exit is restricted because only one door is provided. This example gate can also be operated in reverse, allowing people to enter through only one door and leave through many.

Electronic logic gates perform similar functions but with electrical signals. In a circuit, they perform the necessary logic operation by the dictates of Boolean logic. A discussion of their incorporation in logic design is beyond the scope of this text. The point for this text is that gates, both analog and digital, are formed in a logic circuit by wiring together various components. The circuits are designated by the first letters of the components in the gate. The major bipolar analog logic circuit designations are:

RTL	Resistor-Transistor Logic
DTL	Diode-Transistor Logic
TTL	Transistor-Transistor Logic
ECL	Emitter-Coupled Logic
DCTL	Direct-Coupled Transistor Logic
I^2L	Integrated Injected Logic

Integrated injected logic is another approach to constructing a logic gate. In this arrangement, a bipolar transistor is operated in the reverse mode, with the emitter as collector and collector as emitter. An advanced I^2L gate design is formed with the bipolar transistor constructed with a Schottky diode (sometimes called a clamp) between the base and collector. This arrangement results in a faster transistor, and hence a faster circuit.

Custom-semicustom logic.   Using any of the logic gate approaches listed, hundreds of thousands of different logic circuits can be constructed. They vary from custom small-volume circuits to off-the-shelf standards. Most logic circuits require some degree of customization. Several design and fabrication approaches are used to deliver custom and semicustom circuits to the customer at reasonable costs. The approaches are:

1. Full custom

2. Standard circuit—custom gate pattern

3. Standard circuit—selective wiring gate

4. Programmable array logic

Full custom. A full custom-designed logic circuit is specified by the customer who pays for design and mask-making fees along with the fabrication costs. This approach is expensive and lengthy, and is not geared for experimenting with different circuits in the design stage of a project. Custom-designed circuits are not cost-effective in quantities of less than 100,000.

Standard circuit—custom gate pattern. This fabrication process starts with a standard logic circuit design, but only the gates required for the particular application are formed during the fabrication process. The input, output, and other circuit sections are standard for a family of circuits.

Standard circuit—selective wiring gate arrays. This system is similar to the custom gate approach but based on a standard circuit design for most of the fabrication process. These circuits are built with a standard number of gates. This gate section is called the array and the circuit is known as a *gate array*. Working with the basic design, the customer can instruct the fabrication department to wire together only the gates required to produce the custom circuit logic function.

The result is a faster turnaround time than full custom processes can achieve, and moderate cost. The cost per logic function of gate arrays is higher than that of a custom circuit produced in production quantities. The larger gate section required to allow many different circuit variations results in a larger chip. This larger chip size leads to a higher manufacturing cost per chip and/or a lower yield.

The wafers receive a common process up to the contact mask. The contact mask is customized to form contacts only to the gates required for the particular circuit. After metal deposition, only the gates with contacts are wired into the circuit. A variation of this process is to open the contacts in all the gates but use a customized metal mask that wires in only the wanted gates.

Programmable array logic (PAL). Each of the three systems described requires the chip manufacturer to do the "customizing." This requirement can result in delivery or scheduling problems and generally forces the user to buy a minimum quantity of parts. Monolithic Memories, Inc. (MMI), addressed this problem with the introduction of its PAL line of circuits in 1978. PAL stands for programmable array logic. MMI applied the programmable fuse technology used in memory products to logic circuits. The result was a field programmable (custom) logic circuit.

The concept is similar to the standard gate array, but within the circuit, a fuse connects each logic gate into the circuit. The user programs the circuit by blowing the fuses at the unneeded gates, thus removing them from the circuit. The customer buys a quantity of standard PAL circuits and can program them as needed for the particular applications. In his book *Soul of a New Machine,* Tracy Kidder related the story of a Data General project team's decision to design a 16-bit computer using the newly introduced MMI circuits. The ease of programmability and shortened delivery time is credited with allowing the computer to be brought to market ahead of that of another digital design team using custom circuits.

Memory circuits

Around 1960, industry forecasters began predicting that solid-state memory circuits would overtake the traditional core memory. The advantages of solid circuits were their reliability, small size, and faster speed. This prediction was made every year until the early 1970s, when solid-state memories finally did surpass core memory. The major factor that prolonged the life of core memory was lower cost.

MOS is the favored transistor structure for memory circuits. During the 1960s, however, the cleanliness requirement for MOS processing was not reliably available. High-yield MOS processing also requires accurate alignment and clean thin gate oxides. These processes were not fully developed in those years. The resultant low process yields kept MOS memories more expensive than core memories.

With process improvements and improved silicon gate structures, CMOS is the memory technology of choice. Some bipolar memories are favored for their fast speed and switching capabilities. While logic circuits can be (and are) made with MOS technology, most MOS circuits produced are memories, with the majority incorporated into computers. They are also used in microprocessor-based products, which require auxiliary memory chips. There are two principal types of memory circuits: volatile and nonvolatile (Fig. 17.6).

Nonvolatile memories. A nonvolatile memory device is one that does not lose its stored information when it loses its power. An example of this is a compact disk (CD), which is an information-storage device. If

Figure 17.6 Memory circuit types.

Figure 17.7 Nonvolatile memory types.

power to the CD player is lost, the songs are not lost from the disk itself. A number of nonvolatile memory circuits are listed in Fig. 17.7.

ROM. In integrated circuits, the ROM design is the principal nonvolatile circuit. ROM stands for *read only memory.* The sole function of this type of circuit is to give back precoded information. The information required in the circuit is specifically designed into the chip memory array section during fabrication. Once the chip is made, the stored information is a permanent part of the circuit.

Other memory types have *read and write* capability. That is, they can receive and store information from an input device (keyboard, magnetic tape, floppy disk, etc.). A compact disk is a nonvolatile ROM device. A magnetic tape is an example of a nonvolatile device with both read and write capabilities because information can be erased and re-recorded.

In a calculator, the constants and the rules required for the math operations are available in the ROM section. ROM circuits, like logic circuits, number in the hundreds of thousands. Although there are many standard types, the industry also produces many custom ROM circuits. The choices offered to the user in selecting a standard or custom chip are similar to those available with logic circuits. The user can buy a standard circuit; specify a variation on a standard basic circuit; design a total custom circuit; or buy a PROM, EPROM, or EEPROM.

PROM. PROM stands for programmable read only memory. A PROM is the memory equivalent of a PAL. Each memory cell is connected into the circuit through a fuse. Users program the PROM to their own memory circuit requirements by blowing fuses at the unwanted memory cell locations. After programming, the PROM is changed to a ROM, and the information is permanently coded in the chip, where it becomes a read only memory.

EPROM. For some applications, it is convenient to change the information stored in the ROM without having to replace the whole chip. EPROM chips (erasable programmable ROM) are designed for this use. The erasable feature is built in with the use of MMOS (memory MOS) transistors (detailed in Chap. 16). The transistors can be selectively charged (or programmed) and they hold the charge for a long time, which holds the information in a nonvolatile fashion. Programming is by the mechanism of hot electron injection. When reprogramming is required, the charge in the transistors can be drained off (erasing the memory) by shining ultraviolet light on the chip. Reprogramming of the chip takes place by removing it from the circuit and putting in new memory information with an external programming machine. A typical EPROM can be reprogrammed up to 10 times.

EEPROM. The next level of convenience in memory design is the ability to program and reprogram the chip while it is in a socket in the machine. This convenience is available with the EEPROM (or E^2PROM), standing for electronically erasable PROM. Programming and erasing take place by pulses from the outside that place charges in selected memory cells or drain the charges away. Programming is by the same mechanism used for EPROMs, hot electron injection. Charge is drawn from the memory cell by a mechanism called Fowler-Nordheim tunneling. This convenience comes at the expense of a larger memory cell size and a commensurate reduction in chip density.

Flash memory. Flash memory is a form of EEPROM. It is based on a one-transistor cell design,[1] like an EPROM, but has the convenience of in-socket programming and erasure. Also, blocks, or the entire array, can be erased at one time.

Volatile memories. Semiconductor circuit and computer design involves the constant evaluation of tradeoffs. In the case of memory, nonvolatile memory provides protection against power loss, but these memories are frequently slow and not very dense. More importantly, none of the circuits described previously has a write capability, an essential feature in operating a computer. New information, such as a change in pay status, must be conveniently entered into the computer and stored temporarily while the new check is being written. Memory must also be easily erasable so the computer can quickly process new information or accept a completely new program. There are several memory circuit designs used to produce fast, high-density memory circuits. Both are of the volatile type; that is, when power is lost to the chip all the stored information is lost; information presented on a computer screen, and not saved, is eligible for loss if the power to the computer goes off.

Figure 17.8 RAM cell designs.

RAM. One type of circuit used for high-density memory storage (Fig. 17.8) is random-access memory, or RAM. "Random" refers to the ability of the computer to directly retrieve any information stored in the circuit. Unlike a serial memory, the RAM design allows the chip to find the exact information asked for wherever it is located in the computer memory. This feature allows faster information retrieval and makes the RAM the principal memory circuit in computers.

Dynamic random access memory (DRAM). RAMs come in two principal designs: dynamic and static (Fig. 17.8). A dynamic memory design called a DRAM, for dynamic RAM, is used in great quantities in computer memories. The memory cell design is based on only one transistor and a small capacitor (Fig. 17.9). The information is stored in it by a charge built up in the capacitor. Unfortunately, the charge drains away very rapidly. To combat this problem, the memory information must be re-inputted to the circuit on a constant basis. The term for this function is *refresh*. The refreshing of the circuit occurs many thousands of times per second. Dynamic RAMs are vulnerable to both power loss and interruption, or problems with the refreshing circuit.

The goal of DRAM design is small-cell design for high-density and closely spaced components with small and thin parts for speed. These requirements have driven DRAM design and processing to the highest levels of the technology. All the advantages available by advanced, state-of-the-art equipment and processing are applied to DRAM circuits. This fact makes them the industry's leading-edge circuit.

Static random access memory (SRAM). Static random-access ram (SRAM) memories are based on a cell design that does not need a

Figure 17.9 Dynamic RAM cell schematic.

refresh function. Once the information is put into the chip, it will stay as long as the power remains on. This is accomplished with a cell containing several transistors and capacitors (Fig. 17.10). The information is stored as conditions of the transistors are alternately on or off. Information can be read and written with an SRAM cell much faster than with a DRAM design, since transistors can be switched faster than capacitors can be charged and drained. The penalty paid for this lesser degree of volatility and speed is loss of space. The larger cell design makes static memories less dense than DRAMs.

Memory capacity is measured by the number of bits that can be stored. A 1K RAM has a capacity of 1024 bits of information; 1024 is a power of 2. A 64K RAM actually has a capacity of 65,536 bits of information. RAM capacity has expanded rapidly, with larger megabits memories (64 and higher) expected to be produced in quantities with presently identified technology. Each step upward in RAM capacity places greater pressure on wafer processing and yield improvement. The nature of the semiconductor chip business is exemplified by the 64K RAM, introduced by IBM in 1977. The chips were soon available in the merchant market, priced at over $100 per circuit. By 1985 competition and yield improvements had lowered the price to under $1 per circuit!

Redundancy

Redundancy is the inclusion of extra circuit components in the design. If one or more of the components don't work, there are others available that do. The tradeoff for redundancy is larger chip size. Also, extra circuitry is required within the main circuit to detect the functioning and nonfunctioning components and direct the selection of a

Figure 17.10 Static RAM cell schematic.

functioning component. Although this approach to higher yield has been discussed for years, it hasn't yet become a mainstay of circuit design. This is due to the problem of locating the working and non-working components and wiring the working ones into the circuit.

Wafer Scale Integration

Wafer scale integration (WSI) is a novel approach to integration aimed at fuller utilization of all the functioning areas on a wafer. The surface of a wafer using WSI is not covered by individual die. Instead, the wafer contains sections that are the functional subsegments of a circuit.

Failed chips on a conventional wafer usually fail from a single defect (or several) that prevents only one of the subsections of the circuit from performing. Even failed chips have many areas or subsections that *do* function. In the conventional chip approach, vast areas of the chip that contain working subsections are thrown away.

A WSI wafer is not covered by individual chips. Part of the wafer contains input sections, part has output sections, and so on. At the end of the processing, each of the subsections is electrically probed and tested. The locations of the working subsections are recorded. After testing, a "super" circuit is wired together from the working subsections. For example, consider a circuit board requiring 10 logic chips and 20 memory chips. The conventional approach requires manufacturing enough wafers to deliver the required number of individual chips.

With WSI, the wafer surface is covered with individual circuit subsections rather than complete circuits. In the supercircuit sample just given, the WSI wafer would have all the subsections plus extras required to make the 10 logic and 20 memory sections. The result is a supercircuit that uses all the functioning parts of the wafer surface. Other benefits include faster speed due to the lower number of package connections and lower cost resulting from eliminating individual packages. The drawbacks to WSI include longer processing time to form both logic and memory functions on the same wafer, reliability problems stemming from high heat levels which the more dense circuit generates, and the need for a more flexible metallization patterning than is currently available.

The Next Generation

Since the 1950s the semiconductor industry has maintained a constant rate of product evolution. In general, there has been a new product generation every three years.[2] Within each generation, the density of memory chips has increased four times and logic chips 2 to 3 times. Every two generations (six years), the feature size has decreased by a factor of 2 and chip area and package pin count have increased by a factor of 2.

Projecting the future is always difficult but there are identified end points for chip circuits as we now know them. The Semiconductor Industry Association (SIA) technology roadmap[3] profiles development to the 0.10-μm feature size level by the year 2007. That size will require chip densities of 16G (billion) bytes for memory and 20MB (million) logic gates. Wafer diameters will grow to 200–400 mm and die sizes will be in the 1000-mm^2 range (1.2 inches per side).

Reaching this level will require development of x-ray lithography, low-resistance metals, low-leakage contacts, and ultraclean materials and processes. Planarization techniques will be strained as the number of metal interconnect levels climbs to 6 or 7. Perhaps the major challenge will be reliable, stable gate oxides in the 35-Å range. Below 35 Å, carriers can tunnel through the layer, causing leakage problems. These technical problems must be overcome in a manufacturing atmosphere that is becoming increasingly expensive and requiring more automation and extensive measurement and process control programs.

Circuits developed by advanced technology will be a mixture of existing circuits with increased performance and new circuits. Certainly DRAMs will continue to increase in memory capacity and microprocessors will become more powerful and faster. New areas for chip development are speech recognition, expert systems, and the continuing "microchipization" of automobiles and most power-consuming products (control and conservation).

Key Concepts and Terms

Analog	RAM, DRAM, SRAM
Binary	ROM, PROM, EPROM, EEPROM
Digital	Volatile and nonvolatile memories
Flash memory	Wafer scale integration
Memory/logic	

Review Questions

1. Name the two principal solid-state circuit types.

2. Name the two principal logical circuit types.

3. Give an example of a digital electrical device.

4. Which of the following describes a PAL circuit?

   a. Factory programmable
   b. Field programmable
   c. *a* and *b*
   d. Not programmable

5. How many times can a thin-film fuse PROM be programmed?

   a. Once
   b. Five times
   c. Many times

6. What does RAM stand for?

7. What is the difference between a DRAM and an SRAM?

8. How does a serial memory device read back information? Give an example.

9. Why are DRAM circuits considered the leading-edge ULSI circuits?

10. How does the WFI approach to design and processing differ from conventional approaches?

References

1. M. McConnell, et al., "An experimental 4-Mb Flash EEPROM with Sector Erase," *IEEE Journal of Solid State Circuits,* vol. 26, no. 4, April 1991.
2. C. Hu, "MOSFET Scaling in the Next Decade and Beyond," *Semiconductor International,* June 1994, p. 105.
3. Semiconductor Industry Association, Technology Roadmap, Palo Alto, CA, 1992.

18

Packaging

Overview

After wafer sort, the chips on the wafer surface are completed and the electrically functioning ones are identified. The chips will be incorporated into an individual protective package, mounted with other components in hybrid or multichip modules, or connected directly onto a printed circuit board. This chapter will describe the packages and processes used for chip protection.

Objectives

Upon completion of this chapter, you should be able to:

1. List the four functions of a semiconductor package.
2. List the five common parts of a package.
3. Recognize and identify the major package designs.
4. List and describe the major packaging process flows.

Introduction

After wafer sort, the chips are still part of the wafer. For use in a circuit or electronic product they must be separated from the wafer and, in most cases, put in a protective package. They may also be mounted onto the surface of a ceramic substrate as part of a hybrid circuit, put into a large package with other chips as part of a multichip module (MCM) or connected directly onto a printed-circuit board (Fig. 18.1).

Figure 18.1 Chip packaging options.

All three options share some common processes. In this chapter, the basic packaging processes and the types of packages used to protect the chip will be explained. This series of processes is known variously as *packaging, assembly,* or the *back-end process*. In the packaging process, the chips are called *dies* or *dice* and in certain parts of the country they are referred to by the Texas Instruments' term, *bars*. In this chapter they will be referred to as *chips* or *die*.

Over the years, semiconductor packaging has lagged behind wafer fabrication in process sophistication and manufacturing demands. The advent of the VLSI–ULSI era in chip density has forced a radical upgrading of chip packaging technology and production automation. Higher-density chips result in more bonding pads, requiring more electrical connections and cleaner packages and processes. The SIA technology roadmap is projecting packages with up to 1000 chip connections to package very large memory chips. As solid-state circuits have found more applications, the need for special package designs has increased. The chips may end up in either consumer products or the demanding and harsh environments of space, automotive, and military hardware. These harsh environments require special packaging, processing, and testing to ensure high reliability in the field. These packages, processes, and tests are referred to as *hi-rel* (high-reliability). The other chips and packages are referred to as *commercial* parts.

No longer is packaging the stepchild of the semiconductor industry. Many feel that eventually packaging will be the limiting factor on the growth of chip size. For the time being, however, much effort is going into new package designs, new material development, and faster and more reliable packaging processes.

Chip Characteristics

Throughout this text many characteristics of discrete devices and integrated circuits have been mentioned. Several of them have a direct bearing on package design and the packaging processes (Fig. 18.2). The chip density (integration level) determines the number of connections required, with higher-density chips having larger surface areas and more bonding pads. The trend to larger chips has resulted in the need for thicker and larger-diameter wafers. These factors have caused changes in die separation processes and package design and have created the need for wafer thinning.

In previous chapters it was pointed out that the functioning of the chip components (transistors, diodes, capacitors, resistors, and fuses) can be altered by various contaminants. Primary among them are chemicals such as sodium and chlorine. Additionally, other chemicals can attack the chip layers, and environmental factors, such as particulates, humidity, and static can ruin chips or change their performance. Other concerns are the influence of light and radiation impinging on the chip surface. Some chips are extremely light or radiation sensitive. This factor is considered in the selection of package materials and processing. A dominant chip characteristic is the extreme vulnerability of its surface to physical abuse. The surface components are only a small distance down into the wafer surface, and the surface wiring can be as thin as 1 μm.

These environmental and physical concerns are addressed in two ways. First is the passivation layer deposited near the end of the fabrication process. This layer is known by several different names as listed in Fig. 18.3. The two materials used for the passivation layer are silicon dioxide and silicon nitride. Often they are doped with boron, phosphorus, or both to increase their protective properties. The second method of protecting the chip is provided by a package.

Another chip characteristic of importance to the packaging process is heat generation. Chips used in high-power circuits and

- Integration Level
- Wafer Thickness
- Dimensions
- Environmental Sensitivity
- Physical Vulnerability
- Heat Generation
- Heat Sensitivity

Figure 18.2 Chip characterizations affecting packaging process.

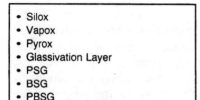

- Silox
- Vapox
- Pyrox
- Glassivation Layer
- PSG
- BSG
- PBSG

Figure 18.3 Passivation layer names.

highly integrated circuits can generate enough heat to actually damage themselves and the circuit. Package design includes heat dissipation factors. Heat is also an important parameter in packaging processes, with packaging process temperatures limited to 450°C. Above this temperature the aluminum and silicon contacts on the chip can form an alloy in the wafer surface that causes electrical shorts.

Package Functions

There are four basic functions performed by a semiconductor package. They are

1. A substantial lead system
2. Physical protection
3. Environmental protection
4. Heat dissipation

Substantial lead system

The primary function of the package is to allow connection of the chip to a circuit board or directly to an electronic product. This connection cannot be made directly due to the thin and fragile metal system used to interconnect the components on the chip surface. The metal leads are typically less than 1.5 μm thick and often only 1 μm wide. The thinnest wires available are 0.7 to 1.0 mils in diameter, which is many times larger than the surface wiring. This difference in wiring sizes is the reason that the chip wiring terminates in the larger bonding pads around the edge of the chip.

Even though the wires are larger, at 1 mil in diameter they too are very fragile. This fragility is overcome by a more substantial electrical lead system that serves as the connection of the chip to the out-

Figure 18.4   Dip through-hole
assembly.

side world (Fig. 18.4). The lead system is an integral part of the
package.

Physical protection

The second function of the package is the physical protection of the
chip from breakage, particulate contamination, and abuse. Physical
protection needs vary from low, as in the case of consumer products,
to very stringent, as in the case of automobile circuits, space rockets,
and military uses. The protection function is accomplished by securing the chip to a die-attachment area and surrounding the chip, wire
bonds, and inner package leads with an appropriate enclosure. The
size and eventual use of the chip dictate the choice of materials for
the enclosure and the design and size of the package.

Environmental protection

Environmental protection of the chip from chemicals, moisture, and
gases that may interfere with the chip functioning is provided by the
package enclosure.

Heat dissipation

Every semiconductor chip generates some heat during operation.
Some generate large quantities. The package enclosure materials
serve to draw the heat away from most chips. Indeed, one of the factors in choosing a package material is its thermal dissipation property. The chips that generate large quantities of heat require additional
consideration in the package design. This consideration will influence
the size of the package and will often require the addition of metal
heat-dissipating fins or blocks on the package.

Packaging Operations

In this section a simple overview of the primary packaging process is presented.

Cleanliness and static control

The vulnerability of the chips to contamination exists during their entire lifetime. While assembly areas are not generally required to maintain the same cleanliness levels as fabrication areas (see Chap. 4), cleanliness is important. The table in Fig. 18.5 lists the common contamination control practices used in packaging areas. Hi-rel areas, particularly, demand higher cleanliness levels. In fact, many companies are finding that half-way contamination control programs are doomed to failure. Consequently, more assembly areas are practicing very stringent controls, especially for people-generated particles and chemicals.

An environmental danger that is most serious in packaging areas is static. In the fabrication clean rooms, static is controlled primarily to prevent the attraction of particles to the wafer surface. This is also a concern in a packaging area. But the greatest concern is *electrostatic discharge,* or *ESD.* Static charge can build up to levels of tens of thousands of volts. If this voltage is suddenly discharged onto a chip surface, it can easily destroy a portion of the circuit. MOS gate structures are particularly vulnerable to ESD.

Every packaging area making high-density chips should have an active antistatic program (Fig. 18.6). Those assembling military parts

* HEPA Filters/VLF Air
* Smocks, Hats, Shoe Covers
* Finger Cots or Gloves
* Filtered Chemicals
* Sticky Floor Mats
* Static Control

Figure 18.5   Contamination control practices.

* Wrist Ground Straps
* Nonstatic Garments
* Antistatic Materials
* Grounded Equipment
* Grounded Work Surfaces
* Grounded Floors or Floor Mats

Figure 18.6   Static control practices.

will need to have one as a condition of getting the contract. The anti-static program is based on operators wearing grounding straps and nonstatic smocks; the use of antistatic carrier materials; moving work by lifting rather than sliding; and grounded equipment, work surfaces, and floor mats. Static is also reduced by the placement of ionizers in nitrogen and air blow-off guns (Fig. 18.7) and in the path of air coming out of HEPA filters.

Basic processes

In wafer fabrication the wafers pass many times in and out of four basic operations (layering, patterning, doping, and heat processing). In packaging also there are several basic operations (Fig. 18.8). However, packaging is a once-through process; each of the major processes is required only once. As in fabrication, the exact order of the operations is determined by the package type and other factors. Each operation may or may not be used in a particular process and each is customized to the particular chip and package requirements.

The basic operations are *backside preparation, die separation, die pick, inspection, die attach, wire bonding, preseal inspection, package*

Figure 18.7 Static control techniques.

• Backside Preparation	• Preseal Inspection
• Die Separation	• Packaging Sealing
• Die Pick	• Plating
• Die Attach	• Lead Trim
• Inspection	• Marking
• Bonding	• Final Tests

Figure 18.8 Basic packaging operations.

sealing, plating, trimming, marking, and *final tests.* The details of the various operations and their optional processes are explained in more detail in the remainder of this chapter.

Backside preparation.   At the end of the fabrication process some wafers have to be thinned (wafer thinning) to fit in the package and/or to remove backside damage or junctions. Wafers whose die are going to be attached to the package by a gold-silicon eutectic will receive a deposited layer of gold (backside gold).

Die separation.   The wafer is separated into individual die by sawing or scribe-and-break techniques.

Die pick and plate.   The functioning die identified at wafer sort are picked from the separated wafer and placed in carriers (Fig. 18.9).

Inspection.   The plated dies are optically inspected for edge integrity, contamination, and cosmetic defects.

Die attach.   Each die is attached into the die-attach area of the package by either a gold-silicon eutectic layer or an epoxy adhesive material (Fig. 18.10).

After Sort

After Die Separation

Plated Die

Figure 18.9   Die separation to plate.

Figure 18.10 Die attach area of package.

Die Attach
Area

Lead bonding. Thin wires are bonded between the chip bonding pads and the inner leads of the package. Bonding is also done by bumps of metal on the bonding pads or by the TAB technique.

Preseal inspection. The bonded die is optically inspected in the package. Criteria checked for are the alignment of the die in the package, proper bond placement, contamination, die-attach quality, and bonding quality.

Plating, trimming, and marking. The outer package leads are plated with an additional layer of conductive metal to improve their solderability into a printed-circuit board. Near the end of the process, each package with outer leads will go through a trimming operation to separate the leads from supports. The marking operation is performed to code important information (Fig. 18.11) on the outside of the package enclosure.

Final test. A series of final tests, including electrical and environmental, are performed on each packaged chip to ensure quality. Some packages will receive a "burn-in" test to detect early failures.

Figure 18.11 Marked package.

Common package parts

The four functions of a chip package are accomplished through the use of a wide variety of package designs. However, most packages have five common parts: the die-attachment area, the inner leads, the outer leads, the chip to package connection, and the enclosure.

Die-attachment area.   In the center of every package is an area where the chip is securely attached into the package. This die-attachment area may have an electrical connection that serves to connect the back of the chip to the rest of the lead system. A major requirement for this area is absolute flatness in order to intimately support the chip in the package (Fig. 18.10).

Inner and outer leads.   The metal lead system chip is continuous from the die-attach cavity to the printed-circuit board or electronic product. The system inner connections are called the *inner leads, bonding lead tips,* or *bond fingers.* The inner leads are generally the narrowest portion of the lead system. The leads become progressively wider or thicker, finally ending outside the package. These portions of the lead system are called the *outer leads* (Fig. 18.12). Most of the lead systems are composed of one continuous piece of metal. One exception is the side-brazed package. In this package construction method, the outer leads are brazed onto the interior leads. Two different alloys are used for the outer lead system, either an iron-nickel alloy or a copper alloy. The iron-nickel alloy is desirable for its strength and stability, while copper is used for its electrical and heat-conduction properties.

Inner
Leads

Outer
Leads

Figure 18.12   Inner and outer leads.

Bonding Wires

Figure 18.13 Bonding wires.

Chip to package connection

The chip is electrically connected to the lead system of the package with bonding wires, balls, or other on-chip connectors (Fig. 18.13).

Enclosures. The die-attach area and inner and outer leads constitute the electrical parts of the package. The other part is the *enclosure* or *body*. This is the part that provides the protection and heat-dissipation functions. These functions are achieved by several different techniques and package designs as described in the sealing section. The completeness of the seal falls into two categories, hermetic and nonhermetic (Fig. 18.14).

Hermetic sealing results in a package that is impervious to the penetration of moisture and other gases. Hermetic seals are required for chips operating in harsh and demanding environments such as rockets or space satellites. Metal and ceramic enclosures are the preferred materials for hermetically sealed packages.

Nonhermetically sealed packages are adequate for most consumer applications such as computers and entertainment systems. This sealing system provides good and adequate environmental protection of the chip, except in the most demanding situations. A better term

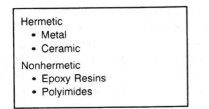

Hermetic
  • Metal
  • Ceramic
Nonhermetic
  • Epoxy Resins
  • Polyimides

Figure 18.14 Package sealing designations.

for this type of enclosure sealing method would be "less hermetic." Nonhermetic packages are composed of epoxy resins or polyimide materials and are generally referred to as "plastic packages."

Packaging Processes

Prepackaging wafer preparation

After the final passivation layer and an alloy step in wafer fabrication, the circuits are complete. However, one or two additional processes may be performed on the wafer before transfer to packaging. These steps, wafer thinning and backside gold, are optional, depending on the wafer thickness and the particular circuit design.

Wafer thinning.   The trend to thicker wafers presents several problems in the packaging process. Thicker wafers require the more expensive complete saw-through method at die separation. While sawing produces a higher-quality die edge, the process is more expensive in time and consumption of diamond-tipped saws. Thicker die also require deeper die attach cavities, resulting in a more expensive package. Both of these undesirable results are avoided by thinning the wafers before die separation. Another situation requiring wafer thinning is electrical in nature. If the wafer backs are not protected as the wafers go through the dopant operations in fabrication, the dopants will form electrical junctions in the wafer back, which may interfere with good conduction in the back contact, which is required for the circuit to operate correctly. These junctions may require physical removal by wafer thinning.

The thinning step generally takes place between wafer sort and die separation. Wafers are reduced to a thickness of 0.2-0.5 mm.[1] Thinning is done by the same processes (mechanical grinding and chemical-mechanical polishing or CMP) used to grind wafers in the wafer preparation stage (see Chap. 3). An alternate method is to protect the front of the wafers and chemically etch material from the back.

Wafer thinning is a worrisome process. In backgrinding there is the concern of scratching the front of the wafer and of wafer breakage. Since the wafer must be held down on the grinder or polishing surface, the front of the wafer must be protected, and once thinned, wafers break more easily. In backetching there is a similar need to protect the front of the wafer from the etchant. The protection can be provided by spinning a thick layer of photoresist on the front side. Other methods include attaching adhesive-backed polymer sheets cut to the wafer diameter. Stresses induced in the wafer by the grinding and polishing processes must be controlled to prevent wafer and die warping. Wafer warping interferes with the die separation process (broken and cracked die). Die warping creates die-attach problems in the packaging process.[2]

Backside gold.   Another optional wafer process is adding a layer of backside gold. A layer of gold is required on wafers that are going to be attached to the package by eutectic techniques (see the "Die attach" section). The gold is usually applied in the fabrication area (after backgrinding) by evaporation or sputtering.

Die separation

The chip-packaging process starts with the separation of the wafer into individual dies. The two methods of die separation are scribing and sawing (Fig. 18.15).

Scribing.   *Scribing,* or *diamond scribing,* was the first production die-separation technique developed in the industry. It requires the alignment of the wafer on a precision stage, followed by dragging a diamond-tipped scribe through the center of the scribe lines. The scribe creates a shallow scratch in the wafer surface. The separation of the die is completed by stressing the wafer with a cylindrical roller. As the roller is moved over the surface, the wafer separates along the scribe lines. The breaks follow the crystal structure of the wafer, thus creating a right-angle edge on the die. Scribing becomes unreliable in wafers over 10 mils thick.

Sawing.   The advent of thicker wafers has led to the development of sawing as the preferred die-separation method. A saw consists of a wafer table with rotation capability, a manual or automatic vision system for orienting the scribe lines, and a diamond-impregnated round saw. Two techniques are used. Both start with the passage of

Figure 18.15 Scribe and saw separation.

the diamond saw over the scribe lines. For thinner wafers, the saw is lowered into the wafer surface to create a trench about one-third of the way through the wafer. The separation of the wafer into die is completed by the stress and roller technique used in the scribing method. The second sawing method is to separate the die by a complete saw-through of the wafer.

Often the wafers undergoing complete saw-through are first mounted on a flexible plastic film. The film holds the die in place after the sawing operation and aids the die-pick operation. Sawing is the preferred die-separation method due to the cleaner die edges and the fewer cracks and chips left on the sides of the die (Fig. 18.16).

Die pick and plate

After sawing, the separated die are transferred to a station for selection of the functioning die (nonlinked). In the manual method, an operator will pick up each of the nonlinked dies with a vacuum wand and place it in a sectioned plate. Wafers that come to the station on the flexible film are first placed on a frame that stretches the film. The stretching separates the die, which aids the die-pick operation.

In the automated version of this operation, a memory tape or disk with the locations of the good die (from wafer sort) is loaded into the tool. A vacuum wand picks up good die and automatically places them in a sectioned plate for transfer to the next operation.

Die inspection

Before being committed to the rest of the process, the die are given an optical inspection. Of primary interest is the quality of the die edge, which should be free of chips and cracks. This inspection also sorts out surface irregularities, such as scratches and contamination. Inspection may be manual with microscopes or automated with a vision system. At this step the die are ready to go into a package.

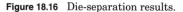

Figure 18.16 Die-separation results.

Die attach

Die attachment has several goals: to create a strong physical bond between the chip and the package, to provide either an electrical conducting or insulating contact between the die and the package, and to serve as a medium to transfer heat from the chip to the package.

A requirement is the permanency of the die-attachment bond. The bond should not loosen or deteriorate during subsequent processing steps or when the package is in use in an electronic product. This requirement is especially important for packages that will be subjected to high physical forces such as those used in rockets. Additionally, the die-attach materials should be contaminant-free and should not outgas during subsequent process heating steps. Lastly, the process itself should be productive and economical.

There are two principal methods of die attach, *eutectic* and *epoxy adhesives*.

Eutectic die attach. The eutectic method is named for the phenomenon that takes place when two materials melt together (alloy) at a much lower temperature than either of them separately. For die attach, the two eutectic materials are gold and silicon (Fig. 18.17). Gold melts at 1063°C, while silicon melts at 1415°C. When the two are mixed together, they start alloying at about 380°C. Gold is plated onto the die-attach area and alloys with the bottom of the silicon die when heated.

The gold for the die-attach layer is actually a sandwich. The bottom of the die-attach area is deposited or plated with a layer of gold. Sometimes a preformed piece of metal composed of a gold and silicon mixture is placed in the die-attach area. When heated, these two layers, along with a thin layer of silicon from the wafer back, forms a thin alloy layer. This layer is the actual bond forming the die-package attachment.

Eutectic die attach requires four actions. First is the heating of the package until the gold-silicon forms a liquid. Second is the placement

Conducting
- Gold/Silicon Eutectic
- Metal-Filled Epoxy
- Conducting Polyimide

Nonconducting
- Epoxy Adhesive
- Insulating Polyimide

Figure 18.17 Die-attach material matrix.

of the chip on the die-attach area. Third is an abrasive action, called "scrubbing," that forces the die and package surfaces together. It is this action in the presence of the heat that forms the gold-silicon eutectic layer. The fourth and final action is the cooling of the system, which completes the physical and electrical attachment of the chip and package.

Eutectic die attach can be performed manually or by an automated machine that performs the four actions. Gold-silicon eutectic die attach is favored for high-reliability devices and circuits for its strong bonds, heat dissipation properties, thermal stability, and lack of impurities.

Epoxy die attach.   The alternate die-attach process uses thick liquid epoxy adhesives. These adhesives can form an insulating barrier between the chip and package or become electrically and heat conducting with the addition of metals such as gold or silver. Polyimide may also be used as an adhesive. Popular adhesives are silver-filled epoxy for copper lead frames and silver-filled polyimide for alloy 22 metal frames.[3]

The epoxy process starts with the deposit of the epoxy adhesive in the die-attach area by dispensing the adhesive with a needle or screen printing it into the die-attach area. The die, held by a vacuum wand, is positioned in the center of the die-attach area. The second action is to force the die into the epoxy to form a thin uniform layer under the die. The final action is a curing step in an oven at an elevated temperature that sets the epoxy bond.

Epoxy die attach is favored for its economy and ease of processing, in that the package does not have to be heated on a stage. This factor makes the automation of the process easier. When compared to gold-silicon eutectic die attach, epoxy has the disadvantage of potential decomposition at the high temperatures of bonding and sealing operations. Epoxy die-attach films also do not have the bonding power of the eutectics.

Regardless of the attachment method used, there are several marks of a successful die attach. One is the proper and consistent alignment of the die in the die-attach area. Proper placement pays off in faster and higher-yield automatic bonding. Another desired result is a solid, uniform, and void-free contact over the entire area of the chip. This is necessary for mechanical strength and good thermal conduction. One evidence of a uniform bond is a continuous joint or "fillet" between the die edge and the package. The final mark of a good die-attach process is a die-attach area free of flakes or lumps that can come loose during use and cause a malfunction.

Bonding pad and package connection

Once the die and package are attached, they go to the bonding process. This is perhaps the most critical of all the assembly operations. Three techniques provide the critical chip to package connection:

- Wire bonding
- Flip-chip and bump
- TAB

The wire bonding procedure is simple in concept. A thin (0.7- to 1.0-mil) wire is first bonded to the chip bonding pad and spanned to the inner lead of the package lead frame. The third action is to bond the wire to the inner lead. Last, the wire is clipped and the entire process repeated at the next bonding pad. While simple in concept and procedure, wire bonding is critical because of the precise wire placement and electrical contact requirements. In addition to accurate placement, each and every wire must make good electrical contact at both ends, span between the pad and inner lead in a prescribed loop without kinks and be at a safe distance from neighboring wires. Wire loops in conventional packages are 8 to 12 mils, while those in ultrathin packages are 4 to 5 mils.[4] The distance between adjacent wires is called the pitch of the bonding.

Wire bonding is done with either gold or aluminum wires. Both are highly conductive and ductile enough to withstand deformation during the bonding steps and still remain strong and reliable. Each has its advantages and disadvantages and each is bonded by different methods.

Gold wire bonding. Gold has several pluses as a bonding wire material. It is the best-known room-temperature conductor and is an excellent heat conductor. It is resistant to oxidation and corrosion, which translates into an ability to be melted to form a strong bond with the aluminum bonding pads without oxidizing during the process. Two methods are used for gold bonding. They are *thermocompression* and *thermosonic.*

Thermocompression bonding (also known as *TC bonding*) starts with the positioning of the package on the bonding chuck and the heating of the chip and package to between 300 and 350°C. Chips that are going to be enclosed in an epoxy molded package are processed through die attach and bonded with the chip on the lead frame only. The bonding wire is fed out of a thin tube called a *capil-*

Figure 18.18 Gold ball bonding.

lary (Fig. 18.18). An instantaneous electrical spark or small hydrogen flame melts the tip of the wire into a ball and positions the wire over the first bonding pad. The capillary moves downward, forcing the melted ball onto the center of the bonding pad. The effect of the heat (thermal) and the downward pressure (compression) forms a strong alloy bond between the two materials. This type of bonding is often called *ball bonding*. After the ball bond is complete, the capillary feeds out more wire as it travels to the inner lead. At the inner lead, the capillary again travels downward to where the gold wire is forced by the heat and pressure to melt onto the gold-plated inner lead. The spark or flame then severs the wire, forming the ball for the next pad bond. This procedure is repeated until every pad and its corresponding inner pad are connected.

Thermosonic, gold ball bonding follows the same steps as thermocompression bonding. However, it can take place at a lower temperature. This benefit is provided by a pulse of ultrasonic energy that is sent through the capillary into the wire. This additional energy is sufficient to provide the heat and friction to form a strong alloy bond.

The majority of production gold wire bonding is done on automatic machines that use sophisticated techniques to locate the pads and span the wire to the correct inner lead. The fastest bonding machines can perform thousands of bonds per hour. There are two major drawbacks to the use of gold bonding wires. First is the expense of the gold. Second is an undesirable alloy that can form between the gold and aluminum. This alloy can severely reduce the conduction ability of the bond. It forms a purplish color and is known as the "purple plague."

Aluminum wire bonding. Aluminum wire, while not having the conduction and corrosion-resistance properties of gold, is still an important bonding wire material. A primary advantage is its lower cost. The second advantage is that the bond with the aluminum bonding pad is a monometal system and thus less susceptible to corrosion. Also, aluminum bonding can take place at lower temperatures than gold bonding, making it more compatible with the use of epoxy die-attach adhesives.

The bonding of the aluminum follows the same major steps as gold wire bonding. However, the method of forming the bond is different. No ball is formed. Instead, after the aluminum wire is positioned over the bonding pad, a wedge (Fig. 18.19) forces the wire onto the pad as a pulse of ultrasonic energy is sent down the wedge to form the bond. After the bond is formed, the wire is spanned to the inner lead where another ultrasonic-assisted wedge bond is formed. This type of bonding is known variously as *ultrasonic* or *wedge bonding*.

After this bond, the wire is cut. At this point in the process, a major difference between the bonding of the two materials occurs. In gold bonding, the capillary moves freely from pad to inner lead, to pad, and so on, with the package in a fixed position. In aluminum wire bonding, the package must be repositioned for every single bonding step. The repositioning is necessary to line up the pad and inner lead along the direction of travel of the wedge and wire. This requirement places an additional difficulty on the designers of automatic aluminum bonding machines. Nevertheless, most production aluminum bonding is done on high-speed machines.

Flip-chip and bump technology

Wire bonding presents several problems. Minimum height limits are imposed by the required wire loops. There is a chance of electrical performance problems or shorting if the wires come too close to each other. The wire bonds require two bonds and must be placed one-by-one and there are resistances associated with each bond. These problems are addressed by replacing wires with a deposited metal *bump*

Figure 18.19 Aluminum wedge bonding.

Figure 18.20 (*a*) Flip-chip joining; (*b*) tape automatic bonding. (*Source: Microelectronics Packaging Handbook by Tummala and Rymaszewski*)

on each bonding pad. Connection to the package is made when the chip is flipped over and soldered. Each bump connects to a corresponding package inner lead (Fig. 18.20*a*). Packages are lower profile, the electrical connection has lower resistance, and the electrical path is much shorter. IBM calls its version of this technology *controlled collapse chip connection* (C4).[5]

Tape automated bonding. Tape automated bonding (TAB) is used to wire up chips when extreme package thinness is required, such as in credit-card-size radios. The TAB process starts with formation of the lead system on a flexible strip of tape. Various methods are used to form the lead system. The metal for the system is deposited on the tape by sputtering or evaporation. Formation of the lead system is either by mechanical stamping or patterning techniques similar to the fabrication patterning process. The result is a continuous tape containing many individual lead systems. For the bonding operation (Fig. 18.20b), the chip is positioned on a chuck and the tape is moved by sprockets until one of these lead systems is positioned exactly over the chip. In this position, the inner leads of the system should be positioned over the bonding pads of the chip.

The bond is completed with a tool known as a *thermode*. The thermode is faced with a flat diamond surface and is heated. The thermode is moved downward, first contacting the inner leads. It continues downward with enough pressure to force the inner leads onto the bonding pads. The heat and the pressure are regulated to cause a physical and electrical bond between the two. Large chips require a larger TAB bonding area. For these chips, the thermode is faced with a synthetic diamond.

TAB bonding is also used in conjunction with package bonding. The advantages of TAB are speed, in that all of the bonds to the chip are made in one action, and the ease of automation offered by the tape and sprocket system.

Preseal inspection

An important step in the chip-packaging process is the preseal or precap inspection, sometimes called *third optical inspection,* which takes place after the bonding step. The inspection is performed to provide feedback on the quality of the operations already performed. It also is performed to reject packaged chips that may represent reliability hazards when the chip is in operation in the field.

While there are many levels of inspection criteria, they fall into two main categories: commercial and high-reliability (hi-rel). The commercial inspections are given to chips and packages destined for use in commercial systems and are derived from the internal quality levels of the producing company in conjunction with its experience and customer specifications. The high-reliability specifications are derived from a set of government standards called "Mil-Standard 883."

Commercial-level inspections screen the bonded chips for die-attach quality; correct placement of the bonds on the bonding pads and inner leads; the shape and quality of the ball or wedge bond; and the general condition of the chip surface in regard to contamination, scratches, and the like.

Mil-Standard 883 covers the same general issues as the commercial inspections, but to more stringent requirements. In particular, this standard also specifies criteria for the chip surface, including pattern alignment, critical dimensions, and surface irregularities, such as small scratches, voids, and small defects. These criteria serve to reject bonded chips that may malfunction in the rigorous conditions encountered in space and military operation.

Sealing techniques

After the bonded chip passes the optical inspection, it is ready for sealing in a protective enclosure. There are several methods used to achieve the enclosure of the chip. The method chosen depends on whether the seal must be hermetic or nonhermetic, and which type of package is to be used. The principal sealing methods use welded seals, soldered seals, glass-sealed lids, CERDIP package construction, and molded epoxy enclosures (Fig. 18.21).

Metal can.   If the package is a metal can type, a hermetic seal is achieved by welding the flanged lid to a matching flange on the base of the package.

Premade packages.   Premade ceramic packages are sealed by one of two methods, metal or ceramic lids (Fig. 18.22). Packages made for metal lids have a ring of gold around the top of the die-attach cavity, called the *seal ring*. Placed on top of the seal ring area is a preformed piece of gold-tin solder. The metal lid is clamped in position over the seal ring and placed on the belt of a conveyor furnace. As the clamped package passes through the furnace, the lid and package are soldered

```
Hermetic
   • Welding
   • Soldered Lid
   • Glass-Sealed Lid or Top
Nonhermetic
   • Epoxy Molding
   • Blob Top
```

Figure 18.21   Package sealing methods.

Figure 18.22   Premade ceramic package.

Lid

together to form a hermetic seal. The sealing takes place at a temperature range of 320 to 360°C in a pure nitrogen atmosphere.

If the package is to receive a ceramic lid, the procedure is similar. The part of the ceramic lid that contacts the base, outside the die-attach area, is coated with a layer of low-melting-point glass. The hermetic seal is completed as the package passes through a conveyor furnace. This sealing takes place at a temperature of about 400°C in an atmosphere of clean dry air.

CERDIP packages. A completed CERDIP package results in a hermetic seal around the chip and bonding wires. This seal is accomplished with glass, similar to the ceramic seal on the premade packages. In the case of the CERDIP package, the inner metal lead system is buried in a glass layer. The ceramic top of the package system has a cavity (Fig. 18.23). The underside of the lid, outside the cavity area, is coated with a layer of low-melting-point glass. The lid is placed over the base and clamped. The seal is accomplished as the assembly is passed through a conveyor furnace or placed in an oven. In the furnace or oven, the glass melts, fusing the base and top together. The CERDIP glass sealing system is used to seal DIP and flat packs. These latter packages are known as *Cerpacks* and *Cerflats*.

Figure 18.23 CERDIP parts.

Molded Plastic

Figure 18.24   Molded CERDIP.

Molded epoxy enclosures.   The fourth major method of enclosure, *epoxy molding,* produces the plastic package (Fig. 18.24). The resultant seal, while protecting the die from moisture and contamination, is not classified as hermetic. However, there exists a considerable amount of research into the development of improved epoxy materials to create better enclosures. The major advantages of epoxy molded enclosures are weight, low material cost, and manufacturing efficiency.

This sealing method follows a different process flow. The die is attached and bonded to a lead frame containing a number of lead systems (Fig. 18.25). After the preseal inspection, the lead frames are transferred to the molding area. The frames are placed on a mold mounted in a transfer molding machine. The molding machine is in turn charged with pellets of the epoxy material, which have been previously softened by a radio-frequency heater. Inside, the pellets are forced by a ram into a liquid state. The ram then forces the liquid around the die on the lead frames, forming an individual package

Figure 18.25   Lead frame.

around each lead frame. After the epoxy sets in the mold, the frames are removed and placed in an oven for final curing.

Lead plating

An important feature of the completed package is the finish on the package leads. Most package leads are coated with lead-tin solder, tin plate, or gold plate. The plating serves several important functions. First is the solderability of the leads into a circuit board. The additional metal finish improves the lead solderability, resulting in a more reliable electrical connection of the package and the printed circuit. The second benefit of the lead finish is that it protects the leads from oxidation or corrosion during periods of storage prior to mounting on the circuit board. The third benefit of lead plating is the protection of the leads from corrosive agents in the packaging and printed-circuit board mounting processes. These agents include solder flux, corrosive cleaners, and even tap water. The plating continues to protect the leads during their lifetime of use.

Electrolytic plating. Plated layers, such as gold and tin, are applied by electrolytic procedures. The packages are mounted on racks with each lead connected to an electric potential. The racks are placed in a tank containing a plating solution. Next, a small current is passed between the packages and an electrode in the tank. The current causes the particular metal in the solution to plate onto the leads.

Tin-lead solder. Tin-lead solder layers are applied either by dipping the packages into a pot of the molten solder or by a wave soldering technique. This latter technique offers the advantage of good control of the layer thickness and provides a shorter exposure of the package to the molten solder.

Plating process flows

Metal cans, side-brazed DIP packs, and pin grid arrays have their leads plated before starting into the packaging process. CERDIP and plastic packages go through the plating process after the sealing steps.

Lead trimming

One of the last steps in the package assembly process is trimming away excess material from the leads. The outer leads of DIP and flat-pack packages are made with a tie-bar (Fig. 18.26). This bar keeps the leads from becoming bent during the packaging process. At the

Tie Bar

Figure 18.26   Tie-bar.

end of the process, the package goes through a simple trimming machine that simultaneously trims away the tie-bar and trims the leads to the proper length.

Plastic package lead frames have an additional piece of material. It is a bridge of metal close to the package body that functions as a dam to prevent the liquid epoxy material from running into the lead area (Fig. 18.27). The dam is cut away from the frame with a series of precise cutting tools. After the dam is cut away, the packages move to another station on the cutter where the frame is separated into individual packages. If the package is a surface-mount type, the leads will be bent into the required shape.

Deflashing

Plastic packages receive an additional process, called *deflashing,* which is required to remove excess molding material from the package enclosure. Deflashing is done by either dipping the packages in a chemical bath followed by a rinse or by a physical abrasion process. Physical deflashing is done in a machine similar to a sand blaster which uses plastic beads as the abrasive.

Package marking

Once completed, a package must be identified with key information. Typical information coded on the package is the product type, device

Figure 18.27 Lead frame dam.

specifications, date, lot number, and where it was made. The main methods of marking are ink printing and laser inscription. Ink printing has the advantage of good adherence to all of the package materials. The composition of the ink is chosen for permanence in the eventual operating environment of the device. The ink is applied by an offset printer followed by a curing step. Curing is done by oven drying, room temperature air drying, or by ultraviolet light.

Laser printing is especially suited for plastic packages. The mark is permanently inscribed in the package surface and can provide good contrast on the dark packages. Additionally, laser marking is fast and noncontaminating since no foreign material is added to the package surface and no curing step is required. A drawback to laser marking is the difficulty of changing the mark if a wrong code was used or the status of the device is changed. Regardless of the marking method, all marks must meet the requirements of legibility, especially on smaller packages, and permanence when exposed to harsh environments.

Final testing

At the conclusion of the packaging process, the completed package is put through a series of environmental, electrical, and reliability tests. These tests vary in type and specifications, depending on the customer and use of the packaged devices. The tests may be performed on all of the packages in a lot or on selected samples.

Environmental tests

Environmental tests are performed to weed out leaking and defective packages. Defects detected are loose chips, contaminants and particles in the die-attach cavity, and faulty bonding. This testing series starts with a stabilization bake to drive off any volatile substances in the package. A typical bake is at 150°C for 24 hours.

The first environmental test is *temperature cycling*. The packages are loaded into a chamber and cycled between two temperature extremes. The number of cycles may reach several hundred. The high and low temperatures of this test vary with the device use. Commercial parts receive a narrower temperature range than hi-rel parts. The hi-rel cycle range is −25 to 125°C. During the cycling any weakness in the seal, die attachment, or bonding will be aggravated and be detected in later electrical tests.

A second environmental test is *constant acceleration*. In this test the packages are accelerated in a centrifuge (Fig. 18.28) that creates a force as high as 30,000 times the pull of gravity on the earth (30,000 g). During the acceleration, loose particles in the package, poorly attached dies, and weakly attached bonds are stressed so that they will be detected at the final electrical tests.

Leaks in the package enclosure are detected by two tests. *Gross leak* testing (Fig. 18.29) is conducted by submerging the packages in a hot liquid. The heated liquid raises the temperature of the package and forces trapped gases in the cavity to escape. The escaping gases are observable as bubbles rising in the liquid. The chamber has a transparent side, allowing an operator to observe the bubbles. Smaller (or fine) leaks are detected by using tracer gases. For this check, helium is pumped under pressure into a chamber containing the packages. If the package has small leaks, the gas will be pumped into the package cavity. The gas is detected as it escapes through the small leaks by a machine known as a mass spectrometer, which can

Units

Up to 30,000 g's

Figure 18.28   Acceleration test.

Figure 18.29 Gross leak bubbles test.

identify the escaping gas. An alternate fine leak test uses the radioactive gas krypton-85. It too is pumped through any leaks into the package under pressure. Detection of any krypton-85 in the package is by a device similar to a Geiger counter.

Electrical testing

The purpose of the wafer-fabrication and packaging processes is to provide to the customer a specific semiconductor device that performs to specific parameters. Thus, one of the last steps is an electrical test of the completed unit to verify that it performs to the specifications. The tests are similar to the wafer-sort tests. The overall objective is to verify that the good chips identified at wafer sort have not been compromised by the packaging process.

First there is a series of parametric tests. These electrical tests check the general performance of the device or circuit and ensure that it meets certain input and output voltage, capacitance, and current specifications. The second part of the final test is called the *functional test*. This test actually exercises; the specific chip functioning. Logic chips are put through logic tests; memory chips are exercised in their data storage and retrieval capabilities.

The equipment used to conduct the final test is electrically similar to that used in the wafer-sort operation. The electrical tests are performed by a computer-controlled tester that directs the sequences and levels of the parametric and functional tests. The packages are connected to the tester through sockets; the socket unit is known as the *test head*. The packages are inserted into the test head manually or by an automatic unit known as a *handler* (Fig. 18.30). This handler may be mechanical or robotic, depending on the speed and complexity of the operation.

Figure 18.30   Final test.

Tester   Test head   Handler

Burn-in tests

The last of the tests is the optional burn-in test(s). The reason it is optional is that although it is required for all high-reliability device lots, it may or may not be performed on commercial devices. The test requires the insertion of the packages in sockets and mounting in a chamber with temperature-cycling capability. During the test, the circuits are temperature-cycled while under an electrical bias.

The burn-in test is intended to stress the electrical interconnection of the chip and package and drive any contaminants in the body of the chip into the active circuitry, thus causing failure. This test is based on data that indicate that chips prone to these types of failures actually malfunction in the early part of their lifetime. By conducting burn-in tests, the early failures are detected. The devices passing the test are statistically more reliable.

Package Process Flows

It should be obvious that there is no universal packaging process flow. The package construction technique and lead-plating requirements dictate the flow. The table in Fig. 18.31 illustrates typical flows for three of the more common package types.

Packaging Strategies

Integrating a chip into an electronic product or system requires an electrical subsystem between the chip and the system. Two main approaches have emerged, individual packaging and bare die techniques (Fig. 18.32).

Individual package

With individual packaging, there are three techniques used to connect the package to a printed-circuit board (PCB). They are through-

Process Step	Premade Ceramic	CERDIP	Epoxy Molded
Die Separation	X	X	X
Pick and Plate	X	X	X
Die Attach	X	X	X
Wire Bonding	X	X	X
Preseal Inspect	X	X	X
Lid Sealing	X	X	
Epoxy Molding	X		
Deflashing	X		
Lead Plating and Trim		X	X
Marking	X	X	X
Environmental Test	X	X	X
Final Electric Test	X	X	X
Burn-In	X	X	X

Figure 18.31 Table of packaging process flows.

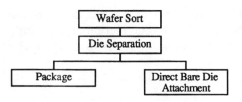

Figure 18.32 Post die-separation options.

Figure 18.33 Dip through-hole assembly.

hole, surface mount, and TAB. Through-hole connections feature straight pins on the package, which are inserted into corresponding holes in the printed-circuit board. (Fig. 18.33). A newer method is *surface mount*, also referred to as *SMD*. This method features packages that have their leads bent into a J shape or bent outward to allow direct soldering to the board surface (Fig. 18.34).

Some SMD packages do not have leads—instead, they terminate in metal traces hugging the package body. They are known as *leadless*

Figure 18.34   Surface-mount device.

packs. For inclusion on a circuit board, they are inserted into chip carriers, which in turn have leads that attach to the printed-circuit board.

Tape automated bonding (TAB) has two uses. One is bonding the chip bonding pads directly to the lead frame (See section on bonding). TAB is also a technique for bonding the outer leads directly to the PCB.

Bare die techniques

More reliable, higher-density, and faster circuits are an ongoing goal and challenge. Reliability is addressed in hybrid circuits. Speed and higher density comes with the elimination of the individual die package. Fewer links in the chain reduce resistances and (in some cases) shorten the length carriers have to travel, hence increasing speed. This strategy, called bare die, is used in hybrid circuits, multichip modules, and chip-on-board technology.

Package Design

Up to the early 1970s, most chips ended up in either a metal package, known as a "can," or in the familiar dual in-line package (DIP). The trends in chip size and integration levels and new electronic products with special packaging requirements (smart cards) have driven the development of new packages and strategies.

Metal cans

Metal cans are cylindrically shaped packages with an array of leads extending through the base (see Fig. 18.35a). The chip is attached to the base and wire-bonded to posts that are connected to the leads. The lid and base have matching flanges that are welded together to create a hermetic seal. These packages are designated by numbers, with the T0-3 and T0-5 being the most common. Metal cans are used to package discrete devices and small-scale integrated circuits.

Dual in-line packages (DIPs)

The DIP is probably the most familiar package design. It features a thick, sturdy body with two rows of outer leads coming out of the side and bending downward. DIPs are constructed by three different techniques (Fig. 18.35b). Chips designed for high-reliability use will be packaged in a premade ceramic DIP. The package is formed with a solid body of ceramic with the leads buried in the ceramic. The die-attachment area is a cavity recessed into the body. The hermetic seal is completed by a soldered metal lid or a glass-sealed ceramic lid.

Another approach to the DIP is the CERDIP, which stands for ceramic DIP. This type of package is composed of a bottom ceramic base with the lead frame held firm in a glass layer. The chip is attached to the base and wire-bonded to the lead frame. The hermetic seal is completed with a ceramic top glass sealed to the base. CERDIP

Figure 18.35   (a) Metal can; (b) DIP packages.

construction is used for a number of package types. The vast majority of DIPs are made by the epoxy molding technique. In this technique the chip is attached to a lead frame and then wire-bonded. After bonding the frame is placed in a molding machine and the package is formed around the chip, wire bonds, and inner leads.

Pin grid arrays

Larger chips, with more leads, have outgrown the DIP configuration. The pin grid array (PGA) is a package designed for larger chips. It features a premade "sandwich" with the outer leads coming out of the bottom of the package in the form of pins (Fig. 18.36). The chip is attached in a cavity that is formed in either the top or bottom of the body. Connections to the chip cover the entire chip area (unlike most chips, which have connections restricted to bonding pads around the periphery of the chip). Ceramic PGAs are hermetically sealed with a soldered metal lid.

Ball grid arrays (BGAs)

BGAs have the same body shape as PGAs. Instead of pins on the package bottom, there is a series of solder bumps (balls) used to connect the package to the PCB (see "Flip-chip and bump technology"). The effect is to lower the package profile and weight as well as provide higher pin counts by using the whole chip surface for bonding pads.

Quad packages

While pin grid array packs are a convenient design for larger chips, their ceramic construction is expensive compared to molded epoxy packages. This consideration led to the development of the quad package. A quad (short for quadrant) pack is constructed by the epoxy-

Figure 18.36 Pin grid array.

molded technique but has leads coming out of all four sides of the package (Fig. 18.37).

Thin packages

New products such as smart cards require thin packages. There are several techniques used to make thinner packages. They are called flat packs (FPs), thin small outline packages (TSOPs), small outline IC (SOIC), and ultrathin packages (UTPs). Flat packs are constructed by the same techniques used to form DIP packages. These packages are designed with flatter height profiles and have leads bent out to the side of the package (Fig. 18.38). Ultrathin packages have total body heights in the 1-mm range. There are also quad flat packs (QFPs).

Lead on chip (LOC)

The LOC package, intended for large die, have the bonding pads arranged down the middle of the chip (Fig. 18.39). Package leads sit on a cushion over the chip surface.

Hybrid circuits

Hybrid circuitry is an old technology long favored for use in military and harsh environments. Hybrid circuits consist of a substrate on which standard electrical and semiconductor devices are mounted. The devices are connected by conductive or resistive thick-film paths

Figure 18.37 Quad package.

Figure 18.38 Flat pack.

Figure 18.39   Lead on chip (LOC). (*Source: Semiconductor International, May 1993*).

on the surface of an insulating substrate. These paths are formed by silk-screening inks containing a proper filler or from thin films evaporated or sputtered onto the surface and patterned by photolithography techniques. The term *hybrid* refers to the mix of solid-state and conventional passive electrical devices (resistors, capacitors) present in the same circuit. Most hybrid substrates are ceramic. High-performance hybrids may have aluminum nitride (AlN), silicon carbide (SiC), or diamond substrates used alone or in combinations.[6]

Hybrid circuits offer the advantages of structural rigidity and low leakage between devices due to the insulating property of the ceramic. They can provide a circuit with a mix of CMOS, bipolar, and other components and offer functions not available in ASIC circuits.[7]

On the downside, they generally have a much lower density than integrated circuits at a higher cost.

Multichip modules (MCMs)

Mounting individual chip packages on a PCB presents several problems. A chip package is several times the area of the chip, and so takes up space on the board. Circuit resistance is increased by the individual resistances of all the package pins, and the electrical path lengths are multiplied by the number of chips and package leads. Each of these problems can be reduced by mounting several chips on the same substrate. This technology, multichip module (MCM), is similar to that used by hybrid circuits, but thick-film screened components are rarely used.

Three types of MCMs have evolved. MCM-L (laminated) is similar to advanced, laminated printed-circuit boards, using copper foil conductors on plastic dielectrics. MCM-C (ceramic) are more like hybrid circuits. The substrate is a cofired ceramic with thick-film conductors. MCM-D (deposited) uses ceramic, metal, or silicon substrates with deposited thin metal conductors.

Chip on board (COB)

The most direct use of bare die is to bond them directly to the printed-circuit board. Bonding techniques include all those used to bond chips into packages. After connection, the chip is protected by a technique called *blob top protection*. The protection is accomplished by covering the chip and bonds with a blob of epoxy resin material (Fig. 18.40). The material is similar in properties to that used to mold the plastic packages. Blob top coverings are used with TAB and other packaging schemes.

The known good die (KGD) problem

In the individual package process, a final test assures quality of the completed product. If the chip had gone bad or the process was faulty, the entire chip and package is discarded. But the cost is much higher when putting bare die into hybrid, MCM, and printed-circuit boards. These vehicles are more expensive to build and carry other expensive chips or components.

One option is to rely on the results of the wafer-sort test to certify die performance. Unfortunately, wafer sort does not include environmental tests or long-term reliability tests. However, performing a final test on a bare die is difficult, if not impossible. Consequently, before the bare-die step, some temporary assembly of the die to a lead frame must be done to facilitate a full final test.

Summary

There are thousands of individual package types and there is no uniform system of identifying them. Some are named by their design (DIP, flat packs, etc.), some are named by their construction technique (molded, CERDIP, etc.), and others are named by their use (e.g.,

Figure 18.40 Blob top.

SMDs). When trying to understand a package type, keep in mind the three considerations: design type, construction technique, and use.

Key Concepts and Terms

Bare die strategies	Package designs
Bonding	Package functions
Cleanliness requirements	Package parts
Die attach	Pick and plate
Final test(s)	Sealing techniques
Hermetic and nonhermetic seals	

Review Questions

1. Name the four functions of a chip package.

2. What is the function of marking dies with ink dots?

3. Name the five parts of a package.

4. Define a hermetic package.

5. Give an example of a hermetic and nonhermetic package.

6. List in order the six major steps in the packaging process.

7. Describe the process and material used for eutectic die attachment.

8. Describe the principal thermosonic and ultrasonic bonding techniques.

9. What is a DIP package and how is it formed?

10. Name two methods of package marking.

References

1. H. Hinzen and B. Ripper, "Precision Grinding of Semiconductor Wafers," *Solid State Technology,* Penwell Publishing, August 1993, p. 53.
2. F. I. Blech and D. Dang, "Silicon Wafer Deformation after Backside Grinding," *Solid State Technology,* Penwell Publishing, August 1994, p. 74.
3. L. Plummer, "Packaging Trends," *Semiconductor International,* Cahners Publishing, January 1993, p. 33.
4. R. Iscoff, "Ultrathin Packages: Are They Ahead of Their Time?" *Semiconductor International,* Cahners Publishing, May 1994, p. 50.
5. R. Tummala and E. Rymaszewski, *Microelectronics Packaging Handbook,* Van Nostrand Reinhold, New York, 1989, p. 156.
6. N. Nguyen, "Using Advanced Substrate Materials with Hybrid Packaging Techniques for Ultrahigh-Power ICs," *Solid State Technology,* Penwell Publishing, February 1993, p. 59.
7. R. Iscoff, "Will Hybrid Circuits Survive?" *Semiconductor International,* Cahners Publishing, October 1993, p. 57.

Glossary

acceptor   An impurity that causes semiconducting materials to accept valence electrons, thereby leaving "holes" in the valence band. The holes act like carriers of positive charge, referred to as P type.

aligner (align and expose)   A process tool used to align wafers and masks or reticles and expose the photoresist with a UV or other radiation source.

alignment   Refers to the positioning of a mask or reticle with respect to the wafer.

alignment marks   Targets on the mask and wafer used for correct alignment.

alloy   (1) A compound composed of two metals. (2) In semiconductor processing, the alloy step causes the interdiffusion of the semiconductor and the material on top of it, forming on ohmic contact between them.

aluminum (Al)   The metal most often used in semiconductor technology to form the interconnects between devices on a chip. It can be applied by evaporation or sputtering.

amorphous   Materials with no definite arrangement of atoms, e.g., plastics are amorphous.

angstrom   A unit of length, an angstrom (Å) is one ten-thousandth of a micron (10^{-4} μm) or 100,000,000 Å = 1 cm.

anistropic   An etch process that exhibits little or no undercutting.

anneal   A high-temperature processing step (usually the last one), designed to minimize stress in the crystal structure of the wafer.

antimony (Sb)   A Group V element that is an N-type dopant in silicon. It is often used as the dopant for the buried layer.

Source: Glossary terms are abstracted from Beverly Griggs, Anne Miller, and Peter Van Zant, *Semiconductor Terminology—Graphic Glossary of Terms,* Semiconductor Services, Redwood City, Calif., 1989.

arsenic   A Group V element that is an N-type dopant in silicon.

assembly   The series of operations after fabrication in which the wafer is separated into individual chips and mounted and connected to a package.

atmospheric oxidation   A process of oxidation of silicon carried out at atmospheric pressure. The equipment used for thermal oxidation is the same as that used for thermal diffusion. It is composed of four subassemblies: a reactant source cabinet, a reaction chamber, a heating source, and a wafer holder.

atomic force microscope (AFM)   A microscope for profiling wafer surfaces by plotting the output of a spring-balanced probe moved over the surface.

atomic number   A number assigned to each element equal to the number of protons (therefore the number of electrons) in the atom.

atomic particles   The parts of an atom: electrons, protons, and neutrons.

base   (1) The control portion of an NPN or PNP junction transistor. (2) The P-type diffusion done using boron that forms the base of NPN transistors, the emitter and collector of lateral PNP transistors, and resistors.

binary notation   A way of representing any number using a power of 2 (the digits 0 and 1).

bipolar transistor   A transistor consisting of an emitter, base, and collector, whose action depends on the injection of minority carriers from the base by the collector. Sometimes called NPN or PNP transistor to emphasize its layered structure.

bi-MOS   A circuit containing both bipolar and MOS transistors.

boat   (1) Pieces of quartz or metal joined together to form a supporting structure for wafers during high-temperature processing steps. (2) A Teflon or plastic assemblage used to hold wafers during wet processing steps.

boat puller   A mechanical arrangement to push a boat loaded with wafers into a furnace and/or withdraw it at a fixed speed.

BOE   See **buffered oxide etch.**

bonding pads   Electrical terminals on the chip (generally around the periphery) used for connection to the package electrical system.

boron (B)   The P-type dopant commonly used for the isolation and base diffusion in standard bipolar integrated circuit processing.

boron trichloride (BCl$_3$)   A gas that is often used as a source of boron for doping silicon.

bubbler   An apparatus in which a carrier gas is "bubbled" through a heated liquid, causing portions of the liquid to be transported with the gas, e.g., a carrier gas (nitrogen or oxygen) is bubbled through deionized water at 98 to 99°C on its way to the oxidation tube.

buffered oxide etch   A mix of hydrogen fluoride (HF) and ammonium fluoride (NH$_4$F) used to allow oxide etching to occur at a slow, controlled rate.

bump connection technology   A structure of metal bumps formed on the bonding pads, allowing connection of the chip to a package when the chip is flipped over.

buried layer   The N+ diffusion in the P-type substrate done just prior to growing the epitaxial layer. The buried layer provides a low-resistance path for current flowing in a device. Common buried layer dopants are antimony and arsenic.

can   A metal package used for connecting a chip to a printed circuit board with from three to five leads.

capacitor   A discrete device which stores electrical charge on two conductors separated by a dielectric.

capacitance   Electrical charge storage capability.

capacitance-voltage plot (C/V plot)   Plot that provides information on the amount of mobile ionic contamination present in the oxide.

carrier gas   An inert gas which will transport atoms or molecules of a desired substance to a reaction chamber.

centistokes   Units used to measure viscosity, centipoise divided by density.

channel   A thin region of a semiconductor that supports conduction. A channel may occur at a surface or in the bulk, essential for the operation of MOSFETs and SIGFETs. In cases where channels are not part of the circuit design, their presence may indicate contamination problems or incomplete isolation.

channeling   A phenomenon in which an ion beam will penetrate into the crystal planes of the wafer. Preventing channeling is accomplished by cutting the wafer "off orientation." The effect is to tilt the crystal planes relative to the beam direction.

charge carrier   A carrier of electrical charge within the crystal of a solid-state device, such as an electron or hole.

chemical etching   Selective removal of material by means of liquid reactants. The precision of the etch is controlled by the temperature of the etchant, the time of immersion and the composition of the acid etchant.

chemical mechanical polishing (CMP)   A wafer flattening and polishing process that combines chemical removal with mechanical buffing. Used for polishing/flattening wafers after crystal growing and wafer planarization during the wafer fabrication process.

chemical vapor deposition   See **CVD.**

chip   Die or device, one of the individual integrated circuits or discrete devices on a wafer.

chrome   A metal often used in mask fabrication to form the layer in which the circuit pattern is generated.

circuit board   See **printed circuit board.**

circuit layout   The calculation of the physical device dimensions required to produce the required electrical parameters. Vertical dimensions determine CVD and doping thickness specifications. Horizontal dimensions determine the wafer pattern dimensions and are the basis for a scaled drawing of the finished circuit (composite drawing).

class number   Number of contaminant particles in a cubic foot of air.

clean room   An area in which semiconductor device fabrication takes place. The cleanliness of the room is highly controlled in order to limit the number of contaminants to which the semiconductor is exposed.

clear field mask   A mask on which the pattern is defined by the opaque areas.

cluster tool   Several process stations or tools served by one loading–unloading chamber and wafer-transport system.

CMOS (complementary field-effect transistor)   N- and P-channel MOS transistors on the same chip.

collector   Along with the emitter and base, one of the three regions of the bipolar type of transistor.

collimated light   Light in which the rays are parallel; used for gross visual inspection of surfaces.

composite drawing   A scaled drawing of the finished circuit.

conductivity   The ability of materials to conduct electricity (measured in siemens for conductance or ohms for resistance).

conductor   A material which has low resistivity and high conductivity.

contact   The regions of exposed silicon that are covered during the metallization process to provide electrical access to the devices.

contact aligner   An aligner tool that clamps the wafer and mask into a tight contact before the resist exposure cycle.

contact mask   The step at which holes are put into the wafer layers to allow the metal layer to reach down to the doped silicon substrate.

contamination   A general term used to describe unwanted material that adversely affects the physical or electrical characteristics of a semiconductor wafer.

critical dimensions (CDs)   The widths of the lines and spaces of critical circuit patterns as well as the area of contacts.

cryogenic pump   A vacuum pump that can produce a vacuum to the 10^{-10} torr range, the same level as the vacuum of space. It does not require forepumps or cold traps and is faster than other types of vacuum pumps.

crystal   A material in which the atoms are arranged in structured groups called unit cells.

crystal defects   Vacancies and dislocations in a crystal which influence the electrical performance of a circuit.

crystal orientation   The orientation of the primary crystal plane, expressed as Miller indices.

crystal planes   The planes in the semiconductor crystal structure along which the die must be aligned in order to prevent "ragged" die edges when the wafer is separated into individual die.

CUM yield   See **fabrication yield.**

current   A measure of the number of charged particles passing a given point per unit time.

curve tracer   A piece of electrical test equipment that displays the characteristics of a device visually on a screen.

CVD (chemical vapor deposition)   A method for depositing some of the layers which function as dielectrics, conductors, or semiconductors. A chemical containing atoms of the material to be deposited reacts with another chemical, liberating the desired material, which deposits on the wafer while byproducts of the reaction are removed from the reaction chamber.

Czochralski crystal grower   A type of crystal grower that uses a seed to pull a crystal from a crucible of molten material.

dark field mask   A mask on which the pattern is defined by the clear portion of the mask.

deep ultraviolet (DUV)   A light wavelength often used to expose photoresist which has the advantage of an ability to produce smaller image widths.

defect density   The density of defects per square centimeter on a chip.

dehydration baking   A heating process by which wafer surfaces are restored to a hydrophobic condition by baking. Surface water is evaporated from the wafer at elevated temperatures.

deionized (DI) water   Process water that is free of dissolved ions. Specification levels are generally 15–18 mΩ of resistance.

depletion layer   The region in a semiconductor where essentially all charge carriers have been swept out by the electric field which exists there.

deposition   Process in which layers are formed as the result of a chemical reaction in which the desired layer material is formed and coats the wafer surface.

design rule   The minimum feature size of a circuit.

develop inspection   The first inspection in the photomasking process consisting of measurement of critical dimensions and inspection for defects. It is done after development or development and hard bake if an automatic baking system is used.

development   A photoresist processing step in which photoresist is removed from areas defined by the masking and exposure step of wafer fabrication.

developer   Chemical used to remove areas defined in the masking and exposure step of wafer fabrication.

device A single-function component such as a transistor, resistor, or capacitor.

DI water Deionized water; purity of this water is measured by its resistivity, with the standard being 18 MΩ.

diborane (B_2H_6) A gas that is often used as a source of boron for doping silicon.

die One unit on a wafer separated by scribe lines; after all of the wafer fabrication steps are completed, die are separated by sawing; the separated units are referred to as *chips*.

die bonding Assembly step in which individual chips are attached to the package with conductive adhesives or metal alloys.

die sort See **wafer sort.**

dielectric A material that conducts no current when it has a voltage across it. Two dielectrics encountered in semiconductor processing are silicon dioxide and silicon nitride.

diffusion A process used in semiconductor production which introduces minute amounts of impurities (dopants) into a substrate material such as silicon or germanium and permits the impurity to spread into the substrate. The process is very dependent on temperature and time.

diffusivity The rate of movement or diffusion of dopants in a semiconductor.

diode Device which enables current flow in one direction but not in another.

DIP (dual in-line package) A rectangular circuit package, with leads coming out of the long sides and bent down to fit into a socket.

discrete device A circuit having a single electrical function. Discrete devices include capacitors, resistors, transistors, diodes, and fuses.

dislocation A discontinuity in the crystal lattice, a type of crystal defect.

DMOS (diffused MOS) A transistor structure which features a narrow (channel length) separation between the source and drain. The channel length is created by two sequential diffusions through the same hole.

donor An impurity that can make a semiconductor N-type by donating extra "free" electrons; electrons carry a negative charge.

dopant An element that alters the conductivity of a semiconductor by contributing either a hole or electron to the conduction process. For silicon, the dopants are found in Groups III and V of the periodic table.

dopant deposition The first step in the diffusion process, in which the dopant atoms are diffused into the wafer surface.

doping The introduction of an impurity (dopant) into the crystal lattice of a semiconductor to modify its electronic properties—for example, adding boron to silicon to make the material P type.

DRAM (dynamic random access memory) Memory device for the storage of digital information. The information is stored in a "volatile" state.

drain Along with the source and gate, one of the three regions of a unipolar or field-effect transistor (FET).

drive-in Stage in diffusion where the dopant is driven deeper into the wafer.

dry etch See **plasma etch.**

dry ox The growth of silicon dioxide using oxygen and hydrogen, which form water vapor at process temperatures, rather than using water vapor directly.

dry oxide Thermal silicon dioxide grown using oxygen.

e-beam (electron beam) An exposure source which allows direct image formation without a mask. An E-beam can be deflected by electrostatic plates and therefore directed to precise locations, resulting in the generation of submicron-size patterns.

e-beam aligner An aligner tool that exposes the resist-coated wafer surface by steering (writing) an electron beam across the wafer surface.

e-beam evaporation (electron beam evaporation) Phase change that uses the energy of a focused electron beam to provide the required energy to change solid metal or alloys from solid to gas.

e-beam exposure system A machine in which the image pattern is stored in a computer memory and used to control the electrostatic plates that in turn direct the E-beam, resulting in the generation of patterns without the use of reticles or photomasks.

edge die The incomplete die located on the edge of the wafer.

EEPROM (electrically erasable PROM) A memory circuit with the capability of data erasure and acceptance of new information by the application of an electrical pulse.

electromigration The diffusion of electrons in electric fields set up in the lead while the circuit is in operation. It occurs in aluminum and is exhibited as a field failure, not as a process defect. The metal thins and eventually separates completely, causing an opening in the circuit.

electron A charged particle revolving around the nucleus of an atom. It can form bonds with electrons from other atoms or be lost, making the atom an ion.

ellipsometer Instruments that use laser light sources to measure thin film thickness.

emitter (1) The region of a transistor that serves as the source or input end for carriers. (2) The N-type diffusion usually done using phosphorus, which forms the emitter of NPN transistors, the base contact of PNP transistors, the N+ contact of NPN transistors, and low-value resistors.

epitaxial (Greek for "arranged upon") The growth of a single-crystal semiconductor film upon a single-crystal substrate. The epitaxial layer has the same crystallographic characteristics as the substrate material.

epoxy package See **molded package.**

EPROM (erasable PROM) A memory circuit with the capability of data erasure and acceptance of new information.

etch   A process for removing material in a specified area through a wet or dry chemical reaction or by physical removal, such as by sputter etch.

evaporation   A process step that uses heat to change a material (usually a metal or metal alloy) from its solid state to a gaseous state with the result of the source being deposited on wafers. Both electron beam and filament evaporation are common in semiconductor processing.

exposure   Method of defining patterns by the interaction of light or other form of energy with photoresist that is sensitive to the energy source.

fabrication   Integrated circuit manufacturing processes.

fabrication yield   The percent of wafers arriving at wafer sort compared with the number started into the process.

feature size   The minimum width of pattern openings or spaces in a device.

FET (field-effect transistor)   A transistor consisting of a source, gate, and drain, whose action depends on the flow of majority carriers past the gate from the source to drain. The flow is controlled by the transverse electric field under the gate. See **unipolar transistor.**

field oxide   The region on an electrical device where the oxide serves the function of a dielectric.

final test   The final assembly step in which the packaged die is put through its last electrical test.

flash memory   An EPROM or EEPROM with the capability of block erasure of data in the memory array.

flat zone   The highly temperature controlled region of a tube furnace.

flip-chip joining   A chip or package connection process where "bumps" of connecting metal are formed on the chip surface and the chip is "flipped" over for soldering to the package.

four-point probe   A piece of electrical test equipment used to determine the sheet resistivity of a wafer.

furnace   A piece of equipment containing a resistance-heated element and a temperature controller. It is used to maintain a region of constant temperature with a controlled atmosphere for the processing of semiconductor devices.

fuse   A circuit component which can be blown to allow a desired memory cell or logic gate to be programmed.

gallium arsenide (GaAs)   Most common of compound semiconductor materials. It has the advantage of producing higher-speed devices than those produced using silicon as a substrate.

gate   Along with the source and drain, one of the three regions of the unipolar or field-effect transistor (FET or MOS).

gate array   Type of integrated circuit made up of an arrangement of interconnected gates used to provide custom functions.

gate oxide (gate ox) The thin oxide which causes the induction of charge, creating a channel between source and drain regions of a MOS transistor.

germanium Semiconducting material used in the manufacture of crystal diodes and of early transistors.

HEPA filter (high-efficiency particulate attenuator) A filter constructed of fragile fibers in an accordion-folded design which allows a larger filtering area at an air velocity low enough for operator comfort. This filter permits a filtering efficiency of 99.99 percent.

hexamethyldisilizane (HMDS) Primer used to promote photoresist adhesion.

high-pressure oxidation Oxidation carried out at high pressure (10 to 20 atm) in order to reduce the amount of heat or time required. The reaction chamber for this process must be constructed of stainless steel to safely contain the pressure.

hole (1) The absence of a valence electron in a semiconductor crystal. Motion of a hole is equivalent to motion of a positive charge. (2) A "hole" in a surface layer created by the photomasking process.

hybrid integrated circuit A structure consisting of an assembly of one or more semiconductor devices and a thin-film integrated circuit on a single substrate, usually of ceramic.

hydrofluoric acid (HF) An acid used to etch silicon dioxide; often diluted or buffered before it is used.

hydrogen (H_2) A gas used in semiconductor processing primarily as a carrier gas for high-temperature reaction steps such as epitaxial silicon growth.

hydrophilic Affinity toward water (water-loving); a hydrophilic surface is one that will allow water to spread across it in large puddles.

hydrophobic Aversion to water; a hydrophobic surface is one that will not support large pools of water. The water is pulled into droplets on the surface. These surfaces often are termed "dewetted."

hydroscopic Attracts and absorbs water.

integrated circuit A circuit in which many elements are fabricated and interconnected on a single chip of semiconductor material, as opposed to a "nonintegrated" circuit, in which the transistors, diodes, resistors, etc., are fabricated separately and then assembled.

integration level The range of total component count in a die. Varies from SSI (small-scale integration, less than 50 components) to ULSI (ultralarge-scale integration, over 1,000,000 components).

intrinsic semiconductor An element or compound that has four electrons in its outer ring (i.e., elements from Group IV of the periodic table or compounds of Group III and V).

ion An atom that has either gained or lost electrons, making it a charged particle (either negative or positive).

ion beam milling A dry etching method which uses an ion beam. Argon atoms are ionized and accelerated toward a wafer. The exposed areas are removed through a sputtering action.

ion implantation Introduction of selected impurities (dopants) by means of high-voltage ion bombardment to achieve desired electronic properties in defined areas.

interconnect See **lead.**

isolation diffusion Diffusion step resulting in P-N junctions surrounding the areas to be separated.

isotropic etching Refers to the etching of the photoresist both downward and to the side.

JFET (junction field-effect transistor) Device in which voltage is applied to a terminal to control current between the source and drain regions.

junction The interface at which the conductivity type of a material changes from P type to N type or vice versa.

killer defect A defect that causes the failure of a device or circuit.

lateral diffusion The diffusion of dopants from side to side every time the wafer is heated near the diffusion temperature range.

layering A process by which thin layers of different materials are grown on, or added to, the wafer surface.

lead A metal strip on the wafer surface.

LED (light-emitting diode) A semiconductor device in which the energy of minority carriers in combining with holes is converted to light. Usually, but not necessarily, constructed as a P-N junction device.

lift-off process A material removal process in which the material is deposited into a hole in a photoresist layer and the pattern defined where the photoresist layer is removed (lifted off) the surface.

light field mask See **clear field mask.**

lithography Process of pattern transfer; when light is utilized, it is termed photolithography; and when patterns are small enough to be measured in microns, it is referred to as microlithography.

local oxidation of silicon (LOCOS) A MOS surface isolation scheme where silicon dioxide is grown around islands of silicon nitride, which is in turn removed to leave oxide-free areas for the formation of the circuit components.

low-pressure CVD (LPCVD) A CVD system and process performed in a low-pressure range.

LSI (large-scale integration) Refers to chips with between 5000 and 100,000 components each.

majority carrier The mobile charge carrier (hole or electron) that predominates in a semiconductor material—for example, electrons in an N-type region.

mask A glass plate covered with an array of patterns used in the photomasking process. Each pattern consists of opaque and clear areas that respectively prevent or allow light through. Masks are aligned with existing patterns on silicon wafers and used to expose photoresist. Mask patterns may be formed in emulsion, chrome, iron oxide, silicon, or a number of other opaque materials.

masking   See **patterning.**

memory   The storing of data or information.

metal mask   The step at which an island of conducting material is left on the wafer surface.

metalorganic CVD (MOCVD)   A VPE process that uses halides and metalorganic sources.

micrometer   One-millionth of a meter (10^{-6} m); symbol is μm.

micron   Same as micrometer.

microchip   See **chip.**

Miller indices   A numerical system of three numbers used to identify the orientation of planes in a crystal.

minienvironment   An environment that maintains wafer cleanliness by storing, transporting, and loading or unloading wafers in small, clean enclosures.

minority carrier   The nonpredominant mobile charge carrier in a semiconductor—for example, electrons in a P-type region.

MMOS (memory MOS)   A nonvolatile memory device structure that enables information to be retained during power shutdown.

mobile ionic contaminant   Wafer contaminants that are electrically charged particles, which can cause electrical failures in a wafer or device.

molded package   A package of epoxy or other polymer material that is molded around the chip and package lead system.

molecular beam epitaxy (MBE)   An evaporative deposition process capable of extreme control of the deposition process.

molecule   Smallest quantity of a substance that retains the properties of that substance.

monochromatic light   Light of a single wavelength.

MOSFET   A field-effect transistor containing a metal gate over thermal oxide over silicon.

MSI (medium-scale integration)   Refers to chips with between 50 and 5000 components each.

multichip module (MCM)   A package containing two or more IC chips connected by a thin-film metal system.

multilayer resist process   An image-resolution process that uses two or more layers of photoresist.

negative resist   Photoresist that remains in areas that were not protected from exposure by the opaque regions of a mask while being removed by the developer in regions that were protected. A negative image of a mask remains following the develop process. A "clear" or "light" field mask is most often used with negative resist.

nitric acid (HNO$_3$)   A strong acid often used to clean silicon wafers or etch materials.

nitridation   The formation of silicon nitride by the high-temperature exposure of a silicon surface to nitrogen.

nitrogen (N$_2$)   A gas that seldom reacts with other materials. It is often used as a carrier gas for chemicals in semiconductor processing.

NMOS   N-channel MOS; type of MOSFET in which the channel is negative during conduction.

nonvolatile memory circuit   A memory circuit that retains its data when power to the chip is lost.

NPN transistor   A transistor which has a base of P-type silicon sandwiched between an emitter and a collector of N-type silicon.

N-type   A semiconductor material in which the majority of carriers are electrons and therefore negative. N-type dopants in silicon are Group V elements, in which the fifth outer electron is free to conduct current.

Ohm's law   A relationship between resistance, voltage, and current; $R = V/I$.

oil diffusion pump   A type of high-vacuum pump that uses evaporated hot oil particles to "push" chamber particles out of the system.

overall yield   The percentage of functioning packaged chips from a wafer related to the number of die mapped onto the wafer. Overall yield is the product of fabrication yield, sort yield, and assembly yields.

oxidation   The growth of oxide on silicon when exposed to oxygen. This process is highly temperature dependent.

oxidation reaction chamber   A chamber in which oxidation takes place. Quartz or silicon carbide tubing is used to make an oxidation reaction chamber due to their thermal resistance and purity.

oxide   See **silicon dioxide.**

oxide etching   An etching process which uses acid [usually hydrofluoric acid (HF)]. The acid must be buffered in order for the reaction to proceed at a rate slow enough to be controlled. Buffered oxide etch (BOE) is often used.

oxygen (O$_2$)   Gas used to combine with silicon to form silicon dioxide.

package   Protective container for semiconductor chip having electrical leads.

packaging yield   The percentage of packaged die passing the final tests compared to the number of good die that entered the packaging process.

passivation   Sealing layer added at the end of the fabrication process to prevent deterioration of electronic properties through chemical action, corrosion, or handling during the packaging processes. The passivation layer, usually silicon dioxide or silicon nitride, protects against moisture or contamination.

patterning   A process in which the pattern in a reticle or photomask is transferred to a wafer resulting in the identification of areas to be doped or selectively removed.

pellicle   A thin film of an optical-grade polymer that is stretched on a frame and secured to a mask or reticle. This solves the problem of airborne dirt collecting on the mask and acting as an opaque spot. During the exposure, the dirt is held out of the focal plane and does not "print" onto the wafer.

phosphine (PH$_3$)   A gas that is often used as a source of phosphorus for doping silicon.

phosphorus (P)   The N-type dopant commonly used for the sinker and emitter diffusions in standard bipolar integrated circuit technology.

phosphorus oxychloride (POCl$_3$)   A liquid that is often used as a source of phosphorus for doping silicon.

photomasking   See **patterning.**

photoplate   A coated mask blank before imaging.

photoresist   The light-sensitive film spun onto wafers and "exposed" using high-intensity light through a mask. The exposed (or unexposed depending on its polarity) photoresist is dissolved with developers, leaving a pattern of photoresist which allows etching to take place in some areas while preventing it in others.

pinhole   A small undesired hole in the photoresist or in the opaque region of a mask or reticle.

pin grid array (PGA)   A large chip package with many leads coming out of the entire bottom surface of the package.

planar structure   A flat-surfaced device structure fabricated by diffusion and oxide masking, with the junctions terminating in a single plane.

plasma   High-energy gas made up of ionized particles.

plasma-enhanced CVD (PECVD)   A CVD system and process using plasma energy to drive the deposition.

plasma etch   A dry-etch process using reactive gases energized by a plasma field.

plastic package   See **molded package.**

plug (via plug)   A metal (generally a refractory metal) deposited in a connecting via hole between the conducting layers of a multilayer metal system.

PMOS (P-channel MOS)   Type of MOSFET in which the channel is positive due to conduction achieved by holes.

PNP   Semiconductor crystal structure consisting of an N-type region sandwiched between two P-type regions, as commonly used in bipolar transistors.

polycide MOS gate   A MOS gate structure composed of a sandwich of silicon dioxide topped with a layer of polysilicon that is covered with a refractory metal silicide.

polycrystalline silicon (Poly)   Silicon composed of many crystal unit cells randomly arranged.

polymer   A complex organic chemical compound made up of repeating units.

positive resist   Photoresist that is removed in areas that were not protected from exposure by the opaque regions of a mask while remaining after develop in regions that were protected from exposure. A positive image of the mask remains following the develop process. A "dark field" mask is used most often with positive resist.

post exposure bake (PEB)   A baking step performed after resist exposure to reduce standing wave effects.

predeposition (predep)   The process step during which a controlled amount of a dopant is introduced into the crystal structure of a semiconductor.

primer chemical   A chemical which enhances the adhesion of a desired layer (in semiconductor technology, the layer is usually photoresist).

programmable read-only memory   See **PROM.**

projection alignment   An exposure system in which the image on the mask is projected onto the wafer. This results in little mask or photoresist damage and has about the same productivity as the contact method. For LSI and VLSI, projection alignment is standard.

projection aligner   An aligner tool that projects the mask or reticle image over a distance onto the wafer.

PROM (programmable read-only memory)   A technology in which fuses are used in every memory cell and selected fuses are blown in order to program the chip with user-specified information.

proximity aligner   An aligner tool that holds the wafer and mask or reticle a small distance apart during the resist exposure cycle.

P-type   Semiconductor material in which the majority carriers are holes and therefore positive. P-type dopants in silicon are Group III-A elements.

quartz   Commercial name for silicon dioxide formed into glass products. Because of its high temperature resistance, quartz is used in many processing steps in integrated circuit fabrication.

RAM (random access memory)   Device that temporarily stores digital information.

rapid thermal oxidation (RTO)   An oxidation process performed in a rapid thermal processing (RTP) tool.

rapid thermal processing (RTP)   A single-wafer processing tool that uses high intensity lights or other sources to heat and cool the wafer in milliseconds.

RCA clean   A multiple-step process to clean wafers before oxidation; named after RCA, the company that developed the procedure.

reactive ion etching (RIE) An etching process that combines plasma and ion beam removal of the surface layer. The etchant gas enters the reaction chamber and is ionized. The individual molecules accelerate to the wafer surface. At the surface, the top layer removal is achieved by the physical and chemical removal of the material.

reactor (1) A piece of equipment used for the deposition of a layer of material used in semiconductor processing. Common types of reactors are epitaxial reactors, vapox reactors, and nitride reactors. (2) See **plasma etcher.**

resistivity A measure of the resistance to current flow in a material. A function of the attraction between the outer electrons and inner protons of a material. The more tightly bound the electrons, the greater the resistivity.

resolution capability The minimum feature size capability of a photolithography process or tool.

reticle An exposure mask with only a portion of complete die pattern.

rinse The removal of wet etchants or developers from the wafer. This process results in stopping the etching or developing processes and removing the active chemical from the surface. There are several different methods of rinsing: overflow rinsing, spray rinsing, dump rinsing, and spin-rinse dryers.

ROM (read-only memory) A memory circuit containing permanent data with no capability of accepting new information.

salicide MOS gate A polycide MOS gate structure sequenced in the process to self-align to the source or drain. See **polycide MOS gate.**

scanning electron microscope (SEM) Microscope used to magnify images as much as 50,000 times by means of scanning with an electron beam. The impinging electrons cause electrons on the surface to be ejected. The ejected electrons are collected and translated into a picture of the surface.

scribe lines Lines used to separate die on a wafer. The wafer will be sawed along the scribe lines, resulting in individual chips.

self-aligned gate A MOS structure that allows the direct alignment of the source or drain to the gate without a photoresist alignment step.

semiconductor An element such as silicon or germanium, intermediate in electrical conductivity between the conductors and the insulators, in which conduction takes place by means of holes and electrons. Common single-element semiconductors are Si (silicon) and Ge (germanium); a common compound semiconductor is GaAs (gallium arsenide).

sheet resistance A measurement of resistance with dimensions of ohms per centimeter squared that shows the number of N-type or P-type donor atoms in a semiconductor.

side diffusion See **lateral diffusion.**

silicon (Si) The Group IV element used for fabricating diodes, transistors, and integrated circuits.

silicon dioxide (SiO$_2$) A nonconducting layer that can be thermally grown or deposited on silicon wafers. Thermal silicon dioxide is commonly grown using either oxygen or water vapor at temperatures above 900°C.

silicon gate MOS   A MOS gate structure with a layer of polysilicon on top of a thin layer of silicon dioxide.

silicon nitride (Si_3N_4)   A nonconductive layer chemically deposited on wafers at temperatures between 600 and 900°C. When it is the final layer in the process, it protects devices against contamination.

single crystal   Refers to substances which have all unit cells arranged in a definite and repeated fashion as opposed to polycrystalline materials, which have unit cells randomly arranged.

slope etching   Controlled undercutting; an etch strategy in which the sides of the contact holes are purposely overetched so as to reduce the shadow effect of the sidewall and the resultant thinning of the film.

soft baking   A heating process used to evaporate a portion of the solvents in resist. The term "soft" describes the still soft resist after baking. The solvents are evaporated to achieve two results: to avoid retention of the solvent in the resist film and to increase the surface adhesion of the resist to the wafer.

solid-state electronics   Designation used to describe devices and circuits fabricated from solid materials such as semiconductors, ferrites, or thin films, as distinct from devices and circuits making use of electron tube technology.

source   Along with the gate and drain, one of the three regions of a unipolar or field-effect transistor (FET).

spinning   A technique in which the photoresist is spun onto the wafer, resulting in a typical photoresist layer 0.5 μm thick with an allowable thickness variation of 10 percent.

spin rinse dryer   A machine that automatically rinses and dries wafers by spinning them in cassettes around a central axis.

sputtering   A method of depositing a thin film of material on wafer surfaces. A target of the desired material is bombarded with radiofrequency-excited ions which knock atoms from the target; the dislodged target material deposits on the wafer surface.

spectrophotometer   An analytical instrument used to collect interference measurements which are used to calculate film thickness.

spreading resistance   A technique used for measuring the dopant concentration profile in a wafer.

SSI (small-scale integration)   Refers to chips with between 2 and 50 components each.

standard mechanical interface (SMIF)   A system that allows the mating of portable clean wafer boxes (called pods) to the clean microenvironment loading stations of process tools.

standing wave effects   A vertical resist exposure pattern that follows standing waves set up in the resist layer by constructive interference of the exposure light reflecting off the wafer surface.

static RAM (static random access memory) Fast read-write memory cell based on transistors.

steam oxide Thermal silicon dioxide grown by bubbling a gas (usually oxygen or nitrogen) through water at 98 to 100°C.

step and repeat An operation in which the pattern on the reticle is tranferred to the mask or wafer. The photoresist-coated mask blank (chrome, emulsion, or iron oxide) or wafer is placed on an x-y stage and the reticle pattern is repeatedly imaged until the entire surface is filled with the reticle pattern.

step coverage The ability of new layers to evenly cover steps formed in the existing wafer layers.

stepper An aligner tool that aligns and exposes one (or a small number) of die at a time. The tool "steps" to each subsequent die on the wafer.

stripping Removal process; usually refers to photoresist.

subcollector See **buried layer.**

substrate The underlying material upon which a device, circuit, or epitaxial layer is fabricated.

sulfuric acid (H_2SO_4) A strong acid often used to clean silicon wafers and to remove photoresist.

susceptor The flat slab of material (usually graphite) on which wafers are held during high-temperature deposition processes such as epitaxial growth or nitride deposition.

tape automated bonding (TAB) A chip-to-package connection process in which the package leads are formed on a flexible tape and all the lead fingers are bonded to the chip in one action.

target The material to be sputtered during the sputtering process.

test die Die on a wafer that appear to have a different pattern from most others. These contain test devices created by the same processes as the regular die; however, the devices on these die are designed on a larger scale to allow in-process quality control.

TCE (trichloroethylene) A solvent used for wafer and general cleaning.

tetraethylorthosilicate (TEOS) A chemical source for the deposition of silicon dioxide.

thermal diffusion A process by which dopant atoms diffuse into the wafer surface by heating the wafer in the range of 1000°C and exposing it to vapors containing the desired dopant.

thermal oxide On silicon semiconductor devices, an oxide fabricated by exposing the silicon to oxygen at high temperatures. The resulting interface has low levels of ionic impurities and defects (surface states).

thermocouple A device to measure the temperature in a furnace of a reactor. It is made by welding two wires together at a point. Heat generates a voltage between the two materials that is proportional to the temperature.

torr   Pressure unit; international standard unit replacing the English measure, millimeters of mercury (mmHg).

transistor   A semiconductor device that uses a stream of charge carriers to produce active electronic effects. The name was coined from the electrical characteristic of "transfer resistance."

tube   (1) See **furnace.** (2) A cylindrical piece of quartz with fittings on one or both ends. It is placed in a furnace to provide a contamination-free and controlled atmosphere.

undercutting.   See **isotrophic etching.**

ULSI (ultra-large-scale integration)   Refers to chips with over 1,000,000 components each.

ultraviolet (UV) light   A portion of the electromagnetic spectrum from 250 to 500 nm. High-pressure mercury sources emit UV light for photoresist exposure.

unipolar transistor   A transistor such as an FET whose action depends on majority carriers only.

vacancy   (1) A position in the crystal for an atom which is empty. (2) A type of crystal defect.

vacuum   A low-pressure condition.

vapor phase epitaxy (VPE)   An epitaxial deposition system that can combine several source gases to deposit compound semiconductors.

vapor priming   A technique in which primer is applied in a vapor state such that the wafer never comes in contact with any possible contamination in the liquid or, in the case of HMDS, any particles of hydrolyzed HMDS.

via   Vertical opening filled with conducting material used to connect circuits on various layers of a device to one another and to the semiconducting substrate; serves same purpose as "contacts."

viscosity   The qualitative measure of liquid flow. Viscosity measurements are made by measuring the force required to move an object through the liquid. It is a measurement of "internal friction."

VLF hood   A work station with vertical laminar airflow to keep particulate levels low.

VLSI (very-large-scale integration)   Refers to chips with between 100,000 and 1,000,000 components.

volatile memory circuit   A memory circuit that loses its data when power to the chip is lost.

voltage   The force applied between two points causing charged particles (and hence current) to flow.

wafer   A thin, usually round slice of a semiconductor material, from which chips are made.

wafer fabrication The series of manufacturing operations in which the circuit or device is put in and on the wafer.

wafer flat Flat area(s) ground onto the wafer's edges to indicate the crystal orientation of the wafer structure and the dopant type.

wafer sort The step after wafer fabrication during which the electrical parameters of integrated circuits are tested for functionality. Probes contact the pads of the circuit to conduct the test leading to the name "prober" for the equipment that performs electrical tests on each die site of completed wafers.

wafer sort yield The number of functioning die at wafer sort as compared to the total number of die started; typically, the lowest major yield point for integrated circuits.

wire bonding An assembly step in which thin gold or aluminum wires are attached between the die bonding pads and the lead connections in the package.

x-ray aligner An aligner tool that uses x rays and a mask to expose resist-coated wafers.

x-ray exposure system Imaging system using x rays as the exposure source. Due to their short wavelengths, x rays exhibit no detrimental diffraction effects.

yield A percentage used in the semiconductor industry which indicates the amount of finished products leaving a process as compared to the amount of product entering that process.

Index

Note: Page numbers in *italics* indicate illustrations. Page numbers followed by *t*s indicate tables.

acids vs. alkalis, 45–46
acoustic wave devices, 525
adiabatic cooling, 250–251, 413–414
aerial image: control methods, 286
air filters, 75–78
air quality, 72–73, *74*, *75*
airborne particulates, 68–69, 72–73
aligners, 226–228, *229*, 234–242, 243*t*, 587
alignment, 192–193, *197*, 226–242, 587
alignment marks, 101, *115*, 232–233, 587
alkalis vs. acids, 45–46
alloy (process), 105, 112, 390, 587
alloy junctions, 15–16, 312
alloys: deposition, 401, 406
aluminum:
    conduction, 391–394, 397, 587, 593
    deposition, 400–401, 408
aluminum film wet etching, 263–264
aluminum wire bonding, 566–567
AMLCDs, 526, *527*
ammonium persulfate, 176, 274
amorphous silicon deposition, 379–380
amorphous structure, *356*
amplification, 3
    bipolar transistors, 431–432, 499–500
    MOS transistors, 433, 502–503
amplifiers: analog circuits, 537
analog circuits, 536–538
analyzers: ion selection, 335
angstrom: definition, 143, 587
anisotropic etch, 260, *261*, 307, *308*
annealing process, 105, 166, 343–344, 587
annular ring illumination, 289–290
anodic oxidation, 185–186, 426
antimony-based dopants, 324*t*, 333, 587, 589
antireflective coatings, 292–293, *294*
APCVD (*see* atmospheric pressure CVD)
argon gas, 405–406, 406
arsenic-based dopants, 324*t*, 325, 333, 342–343, 347–348, 588, 589
aspect ratio (photoresists), 202, 205, 270

assay numbers (chemicals/gases), 91, 92
assembly (*see* packaging)
atmospheric oxidation, 150–170, 588
atmospheric pressure CVD, 358, 359–362
atomic force microscope, 446, 456–457, 588
atomic number, 27, 588
atomic structure, 26–29
Auger electron spectroscope, 458–459
autodoping: in epitaxial layers, 376–377
automation, 19–20, 472–478

backside processes (*see* wafer backsides)
bacterial contaminants, 70–71, 89
baffles (in tube furnaces), 324, *325*
baking (*see* heat treatments)
ball bonding, 565–566
ball grid arrays (BGAs), 582
band gap, 38
bare die, 580, 585
barrel plasma etching, 267–268
barrier metals, 392, 393
bars, 6, *100*, 101
base region, 432, 499–501, 588
bases vs. acids, 45–46
batch vs. single-wafer processing, 478–479
BGAs (ball grid arrays), 582
Bi-MOS circuits, 520–521, *522*, 588
biasing: in diodes, 497–498
binary notation, 532–534, 588
bipolar circuits, 512–516
    analog, 537, 538
    layers, 351–352
    vs. MOS circuits, 510–511, 540
bipolar transistors, 15–16, *17*, 431–432, 499–502, 588
    dopant concentration profile, *501*
    electrical measurements, 431–432
    epitaxial layer development, 18, 372
    ion implantation uses, 347
bird's beak problem, 171, 516–517

ABOUT THE AUTHOR

Peter Van Zant's career spans over thirty years of professional accomplishments as an engineer, business consultant, author, and trainer. His experience includes semiconductor chip manufacturing, small company management and marketing, business consulting, and training program design and presentation. He is the author of several books and a number of professional articles.